# THE GLYCOME

## Understanding the Diversity and Complexity of Glycobiology

# THE GLYCOME

## Understanding the Diversity and Complexity of Glycobiology

*Edited by*
**Adeel Malik, PhD**
**Tanveer A. Dar, PhD**

First edition published 2022

**Apple Academic Press Inc.**
1265 Goldenrod Circle, NE,
Palm Bay, FL 32905 USA
4164 Lakeshore Road, Burlington,
ON, L7L 1A4 Canada

**CRC Press**
6000 Broken Sound Parkway NW,
Suite 300, Boca Raton, FL 33487-2742 USA
2 Park Square, Milton Park,
Abingdon, Oxon, OX14 4RN UK

**Library and Archives Canada Cataloguing in Publication**

Title: The glycome : understanding the diversity and complexity of Glycobiology / edited by Adeel Malik, PhD, Tanveer A. Dar, PhD.

Names: Malik, Adeel (Research professor), editor. | Dar, Tanveer Ali, editor.

Description: First edition. | Includes bibliographical references and index.

Identifiers: Canadiana (print) 20210119934 | Canadiana (ebook) 20210120002 | ISBN 9781771889971 (hardcover) | ISBN 9781774638279 (softcover) | ISBN 9781003145394 (ebook)

Subjects: LCSH: Glycomics.

Classification: LCC QP702.G577 G69 2021 | DDC 572/.567—dc23

**Library of Congress Cataloging-in-Publication Data**

Names: Malik, Adeel (Research professor), editor. | Dar, Tanveer Ali, editor.

Title: The glycome : understanding the diversity and complexity of glycobiology / edited by Adeel Malik, PhD, Tanveer A. Dar, PhD.

Description: First edition. | Palm Bay : Apple Academic Press, 2021. | Includes bibliographical references and index. | Summary: "This volume, The Glycome: Understanding the Diversity and Complexity of Glycobiology, provides a comprehensive understanding of the enigmatic identity of the glycome, a complex but important area of research, one largely ignored because of its complexity. In this volume, the authors thoroughly deal with almost all aspects of the glycome, i.e., elucidation of the glycan identity enigma and its role in regulation of the cellular process and in disease etiology. It bridges the knowledge gap in understanding the glycome, from being a cell signature to its applications in disease etiology. In addition, it details many of the major insights regarding the possible role of the glycome in various diseases as a therapeutic marker. The book systematically covers the major aspects of the glycome, including the significance of substituting the diverse monosaccharide units to glycoproteins, the role of glycans in disease pathologies, and the challenges and advances in glycobiology. The authors stress the significance and huge encoding power of carbohydrates as well as provide helpful insights in framing the bigger picture. The authors begin with an introduction to the trends and developments in glycobiology and then proceed to discussing its role in disease pathologies, followed by its roles in other organisms, including bacteria, plants, and as stress busters. Chapters cover the involvement of the glycome in congenital and noncongenital disorders, the role of glycans in immunological processes, the role of glycans in neurodegeneration and metastasis, the involvement of glycans in host-pathogen interaction, and their role in microbial infection and the immune evasion. The book also provides insights into the therapeutic aspects of acylation as a vital glycosylational modification of proteins. A chapter highlights the scenario of glycans in plants while another discusses glycans as stress coping agents. The Glycome: Understanding the Diversity and Complexity of Glycobiology details state-of-the-art developments and emerging challenges of glycome biology, which are going to be key areas of future research, not only in the glycobiology field but also in pharmaceutics"-- Provided by publisher.

Identifiers: LCCN 2021002967 (print) | LCCN 2021002968 (ebook) | ISBN 9781771889971 (hardcover) | ISBN 9781774638279 (paperback) | ISBN 9781003145394 (ebook)

Subjects: LCSH: Glycomics. | Glycoconjugates.

Classification: LCC QP702.G577 G5963 2021 (print) | LCC QP702.G577 (ebook) | DDC 572/.56--dc23

LC record available at https://lccn.loc.gov/2021002967
LC ebook record available at https://lccn.loc.gov/2021002968

ISBN: 978-1-77188-997-1 (hbk)
ISBN: 978-1-77463-827-9 (pbk)
ISBN: 978-1-00314-539-4 (ebk)

# About the Editors

---

**Adeel Malik, PhD**
*Institute of Intelligence Informatics Technology, Sangmyung University, Seoul, South Korea*

Adeel Malik (PhD) is working as a Research Professor at the Institute of Intelligence Informatics Technology, Sangmyung University, Seoul, South Korea. He obtained his PhD (2009) from the Dept. of Biosciences, Jamia Millia Islamia (JMI), New Delhi, India. During his PhD work, he developed computational methods for the prediction of carbohydrate-binding sites in proteins using sequence and evolutionary information. This was the first sequence-based method to predict the carbohydrate-binding sites in proteins. Later he joined the School of Computational Sciences, Korea Institute for Advanced Study (KIAS), Seoul, South Korea, to pursue his postdoctoral fellowship. As a part of his research, he investigated plant lectin-carbohydrate interactions via community-based network analysis by using glycan array data. He also worked as an Assistant Professor at the Perdana University Center for Bioinformatics (PU-CBi), Malaysia, from 2014–2016. Dr. Adeel has broad research interests in the area of bioinformatics that include protein-carbohydrate interactions, machine learning, NGS, etc. He also has to his credit several research publications in peer-reviewed international journals as well as book chapters and edited books with international publishers.

**Tanveer A. Dar, PhD**
*Clinical Biochemistry, University of Kashmir, Srinagar, Jammu and Kashmir, India*

Tanveer A. Dar, PhD, is an Assistant Professor in the Department of Clinical Biochemistry, University of Kashmir, India. Dr. Dar received his master's degree in Biochemistry from Hamdard University, New Delhi, India, and a PhD in Protein Biophysics from Jamia Millia Islamia, New Delhi, India. After completing his doctorate, Dr. Dar joined as a postdoctorate at the University of Montana, Missoula, USA, where his main research area involved characterization of protein denatured states.

Dr. Dar's research interests mainly include (i) structural and functional characterization of glycosylated therapeutic proteins from medicinal plants and (ii) role of chemical chaperones/osmolytes in modulating fibrillation/aggregation of proteins. He has authored more than 40 publications in both national and international journals of repute in the field of protein biophysics and proteomics and is also a recipient of an INSA visiting fellowship from the Indian National Science Academy (INSA), India. In addition to this, Dr. Dar has co-authored a number of edited volume books published by internationally reputed publishers, including Springer and Elsevier.

# Contents

# Contributors

**Fasil Ali**
Clinical Biochemistry, University of Kashmir, Hazratbal, Srinagar – 190006, Jammu and Kashmir, India

**Mumtaz Anwar**
Department of Pharmacology, University of Illinois at Chicago, Chicago – 60612, Illinois, USA,
Phone: +1 (312)-877-4320, E-mails: mumtazan@uic.edu; mumtaz_anwar1985@yahoo.co.in

**Ayyagari Archana**
Department of Microbiology, Swami Shraddhanand College, University of Delhi, New Delhi – 110 036,
India

**Sunil K. Arora**
Professor, Department of Immunopathology, Postgraduate Institute of Medical Education and Research
(PGIMER), Chandigarh (UT) – 160012, India

**Jahangir A. Dar**
Division of Plant Biotechnology, Sher-e-Kashmir University of Agriculture Sciences and Technology of
Kashmir, Shalimar, Srinagar, Jammu and Kashmir, India

**Parvaiz A. Dar**
Clinical Biochemistry, University of Kashmir, Hazratbal, Srinagar – 190006, Jammu and Kashmir, India

**Tanveer Ali Dar**
Clinical Biochemistry, University of Kashmir, Hazratbal, Srinagar – 190006, Jammu and Kashmir,
India, Phone: 91-9419639396, E-mail: tanveerali@kashmiruniversity.ac.in

**Durgashree Dutta**
Department of Biochemistry, Jan Nayak Chaudhary Devilal Dental College, Sirsa, Haryana, India

**Muzafar Jan**
Assistant Professor, Department of Biochemistry GDC Dooru, University of Kashmir, Srinagar,
Jammu and Kashmir (UT) – 192211, India, E-mail: muzijan@gmail.com

**Raja Amir H. Kuchay**
Department of Biotechnology, Baba Ghulam Shah Badshah University, Rajouri, Jammu and Kashmir,
India, E-mail: kuchay_bgsbu@yahoo.com

**Abhai Kumar**
Interdisciplinary School of Life Sciences, Institute of Science, Banaras Hindu University, Varanasi,
Uttar Pradesh, India, E-mail: singhabhai2000@gmail.com

**Reetika Mahajan**
School of Biotechnology, Sher-e-Kashmir University of Agricultural Sciences and Technology of
Jammu, Chatha, Jammu and Kashmir, India

**Adeel Malik**
Institute of Intelligence Informatics Technology, Sangmyung University, Seoul, South Korea,
E-mails: adeel@procarb.org

**Usma Manzoor**
Clinical Biochemistry, University of Kashmir, Hazratbal, Srinagar – 190006, Jammu and Kashmir, India

**Rinki Minakshi**
Department of Microbiology, Swami Shraddhanand College, University of Delhi,
New Delhi – 110 036, India; D-3, 3502, Vasant Kunj, New Delhi – 110070, India,
E-mails: rinki.minakshi@hotmail.com; minakshi4050@gmail.com

**Yaser Rafiq Mir**
Department of Biotechnology, Baba Ghulam Shah Badshah University, Rajouri, Jammu and Kashmir,
India

**Abhay Mishra**
Department of Biotechnology, Khandelwal College of Management Science and Technology,
NH 74, Bareilly, Uttar Pradesh – 243122, India

**Zeba Mueed**
Department of Biotechnology, Invertis Institute of Engineering and Technology, Invertis University,
Bareilly, Uttar Pradesh – 243123, India

**Shazia Mukhtar**
School of Biotechnology, Sher-e-Kashmir University of Agricultural Sciences and Technology of
Jammu, Chatha, Jammu and Kashmir, India

**Muntazir Mushtaq**
School of Biotechnology, Sher-e-Kashmir University of Agricultural Sciences and Technology of
Jammu, Chatha, Jammu and Kashmir, India

**Muslima Nazir**
Division of Plant Biotechnology, Sher-e-Kashmir University of Agriculture Sciences and Technology of
Kashmir, Shalimar, Srinagar, Jammu and Kashmir, India

**Jasdeep Chatrath Padaria**
National Research Center on Plant Biotechnology, Indian Agricultural Research Institute,
New Delhi, India

**Nitesh Kumar Poddar**
Department of Biosciences, Manipal University Jaipur, Jaipur-Ajmer Express Highway,
Dehmi Kalan, Near GVK Toll Plaza, Jaipur, Rajasthan – 303007, India,
E-mail: niteshkumar.poddar@jaipur.manipal.edu

**Abid Qureshi**
Biomedical Informatics Centre, Sher-i-Kashmir Institute of Medical Sciences (SKIMS), Srinagar,
Jammu and Kashmir, India, E-mail: abidbioinf@gmail.com

**Safikur Rahman**
Department of Medical Biotechnology, Yeungnam University, Gyeongsan – 712-749, South Korea

**Ritika Rajpoot**
Department of Crop Physiology, University of Agricultural Sciences, Bangalore – 560056, Karnataka,
India, E-mail: ritikaa15@gmail.com

**Susheel Sharma**
School of Biotechnology, Sher-e-Kashmir University of Agricultural Sciences and Technology of
Jammu, Chatha, Jammu and Kashmir, India

**Manali Singh**
Department of Biotechnology, Invertis Institute of Engineering and Technology, Invertis University,
Bareilly, Uttar Pradesh – 243123, India

**Smita Singh**
Center of Advance Studies in Botany, Institute of Science, Banaras Hindu University, Varanasi, Uttar Pradesh, India

**Shafiq A. Wani**
Division of Plant Biotechnology, Sher-e-Kashmir University of Agriculture Sciences and Technology of Kashmir, Shalimar, Srinagar, Jammu and Kashmir, India

**Snober Shabeer Wani**
Clinical Biochemistry, University of Kashmir, Hazratbal, Srinagar – 190006, Jammu and Kashmir, India

**Sajad Majeed Zargar**
Division of Plant Biotechnology, Sher-e-Kashmir University of Agriculture Sciences and Technology of Kashmir, Shalimar, Srinagar, Jammu and Kashmir, India, E-mail: smzargar@gmail.com

# Abbreviations

| | |
|---|---|
| 2-AB | 2-aminobenzamide |
| ABA | abscisic acid |
| AD | Alzheimer's disease |
| ADCC | antibody-dependent cellular cytotoxicity |
| AFEX | ammonia fiber expansion |
| AFM | atomic force microscopy |
| AGIs | alpha-glucosidase inhibitors |
| AGPase | ADP-glucose pyrophosphorylase |
| AHP | alkaline hydrogen peroxide |
| AIDS | acquired immunodeficiency syndrome |
| AIR | alcohol-insoluble residues |
| ALL | acute lymphoblastic leukemia |
| aM-II | a-mannosidase II |
| AP | aminopyridine |
| APCs | antigen presenting cells |
| ASF | asialofetuin |
| AuNP | gold nanoparticles |
| Aβ | amyloid beta |
| B4GALT1 | β1,4-galactosyltransferase I |
| BCR | B-cell receptor |
| BSA | bovine serum albumin |
| CAD1 | cinnamyl alcohol dehydrogenase 1 |
| CBDs | carbohydrate binding domains |
| CCMS | cerebrocostomandibular |
| CD4 | constant domain 4 |
| CDG | congenital disorders of glycosylation |
| CE | capillary electrophoresis |
| CLR | C-type lectin receptor |
| CMV | cytomegalovirus |
| CNS | central nervous system |
| COG7 | component of Golgi complex 7 |
| CoMPP | comprehensive microarray polymer profiling |
| ConA | concanavalin A |
| CRDs | carbohydrate recognition domains |

| | |
|---|---|
| CRP | C-reactive protein |
| CS | chondroitin sulfate |
| CSP | clustered saccharide patches |
| CSS | collision cross section |
| CVD | cardio vascular disease |
| CZE | capillary zone electrophoresis |
| DCs | dendritic cells |
| DGC | dystrophin-glycoprotein complex |
| DGD | digalactosyldiacylglycerol |
| DLB | dementia with Lewy bodies |
| DNA | deoxyribonucleic acid |
| Dol | dolichol |
| DS | dermatan sulfate |
| EAE | encephalomyelitis |
| ECM | cell-extracellular matrix |
| EDS | Ehlers-Danlos syndrome |
| EGF | growth factor receptor |
| EKC | epidemic keratoconjuctivitis |
| ELISA | enzyme linked immunosorbent assay |
| ELLA | enzyme linked lectin sorbent assay |
| EPN | Glu-Pro-Asn |
| EPSs | extracellular polymeric substances |
| ER | endoplasmic reticulum |
| ERGIC | ER-Golgi intermediate compartment |
| ESI | electron spray ionization |
| ESI-MS | electrospray ionization-mass spectrometry |
| ESR | erythrocyte sedimentation rate |
| F | fusion |
| FBGs | fetal brain glycoproteins |
| FCEP | fluorescence capillary electrophoresis |
| FDG | 2-fluorodeoxy-D-glucose |
| FNGs | free N-glycans |
| FRKs | fructokinases |
| FT | fructosyltransferase |
| FTICR | Fourier transform ion cyclotron resonance |
| FT-ICR-MS | Fourier transform ion cyclotron mass spectrometry |
| Fuc | fucose |
| GA | Golgi apparatus |
| GAA | $\alpha$-glucosidase acid |

| | |
|---|---|
| GAGs | glycosaminoglycans |
| Gal | galactose |
| GalNAc | galactose N-acetylgalactosamins |
| GALT | gut associated lymphoid tissue |
| GBPs | glycan binding proteins |
| GC | β-glucocerebrosidase |
| GD | Gaucher disease |
| GIPLs | glycosylinositolphospholipids |
| GlcNAc | N-acetyl-glucosamine |
| GnT-III | N-acetyl-glucosaminyltransferase III |
| GnT-V | N-acetyl-glucosaminyltransferase V |
| GolS | galactinol synthase |
| GPC | gel permeation chromatography |
| GPIs | glycosylphosphatidylinositols |
| GSDs | glycogen storage diseases |
| GTs | glycosyltransferases |
| $H_2O_2$ | hydrogen peroxide |
| HA | hemagglutinin |
| HAS | human serum albumin |
| HBGA | histo-blood group antigens |
| HCT | hydroxycinnamoyl CoA: shikimatehydroxycinnamoyl transferase |
| HCV | hepatitis C virus |
| HD | Huntington disease |
| HDACs | histone deacetylase |
| HHV | human herpesvirus |
| Hib | haemophilus influenzae type b |
| HILIC | hydrophilic liquid interaction chromatography |
| HIV | human immunodeficiency virus |
| HPAEC | high-performance anion-exchange chromatography |
| HS | heparan sulfate |
| HSPGs | heparin sulphate proteoglycans |
| HSV | herpes simplex virus |
| HXKs | hexokinases |
| IAPP | islet amyloid polypeptide |
| IgG | immunoglobulin G |
| IM-MS | ion mobility mass spectrometry |
| InfCHEF | InfC hemagglutinin-esterase-fusion protein |

| INLIGHT | individuality normalization when labeling with isotopic glycan hydrazide tags |
| KLH | keyhole limpet hemocyanin |
| KS | keratan sulfate |
| LAD II | leukocyte adhesion deficiency type II |
| LC | liquid chromatography |
| LCs | Langerhans cells |
| LDL | low-density lipoprotein |
| LLO | lipid-linked oligosaccharide |
| LMWHs | low-molecular weight heparins |
| LOS | lipooligosaccharides |
| LPF | long polar fimbriae |
| LPG | lipophosphoglycan |
| LPS | lipopolysaccharide |
| LSDs | lysosomal storage disorders |
| MAC | membrane attack complex |
| MAG | myelin-associated glycoprotein |
| MALDI-MS | matrix-assisted laser desorption ionization-mass spectrometry |
| MBC | minimal bactericidal concentration |
| MBL | mannose binding lectin |
| MBPs | mannose binding proteins |
| MCE | multiple cartilaginous exostoses |
| MDDG | muscular dystrophy-dystroglycanopathies |
| MEKC | micellar electro kinetic chromatography |
| MHC | major histocompatibility complexes |
| miRNA | microRNA |
| MLV | murine leukemia virus |
| MMP | matrix metalloproteinases |
| MPLA | monophosphoryl lipid A |
| MRT7 | mental retardation-7 |
| MS | mass spectrometry |
| N | neuraminidase |
| N | nitrogen |
| NA | neuraminidase |
| NCAM | neural cell adhesion molecule |
| NeuGc | N-glycosylneuraminic acid |
| NGF | nerve growth factor |
| NGL | neoglycolipid |

| NK | natural killer |
|---|---|
| NLGs | N-linked glycans |
| NMR | nuclear magnetic resonance |
| NP | normal-phase |
| NSAID | non-steroid anti-inflammatory drugs |
| NST | nucleotide sugar transporter |
| *NTHi* | nontypeable *haemophilus influenzae* |
| $O_2$ | oxygen |
| OAT | O-acetyltransferase |
| OGA | O-linked N acetyl glucosamine |
| OPP | oxidative pentose phosphate |
| O-SP | O-specific polysaccharide |
| OST | oligosaccharyl transferase |
| OVA | ovalbumin |
| PA-IL | *P. aeruginosa* lectin-I |
| PAL | phenylalanine ammonia-lyase |
| PAMP | pathogen associated molecular pattern |
| PD | Parkinson's disease |
| PEF | plasmid-code fimbriae |
| PET | positron emission tomography |
| PFKL | phosphofructokinase liver type |
| PFKP | phosphofructokinase platelet type |
| PGC | porous graphitized carbon |
| PMM | phosphomannomutase |
| PMP | 1-phenyl-3-methyl-5-pyrazolone |
| PNGaseF | peptide -N-glycosidase F |
| PRRs | pattern recognition receptors |
| PSA | prostate-specific cancer |
| PSGL-1 | P-selectin glycoprotein ligand-1 |
| QPD | Gln-Pro-Asp |
| QTOF | quad time of flight |
| RA | rheumatoid arthritis |
| Rb | retinoblastoma |
| RFOs | raffinose family oligosaccharides |
| RNA | ribonucleic acid |
| ROS | reactive oxygen species |
| RP | reversed-phase |
| S | serine |
| S/G | syringyl/guayacyl |

| | |
|---|---|
| SEM | scanning electron microscopy |
| SF2a | *Shigella flexneri* 2a |
| SIGLEC-2 | Sia-binding immunoglobulin type lectin CD22 |
| SILAC | stable isotope labeling by amino acids in cell culture |
| SLE | systemic lupus erythematosus |
| SLe$^a$ | Sialyl Lewis A |
| SLe$^x$ | Sialyl-Lewis X |
| SOAT | sialate-O-acetyltransferases |
| SPR | surface Plasmon resonance |
| SSEA3 | stage-specific embryonic antigen 3 |
| STs | sialyltranferases |
| T | threonine |
| T6P | trehalose-6-phosphate |
| TACAs | tumor-associated carbohydrate antigens |
| TBE | tick-borne encephalitis |
| TCR | T-cell receptor |
| TFag | Thomsen-Friedenreich antigen |
| T$_H$ | T helper |
| TIMP | tissue inhibitors of metalloproteinases |
| TLRs | toll like receptors |
| TNF | tumor necrosis factor |
| TNFR1 | tumor necrosis factor receptor 1 |
| TNFα | tumor necrosis factor |
| TOF-MS | time-of-flight mass spectrometry |
| TPS | trehalose-6-phosphate synthase |
| TTox | tetanus |
| TTR | transthyretin |
| UPEC | Uropathogenic *Escherichia coli* |
| UTI | urinary tract infection |
| UV | ultraviolet |
| VEGF | vascular endothelial growth factor |
| VZV | varicella-zoster virus |
| WES | whole exome sequencing |
| WGS | whole genome sequencing |
| ZPSs | zwitterionic polysaccharides |

# Preface

The subject of this book, *The Glycome: Understanding the Diversity and Complexity of Glycobiology*, comprehensively deals with most of the aspects of glycome, such as identification of the glycan identity enigma, its role in maintaining cellular process and disease pathology. This book also discusses the challenges and advances in glycobiology in a quest to provide insight for minimizing the knowledge gap in glycan biology.

The word "glycome" is an analogous term to genome and proteome that represents the complete set of glycans (alternately known as carbohydrates, sugars, or saccharides) generated within an organism/tissue/cells. With being very wide and complex, the cell's glycome is thought to averaging in the range of 100,000–500,000 glycans in size. This large diversity of glycome has been attributed to: (i) large extent of variation in monomer structure of the sugar components (linear or branched form); (ii) inter-linking of saccharide units (different bond types and binding); and (iii) variation in glycan attachment sites.

The process by which the assemblage of monosaccharide units to glycans occurs is named glycosylation. Unlike proteins and nucleic acids, glycans are generated by highly complex interactions between enzymes, ion channels, and other proteins. Glycans are also found attached covalently to proteins (glycoproteins) and lipids (glycolipids) that drive key biological processes such as cell-cell interaction, protein folding, cell signaling, and many others. Moreover, various glycan polymers serve as essential energy storage molecules.

All these diverse glycans play an important role in the majority of the fundamental life-supporting functions of the organisms. Not only this, microorganisms act as if bacteria and viruses on infecting cells detect explicit glycans on specific host cell membranes. In response to this, the immune system of the infected individual/cell produces antibodies against these invading pathogens on the basis of glycans present on these microbes. Additionally, glycans are reported to be not only directly involved in the pathophysiology of almost all key diseases but are of substantial significance in personalized medicine as well as the pharmaceutical industry. Furthermore, these glycans in the form of glycocalyx (a dense layer of glycans attached to membrane proteins and lipids) act as a fingerprint for a cell in distinguishing between self and non-self. In fact, due to the huge diversity in glycan structures, these carbohydrate chains may encompass crucial information for regulating various biological processes. Furthermore, due to the fact that glycans are involved in the majority of the biological processes, defects in glycan synthesis is nowadays recognized as a causal agent of a number of diseases. In fact, glycans are promoted as promising disease markers and, thus, potential therapeutic targets.

Despite glycomes being important components of life, studies dealing with glycobiology are much further behind than those of other major cellular biomolecules, which has been largely argued to be due to the inherent structural complexity of glycans, difficult sequence determination, and the unavailability of a genetic template for predicting their biosynthesis. This "unexplainable micro-heterogeneity/complexity," i.e., the enigma, raised a number of questions about the importance of substituting so many diverse saccharide units to a single polypeptide chain. In an attempt to expound on this enigma, a number of advanced technologies have been developed for exploring the glycan complexity leading to the birth of a new molecular biology frontier called glycobiology. For example, carbohydrate microarrays have been used to investigate the probable glycan markers that are overexpressed in breast cancers. Similarly, chromatographic and mass spectrometry (MS) techniques are presently the utmost suitable methods for high-throughput analysis of glycans. In spite of these advances, limitations in the sequencing methods for carbohydrates have stalled the decryption of complete human glycome, and due to this, the enigma still exists there. In fact, understanding this enigma of glycome characterization and its interactions with carbohydrate-binding proteins still remains a major challenge. Keeping in view the importance and the enigmatic nature of the glycan identity, a thorough insight is important for beginners as well as experts in the glycobiology field.

Therefore, the purpose of compiling this edited volume is to systematically cover major aspects of the glycome vis-à-vis the significance of substituting the diverse monosaccharide units to glycoproteins, the role of glycans in disease pathologies, and the challenges and advances in glycobiology in a quest to minimize the knowledge gap in biology. In fact, despite being important cellular processes, we could not find any widespread collection in the literature regarding the cellular glycome with respect to its characterization and the significance of glycans as future therapeutic agents.

The present book is comprised of 10 chapters organized with the intention of providing an introduction to trends and developments in glycobiology, then discussing their role in disease pathologies followed by their role in other organisms including bacteria, plants, and their role as stress busters. Chapter 1 includes the introduction about glycobiology with a detailed overview of recent trends and developments in the field. Chapter 2 describes the involvement of glycome in congenital and non-congenital disorders. Chapter 3 discusses about the role of glycans in immunological processes, while Chapters 4 and 5 introduce the role of glycans in neurodegeneration and metastasis, emphasizing the concept of glycan remodeling. Chapters 6 and 7 describe the involvement of glycans in host-pathogen interaction and their role in microbial infection and immune evasion. Chapter 8 provides insights and therapeutic aspects of acylation as a vital glycosylation modification of proteins. Chapter 9 highlights the scenario of glycans in plants, while Chapter 10 discusses glycans as stress coping agents.

With developing knowledge about the role of glycobiology in biological processes and the technical instrumental advancements for characterizing the glycome, increased attention has been already generated with respect to characterizing the glycan diversity. *The Glycome: Understanding the Diversity and Complexity of Glycobiology* intends to sensitize both students and the scientific community to the significance and huge encoding power of glycans vis-à-vis providing helpful insights in framing a bigger picture of the cellular glycome. Finally, the book will help readers to keep themselves abreast with state-of-the-art developments and emerging challenges of glycome biology, which are going to be the key areas of future research not only in the glycobiology field but also in pharmaceutics.

# CHAPTER 1

# Trends and Advancements in Glycobiology: Towards Development of Glycan-Based Therapeutics

YASER RAFIQ MIR and RAJA AMIR H. KUCHAY

*Department of Biotechnology, Baba Ghulam Shah Badshah University, Rajouri, Jammu and Kashmir, India,*
*E-mail: kuchay_bgsbu@yahoo.com (R. A. H. Kuchay)*

## ABSTRACT

Carbohydrates, often named glycans, represent the most diverse class of biomolecules with vast structural diversity to form complex branched structures compared to other fundamental biomolecules like proteins and nucleic acids. Glycan diversity ranges from being linear or branched with $\alpha$ or $\beta$ linkages between the monosaccharide units. Due to their structural diversity, glycans perform various biological functions quite distinct from that of proteins and nucleic acids, which mainly involve cellular interactions, cell recognition, various cellular and immunological processes important for the development and functioning of complex multicellular organisms. The process of glycosylation enhances and diversifies the structural and functional ability of sugars. Advances in glycobiology techniques have established a critical role of glycosylation in health and diseases with a special focus on congenital disorders of glycosylation (CDG) occurring due to faulty N- and O-linked glycosylation of proteins. With the advent of new techniques such as glycomimetics, glycan arrays, glycoengineering of metabolic intermediates and glycan therapy, glycobiologists have successfully carried out glycan-based clinical trials for various diseases. The inefficiency of carbohydrates due to low immune response (as they are T-cell independent antigens) has been overcome by use of conjugation technology in which different types of carrier peptide molecules like KLH,

BSA, TT, CRM197 and gold nanoparticles (AuNP) have been attached with carbohydrate antigens to elicit a strong T-cell response and presentation to MHC molecule. Various infectious diseases like influenza, pneumonia, and cancer have been targeted with hopeful human trials by the development of multivalent conjugate vaccines. Glycomimetic molecules, which mimic the carbohydrate skeletons, are quite useful in understanding the role of carbohydrates and other intermediates in biological processes. The development of glycan microarrays, especially lectin arrays, has paved the way for glycome profiling to elucidate carbohydrate-protein, carbohydrate ligand interactions involved in endogenous receptor system host-parasite interactions. Advancements in mass spectrometry (MS)-based techniques and bioinformatic analysis have enabled researchers to investigate the complex cellular glycoproteome to gain a deeper insight into the cellular and physiological activities corresponding to a disease. Owing to these technological advancements, glycobiology is fastly emerging as a primary field of interest for biomolecular and biomedical research with sugars now being widely recognized as vital components of cellular life. The present chapter was designed to provide latest glycobiology trends with a focus on carbohydrate-based therapeutics, glycan arrays, glycomimetics, and advances in MS techniques for characterization of glycans.

## 1.1  INTRODUCTION

Glycobiology deals with the study of the structure, function, biosynthesis, and properties of a category of biomolecules called carbohydrates, often called glycans. On the basis of size, glycans do range from monosaccharides to polysaccharides. Glycans stand at disparity with the linear structure of nucleic acids (DNA, RNA) and protein as they have a notable ability to form complex branched-chain structures or molecules. In other words, as compared to the four fundamental molecules of life, glycans represent a class of biomolecules with vast structural diversity. One of the major reasons for this diversity has been attributed to the large heterogeneity of monosaccharide units [1]. Carbohydrates play a leading role in convocation of organisms and complex multicellular organs, which influence each other through cells and adjoining matrix. Every cell and most of macromolecules exhibit an arrangement of covalently linked sugar moieties (monosaccharides or oligosaccharides) known as "glycans." These glycans carry out crucial activities in the development and function of complex multicellular organisms through their interplay in cell-cell interaction, cell, matrix, and organisms, including host-parasite or pathogen [2]. Additionally, glycans do also play crucial role in majority of the cellular processes and act as their regulatory switches.

Glycans, being regulatory in nature, are attached to the proteins by the process of glycosylation, an important post-translational modification of proteins occurring in the Golgi apparatus (GA) and endoplasmic reticulum (ER). Post-translational modifications involve covalent attachment of glycans to macromolecules like proteins and lipids leading to the formation of glycoproteins and glycolipids, respectively [2, 3]. Glycosylation may be either $N$-linked (linked to proteins via asparagine) or $O$-linked (linked to proteins via serine or threonine). Due to the critical role of glycosylation in health and diseases, this field has attracted significant attention in recent times by the scientific community as well as an industry [4–7]. A fast-growing group of genetic diseases referred to as congenital disorders of glycosylation (CDG) occurs due to faulty N- and O-linked proteins modification. Most of the CDGs are severe multisystem disorders with prominent involvement of nervous system. With the help of serum transferrin isoelectric focusing, about 76 CDGs have been identified. Whole exome and genome sequencing (WES/WGS) has also been employed to carry out diagnostic procedure of patients with some uncharacterized CDGs [8].

Analogous to the other terms like genome, transcriptome, and proteome, the complete repertoire of glycans synthesized by an organism is referred as 'glycome.' Complete functional glycans are more dynamic and complex in structure to the extent that still it has to be fully defined as their composition can't be predicted like that of nucleic acids or proteins [2]. Glycans are secondary gene products as very few genes in human genome code for transporters and enzymes crucial for biosynthesis and assembly of these glycans [9]. DNA and RNA based research in molecular biology follows the chief paradigm that these are information repositories for a cell and information flows in a sequence dependent manner coupled with editing of one molecule based on information contained in another one. However, with the discovery of more than 700 genes involved in glycan related processes and their increasing role in diseases, protein diversity, and immune trafficking, many other biological processes have opened a new frontier in the field of molecular biology research [10–12].

In addition to the presence of carbohydrates as glycoproteins and glyco-lipids in all living forms, these also exist as organic matter on earth, mainly as cellulose in plants and chitin in fungi and arthropods [13]. Glycome pool can differ depending on type of species [14]. In metazoan animals, monosaccharides make up glycome stores whereas in bacteria and archaea, many species-specific carbohydrates are present and some of them are still to be discovered [15]. Glycans show a great diversity in structure and number due to presence of two possible anomers and multiple attachment points which in turn forms a molecular basis of the cellular recognition system [13, 16]. Additional diversity is generated by process of glycosylation which gives structural variations to a given protein and thus produces different glycoforms of the same protein. In this way glycans constitute the "sugar code" of a cell which determines pathological and physiological conditions of a cell [17–19].

## 1.2   ADVANCES IN GLYCAN BASED THERAPEUTICS

One of the important frontiers in glycobiology research is the quest to decipher the glycome structure and function. With the advent of new fields in glycobiology such as glycomimetics, glycan array, glycoengineering, and glycan therapy, the glycan biology has been revolutionized as a promising research area to solve the mystery of CDG disorders. Furthermore, advancements in mass spectrometry (MS) based techniques such as ESI-LC/TOF, MALDI-TOF, MALDI-TOF/TOF, Qq-FTMS, ESI-Ion Trap, Orbitrap Hybrid MS, Fleet Ion Trap LC/MS, Chip based MS, ion mobility-mass spectrometry, gas-phase spectroscopy and ESI-Q-q-TOF systems and bioinformatic data analysis approaches have enabled researchers to investigate the complex cellular glycoproteome so as to gain a better understanding of the cellular and physiological activities corresponding to a disease and to expedite the vigorous attempts focused on recognition of biomarkers specific for a particular disease. Increased attention has been diverted towards the use of glycans for therapeutics as discussed below.

### 1.2.1   GLYCAN BASED THERAPEUTICS

It has been observed that any error in glycosylation leads to a number of abnormalities *in vivo* which in turn demonstrates that glycans contain biological information related to diseases. Functional understanding of glycans is of utmost importance for the establishment of new therapeutic techniques and drugs for the prevention or treatment of these diseases. Prominent role of glycans in signaling and cellular recognition makes them highly desired candidates for diagnostic and therapeutic agents in medical fields like, immunity, oncology, inflammation, infection, and neurodegeneration. Carbohydrate-protein interactions are turning up as central elements for treatment of immune and inflammatory diseases, e.g., interaction between carbohydrates and lectins to form siglecs, selectins, and galectins [20]. Many glycan-based drugs are currently available in market and have been successfully able to ameliorate disease symptoms. Topiramate (Topamax) has been used for epilepsy and also as an anticonvulsant [21]. Dosmalfate has been used in gastric ulceration. Hyaluronic acid (Orthovisc) has been used as viscoelastic supplement and Drotrecogin alfa (Xigris) for treatment of sepsis [21]. Few other examples include Acarbose, Glucobay, Precose, and Prandase for diabetes (Type I and II); GG-167, Zanamivir, and Relenza for influenza, antiviral activity; Dalteparin sodium (Fragmin) in thrombosis as anti-coagulant; GD0039 (Swainsonine) for renal, colorectal, breast cancer and Celgosivir (MDL 28574, DRG-0202, BuCast) for HIV/AIDS [21].

### 1.2.2   NATIVE GLYCAN SKELETONS FOR DESIGN OF THERAPEUTICS

Clinical significance of glycans in therapeutics started with the discovery of blood groups in human beings. Advancements have been brought into use of drugs such as heparin and hyaluronic acid, whose development for therapeutics had been restricted due to synthetic limitation [22, 23]. Recent developments to overcome the difficulties associated with protection of polyhydroxyl groups of monosaccharides and challenges faced with stereo-selective assembly of glycosidic linkages in the process of chemical synthesis of carbohydrates have been initiated [24]. Utility of one-pot multistep protecting-group manipulation technique has eliminated the need for traditional methods which dealt with complex process of extracting and purifying carbohydrate intermediates naturally and were time consuming [24]. D-glucose was used as starting material and substituted/unsubstituted benzyl ethers were chosen as protective groups which could be cleaved to mediate glycosidic bond formation at second carbon atom (C2) of the glycozyl contributor by virtue of the anomeric effect [24]. Due to the one-pot multistep method hundreds of monosaccharides were synthesized and finally assembled into $\beta$-1, 6-glucans. These products have been invaluable in the study of structure, function, antibody specificity and other carbohydrate binding proteins such as lectins. Owing to this method, synthesis of H5N1 avian influenza virus receptor, a trisaccharide unit (SAa2, 6Galb1, 4GlcNHAc), in humans for haemagglutinin binding at the start of viral infection was made possible [24].

Another notable target in carbohydrate therapeutics is to develop a rational blueprint of powerful inhibitors for lectins involved in progression of diseases at various stages such as selectins, adhesins, DC-SIGN, and galectins. Prominent role of selectins in exudating cells from blood vessels during inflammation is not hidden. Diseases like cancer, sickle cell anemia, and inflammation have been targeted by mimicking of interaction between selectin and ligand by Sialyl Lewis x (sLex) [25]. A success in this field has been achieved by reversing vaso-occlusion in mice using the glycomimetics drug GMI-1070 (rivipansel), a powerful pan-selectin inhibitor [26, 27]. Since 2009, this drug has been tested on various human volunteers of sickle cell anemia with good tolerance and no significant adverse effects [27]. A decrease in levels of many biomarkers of cellular adhesion such as E-selectin and P-selectin has also been reported in the blood of patients under clinical trials [27]. Similarly, GMI-1271, in combination with chemotherapy is also being analyzed in the clinical trial for acute myeloid leukemia [28]. Regardless of these successful carbohydrate therapies in lectin inhibition, low affinity of these monovalent glycans for protein targets has hampered their use as candidate drugs. For this, glycomimetics came into play by development of multivalent glycan structures such as glyconanoparticles, glycoclusters, and glycodendrimers as a trial to increase the whole affinity, by mimicking the multivalent type of the interplay between natural carbohydrate receptors and associated lectins. Multivalency approach has been employed to synthesize a multivalent inhibitor for multi-subunit Shiga-like toxin of *Escherichia coli* [29]. This inhibitor known as Dendron-like decameric system (named STAR-FISH) contained 10 copies of a trisaccharide ligand (Pk) and displayed an extremely powerful inhibitory action as compared to the univalent molecule [29].

### 1.2.3   GLYCANS AS MOLECULAR SCAFFOLDS FOR THERAPEUTIC DESIGN

A scaffold can be defined as a structure which provides support to other molecules in executing their function. As compared to other biomolecules, glycans due to certain characteristics features like low cost, presence of highest density of functional groups than any other naturally occurring substance and effortless availability in pure form serve as prime candidates for scaffold designing. Glycan-based scaffolds have proven very useful in pharmaceutical and medicinal chemistry [30, 31]. Most tedious task in this kind of therapeutic approach is proper modification of such scaffolds with specific functional groups which are involved in interaction with other receptors/proteins. Since early 1990's use of carbohydrates as scaffolds from β-D-glucoside to synthesize peptido-mimetic (small protein-like chain designed to mimic a peptide) targeted to the hormone somatostatin receptors has been reported [32]. Application of most of the peptides with potential therapeutic ability is limited by their low stability and poor oral activity [33]. This drawback of peptide drugs can be ruled out after incorporation of a different structure while keeping the proper direction of the amino acidic substituent [34]. Before this research, a non-peptide peptidomimetic had been elucidated by using a unique scaffold consisting of cyclohexane ring attached to side chain [35]. With the utilization of carbohydrate skeleton as templates for peptidomimetics, various bioactive molecules have been synthesized from pyranosidic form of sugar molecules. Mimics of cyclic peptide endothelin antagonist BQ123 have been designed and made from glucose and allose [36]. Similarly, other pharmacologically important peptides have been made utilizing carbohydrate scaffolds. Libraries of bioactive molecules have been synthesized from both glycan and modified sugar amino acid scaffolds. Researchers have been able to generate at least 99 bioactive molecules from 3-deoxy L-xylose scaffold wherein they high-lighted the importance of furanose saccharide amino acids as stereo diverse building blocks [37].

### 1.2.4   GLYCAN BASED VACCINES

Vaccines induce cell-mediated and/or humoral immune response against disease causing organisms. The first step in the development of a carbohy-drate-based vaccine is recognition of potential target epitope. However, this process is still in its infancy in spite of an early (1923) elucidation of relation between bacterial polysaccharide capsule and pneumococcal serotype [38]. Among others, the main obstacle in development of carbohydrate-based vaccines is the poor response of antibody to them which is mainly on account of carbohydrates being T-cell independent antigens. Thus, using only carbo-hydrates for immunization will not produce an immune response. Further class change from IgM to other Ig subclasses and development of affinity maturation will not occur while using pure carbohydrates. T-cell indepen-dent responses are of short life span and weak in contradiction to CD4[+] T-cell dependent response which produces robust, long lived, and class switched antibodies. One main strategy to overcome this limitation of carbohydrate based vaccines is to use carrier proteins like ovalbumin (OVA) and limpet hemocyanin (KLH) covalently linked with synthetic or purified carbohydrate resulting in CD4+ T-cells activation [39, 40]. Such vaccines consisting of carbohydrates linked to proteins are known as glyconjugates or conjugate vaccines. Capsular polysaccharides of bacteria, being zwitterionic in nature, present an exception as they can be processed and displayed on class II MHC molecules like proteins for generating CD4+ T-cell mediated response [41]. Few examples of glycan based vaccines include: ActHiB, OmniHiB (Haemophilus b) for Influenza type b; Typhoid Vi (Typhim Vi) for typhoid fever; Prevnar (pneumococcal conjugate vaccine) for pneumonia caused by *Streptococcus pneumonia*; GMK (GM2 KLH/QS-21 ganglioside conjugate vaccine) for malignant melanoma; IGN 301 (antiidiotypic antibody) for cancer or tumors; Globo H conjugate vaccine for breast, prostate cancer; *Neisseria meningitidis* A, C, Y, and W-135 glycoconjugate-meningococcal polysaccharide with DT; MGV (GM2/GD2 KLH QS21 melanoma vaccine) for colorectal, gastric, small cell lung cancer and Theratope (sialyl-Tn Ag) conjugate vaccine for metastatic colorectal and breast cancer [42].

### 1.2.5 GLYCAN VACCINES FOR INFECTIOUS DISEASES

With the advent of new glycobiology techniques, immunotherapy has progressed to battle against microbial infectious diseases. Conjugate vaccines for bacterial pathogens like *Salmonella typhi*, type b *Haemophilus influenzae* (Hib), *Neisseria meningitides* and *Streptococcus pneumoniae* after clinical trials are now available for commercial use [43]. Carbohydrate vaccines used on routine basis for treatment of infectious diseases includes *Haemophilus influenzae* type B (Hib vaccine) and *Streptococcus pneumoniae*. Before the use of Hib conjugate vaccine in 1988, *Haemophilus influenzae* accounted for main cause of bacterial meningitis in US children. Hib vaccine was designed by linking fully synthetic capsular polysaccharide from Hib to a carrier protein and has reduced the causative disease by more than 95% [44–46]. Synthesis of carbohydrate-based vaccines targeted to bacterial pathogens has been challenging due to substantial diversity of capsules and their complexity. Conjugate vaccine for *Streptococcus pneumonia* includes only 23 capsules out of more than 90 reported types [47, 48]. This issue has been sorted out by development of multivalent conjugate vaccines. In this method, capsular polysaccharides have been isolated from clinically most pertinent serotype, proceeded by their lysis into fragments of small size for activation and finally integrated with immuno-stimulant carrier, i.e., proteins to design multivalent conjugate vaccines. Similarly, vaccines for shigellosis (bacterial dyzentery), malaria, etc., have been developed.

### 1.2.6   GLYCAN VACCINES FOR ONCOTHERAPY

Cancer is one of the leading causes of mortality and is estimated to account for about 9.6 million deaths in 2018 all over the world [49]. At present available cancer therapies comprise of radiotherapy, surgery, palliative care, chemotherapy, and immunotherapy. Harmful effects have been reported for most of them [50]. Immune surveillance theory put forward in 1950s by Burnet and Thomas suggested that immune system can identify and remove the defective cells arising in body. It does so by production of abnormal proteins/glycoproteins in the form of tumor specific antigens on the surface of tumor cells. The immune system likely encounters and eliminates cancer cells on a daily basis [51]. Among important features of cancer cells is the change in cell surface glycosylation pattern. These carbohydrate structures are known as tumor-associated carbohydrate antigens (TACAs) and arise due to aberrant glycosylation in cancer cells. These TACAs have been regarded as potent targets for development of onco-therapy.

### 1.2.7   TACAS AND THEIR POTENTIAL AS VACCINES FOR ONCO-THERAPY

The development of cancer in a cell is associated with an altered expression level of glycosylated proteins. Globo-H, sLex, Ley, TF GM3, sLea, sTn, and Gb3 are some of new carbohydrate structures which are displayed on the cell membrane of tumor-associated cells [52]. TACAs are considered as "self-antigens" which confer immune tolerance to body. Expression of TACAs in normal tissues is at low level and their fetal descent has hampered use of TACAs for cancer vaccine development [53]. Cancer vaccine designing must work towards removing laxness of immune system towards the glycans displayed on cancer cells. Incorporation of other particles like lipids [54], nanoparticles [55], carrier proteins [56], and virus-like particles [57] can display these TACAs as more non-native. At present, these techniques are being utilized to break immune tolerance so as to stimulate long-lasting immune responses towards tumor-associated carbohydrate antigens. Overexpression of TACAs may be a sign of malignancy as they are believed to be involved in metastasis and signal transduction of tumor cells [58–61]. Difficult and time-consuming process of obtaining TACAs from tumor tissues was replaced by techniques like one-pot synthesis, solid phase synthesis and chemo-enzymatic synthesis which helped in abundant synthesis of pure TACAs [62–68].

TACAs being T-cell independent antigens are not capable to produce sufficient and effective immune response [69] as described earlier in this chapter. Development of effective cancer vaccine involves combination of antigenic surface carbohydrates of tumor cells to various types of carrier proteins or other particles having T-cell established epitopes as they evoke the T-cell response by binding to MHCs. TACAs alone can only provoke a weak B-cell response and produce only IgM antibodies. However, conjugation with peptides produces a stronger response due to T-cell mediated immunity. It also produces IgG antibodies from B-cells by producing antigen-specific signals through cross-linked B-cell receptors (BCRs) accompanied by a stimulatory signal from $T_H$ cells in the form of cytokines [70–75].

When immunized with conjugate vaccines, APCs will cleave them into peptides followed by peptide epitope presentation in complex with MHC class II molecules on the cell surface. Finally, a cascade will initiate starting with activation of T-cells followed by production of cytokines, B-cell proliferation by $T_H$ cells and producing memory B-cells accompanied by IgG antibody production. Memory B-cells will act immediately upon presentation of same antigen. Additionally, adjuvants (e.g., Toll-like Receptors) are also added to vaccine for immuno-response [76–78].

## 1.2.8  ADVANCED TACA BASED CANCER VACCINES

An ideal TACA-based anti-tumor vaccine should be pure, structurally defined and must address heterogeneity of tumor antigens. In addition to this, the vaccine must also carry some immunogenic molecules in order to produce robust T-cell based response. Carrier proteins like toxin protein from Tetanus (TTox), human serum albumin (HSA), bovine serum albumin (BSA) and OVA have been used in designing synthetic tumor vaccines. Keyhole limpet hemocyanin (KLH) is a high-molecular-weight oxygen carrying glycoprotein of marine origin isolated from *Megathura cranulata* (a giant limpet snail). Two keyhole limpet genes KLH1 and KLH2 share 60% of homology at protein level. It is highly immunogenic due to its structure and phylogenetic distance from humans-it can induce both cell-mediated and humoral responses in animals and humans. KLH is considered as one of the desired antigen carrier for anticancer vaccine development due to its low-grade toxicity and high immuno-stimulation in mammals [79]. KLH is one of the best TACA carriers [80]. Another carrier protein for polysaccharides and haptens known as CRM197, a non-toxic mutant of diphtheria toxin has also become a promising conjugated vaccine carrier.

Based on the number and type of TACAs attached to a carrier protein, conjugate vaccines are mainly of three types, monomeric vaccines (contain one type of TACAs), monovalent clustered vaccines (contain cluster of one type of TACAs) and multivalent vaccines (contain several different types of TACAs) [81]. Till now, most of the studies have been on monomeric vaccines. Several clinical trials with KLH conjugated to Globo H antigen for breast cancer have reported beneficial results [82, 83]. In one study, Globo H was synthesized chemically and used in conjugation with separate proteins like BSA, TT, KLH, and CRM197 [83]. In comparison to Globo-H-KLH/QS21 conjugate, Globo H-CRM197/C34 conjugate evoked more IgG antibodies upon injection in mouse [83]. Produced antibodies had much affinity towards Globo H protein and related epitopes, e.g., stage-specific embryonic antigen 3 (SSEA3) and SSEA4 which have been found overexpressed specifically in mammary gland tumors and its stem cells [83].

Monomeric vaccines either synthesized or isolated with antigens like mucin-linked antigens Tn, STn, and TF, gangliosides GD2, GD3, Globo H and GM2 proved incompetent when tested in humans as compared to the animal trial data with good outcome. For example, Lewis Y-KLH evoked low production of antibodies for ovarian cancer in clinical trials [84–95]. Similarly, cancer vaccine comprising of antigen STn conjugated to KLH protein and adjuvant QS-21 displayed no survival benefit in clinical trial of 1000 women with breast cancer [96]. These results strongly suggested development of vaccines with clustered TACAs as carrier proteins so as to mimic the structure of glycan antigens presented by tumor cells [97]. In contrast to monomeric vaccines, production of clustered vaccines by conjugated Tn, TF or STn antigens in triads with threonine or serine residues of KLH protein with QS21 adjuvant evoked an enhanced immune response in terms of IgM and IgG antibody production. These vaccines were tested in prostate cancer patients [86, 97, 98]. In order to counter heterogeneity of TACAs, multivalent vaccines that may improve the immune response have been synthesized [56]. Mixing TACA-KLH conjugates (using antigens like GD3, MUC1, Lewis-Y MUC2) with subsequent addition of adjuvant QS-21 produced "pooled monomeric vaccine." This mixture was introduced concurrently and antibody production was observed for each class of antigens [56].

### 1.2.9  CONJUGATED VACCINES WITH NANOPARTICLES AS CARRIERS

Metal nanoparticles such as gold nanoparticles (AuNP) have been applied as scaffolds for development of multivalent vaccines against cancer cells. Production of immune response depends upon spatial organization or conformation of antigens on carrier particle. It has been observed that B-cells respond greatly to antigens with highly patterned presentation as compared to antigens with poor presentation [99–104]. With the use of protein carriers like KLH, BSA, TT, and CRM197 it has been observed that accurate and manageable presentation of antigens is very cumbersome. Using nanoparticles as scaffolds provide greater flexibility in designing carriers and increasing antigens density compared to proteins [105–107]. AuNP prepared by a reduction in the thiol driven synthetic neo-glycoconjugate atmosphere have been used at priority because of their small size, stability to enzymes, water solubility, easy improvement, and target specificity [108]. Brinas and his coworkers prepared a vaccine using 3–5 nm core gold particle. They used mucin MUC4 with Thomsen Friedenreich antigen (TFag) (a disaccharide expressed on tumor cells) in conjugation with a 28-residue peptide molecule derived from C3d which activated B-cells (molecular adjuvants). Thiol with 33 atoms was used to link AuNP with antigen. Solid-phase glycopeptide synthesis was used to design vaccine which finally was injected into mice using both construct and control (without antigen). Enzyme immunoassay evaluation of sera from immunized mice targeted towards each antigen presented a small useful immune response with production of both IgG and IgM antibodies [55]. Recently, Biswas and coworkers synthesized AuNP covered with layer of TFag-glycoamino acid conjugates through a combined alkane/polyethylene glycol linker. These nanoparticles were then tested in two cell lines, namely Galectin (Gal)-3 positive and other Gal-3 negative, together with a control having incorporated hydroxyl-terminated linker units. Results depicted that the particles having saccharides specifically restricted tumor formation in Galectin-3 containing cells appreciably more than the Galectin-3 lacking cells. Further, it was found that serine-attached conjugates were less effective than threonine attached TF particles. These outcomes favor the utility of AuNP as antitumor therapeutic agents targeted against cell lines that produce and display specific Gal-3 lectins which show activity to TFag [109].

### 1.2.10   ZWITTERIONIC POLYSACCHARIDES (ZPSS) AS CONJUGATE VACCINES

Preferential use of zwitterionic polysaccharides (ZPSs) over peptide molecules presents an advantage in designing therapy vaccines. This is due to the fact that presence of both positive and negative charges on adjoining monosaccharide components in zwitterionic compounds makes them candidate molecules to be handled by APCs and displayed to $T_H$ cells in the form of MHC-II-ZPS complexes. All this in turn provoke IgM to IgG immunoglobulin class switching [110].

Various zwitterionic compounds have been isolated from bacteria such as Sp1, PSA1, PSB, PSA2, and CP5/CP8 from *Salmonella pneumoniae, Bacteroides fragilis,* and *Staphylococcus aureus* [111–118]. Andreana and his group-immunized mice with conjugate of Tn antigen and chemically modified PSA1 zwitterion in presence and absence of an adjuvant resulting in strong immune response being elicited specific to Tn-antigen. It was concluded that modification by addition of a chemical species produced no change in immunogenicity of PSA1. Injection with adjuvant-free conjugate vaccine produced same result which in turn confirmed that PSA1 polysaccharide not only acts as a carrier but also an adjuvant [119, 120].

### 1.2.11   METABOLIC GLYCOENGINEERING APPROACH AS AN ONCOTHERAPY

Engineering of metabolic intermediates or oligosaccharides by their incorporation into TACAs has been found to stimulate production of unique oligosaccharide on the surface of cancer cells. So, this technique has been used to develop carbohydrate vaccines against cancer. Mammalian cells upon intake of N-acylmannosamine from culture synthesize extraordinary N-acyl neuraminic acid which was introduced into cell surface membrane as glycosphingolipids [121]. Qiu and coworkers used antigen GM3 in conjugation with murine leukemia model FBL3. A vigorous T-cell-dependent immune response was observed against antigen. Cancer cells were glycoengineered for *in-vivo* and *in-vitro* production of N-phenylacetyl-D-neuraminic acid (GM3NPhAc) using its biosynthetic precursor N-phenylacetyl-D-mannosamine (ManNPhAc). Furthermore, it was observed that a powerful cytotoxic response was generated against FBL3 cells treated with ManNPhAc, by injection of antisera and antibodies specific to GM3NPhAc.

Upon injection of GM3NPhAc conjugate vaccine, preceded by ManNPhAc therapy, a notable suppression of tumor growth and extended survival of tumor possessing mouse was observed. This study validated the efficacy of glycoengineering as oncotherapy [122]. Integration of GD3 or N-butyryl GD3 (GD3Bu) with KLH and adjuvant MPLA (monophosphoryl lipid A) produced the maximum IgG concentration without any side reaction to unchanged GD3 when introduced into mice. In the presence of complement mAb and GD3Bu antiserum, it lyzed the human skin melanoma (SK-MEL-28 cell line) cells displaying GD3Bu antigen [123]. Multiple isotype antibodies were produced from modified GM3-KLH conjugates indicating production of vital T-cell-dependent antibodies [124]. Glycoengineering-based cancer vaccines, though promising to treat cancer cells, are associated with some side effects as well.

## 1.3   ADVANCES IN GLYCAN ASSOCIATED TECHNIQUES FOR USEFUL THERAPEUTIC OUTCOMES

### 1.3.1   GLYCAN ARRAYS

Deoxyribonucleic acid (DNA) microarray technology for gene expression and the protein microarray for protein-protein interaction analyzes are some of the quite familiar technologies. With the advances in glycobiology, a new type of arrays called carbohydrate arrays or saccharide array, or glycan arrays have been developed for studying glycome. These arrays are made-up of various defined glycans (oligosaccharides and/or polysaccharides) fixed on a solid platform in a spatially defined orientations. In designing saccharide arrays, a fluorescently labeled glycan-binding protein (GBP) contained in a buffer solution is applied to a microarray and the protein-carbohydrate interaction is identified either by fluorescence emitted by GBP labeled with fluorescent molecule or a secondary reagent that binds to the GBP. This novel glycobiology technique is a powerful tool for glycome profiling so as to elucidate carbohydrate-protein, carbohydrate ligand involved in endogenous receptor system and host-parasite interactions [125–127].

### 1.3.2   HISTORY OF DEVELOPMENT

In 1980s, Feizi and coworkers extended the method of chromatography to study other classes of glycans as initially it was used only for glycolipids.

Neoglycolipids (NGLs) were created wherein chemical modification of native glycans was carried out by attaching lipid linkers to its reducing end [128–130]. Glycan microarrays were introduced in 2002 with use of robotics for immobilization of diverse sugar molecules on array platform [128, 129]. With advanced studies in this direction, multiple methods for fixing of glycans to the slide or platform with multiple wells were produced. In fact, a number of other research laboratories continued to design many other methods for fixing of saccharide molecules to array platforms [131]. This progress was possible due to extension of the already developed DNA arrays in 1995, followed by development of recombinant protein microarray [132]. Continuous efforts led to development of elegant procedures for carbohydrate fixation on surface. Lectin based microarray technologies have emerged as powerful tools complementing to MS techniques [133].

### 1.3.3   FABRICATION AND DETECTION OF GLYCAN ARRAYS

Saccharides used in manufacture of glycan arrays are either extracted from naturally occurring sources or chemically synthesized. Sugar moieties are attached to supporting surface by either covalent or non-covalent methods [133]. Robotics has been used for attaching sugar molecules to the solid surface, mostly a microscopic glass slide. Spots of about 200 μm are made by a standard microarray spotter by using 1 mL of carbohydrate solutions. Different concentrations of glycans are used in replica on array which has almost 1000 fixation points. Immobilization is carried out in a humidity-controlled compartment. Immediately after immobilization, slides are incubated in humid compartments for several hours. Unbound saccharides are removed by washing followed by quenching of active functional groups, then again washed, dried, and stored for a long time. Recent developments in detection of binding to carbohydrate microarrays include MALDI-TOF mass spectrometer, surface Plasmon resonance (SPR) fluorescent assay, and nanoparticles assay. Among these, fluorescent-based approach is most reliable and commonly used. In order to study the binding properties of proteins and antibody to glycan arrays, commonly used molecules include rhodamine, fluorescein isothiocynate, and indodicarbocyanine [134, 135]. Secondary antibodies with tagged probes are also used to enhance sensitivity of arrays. Recently, UK glycoarrays consortium developed a gold microarray approach. Apart from detection of binding interactions by traditional fluorescence method, this technique provides dynamic and fluorescent label free analysis of protein communications by SPR. Additionally, this technique has also helped in determining specificities of carbohydrate associated enzyme by using nano chip MS analysis [136, 137].

### 1.3.4   GLYCAN ARRAY STUDIES

Post-translational modification of proteins with attachment of glycan chains generates functional as well as structural diversity of proteins. In post-genomic era this field is gaining attention across the globe particularly due to advent of new glycobiology techniques such as lectin-based arrays for study of carbohydrate and protein interactions. Hsu and his coworkers studied profile interactions between carbohydrate and immune receptor wherein cloning of extracellular part of 17 receptors as Fc-Fusion proteins was carried out to identify their activity with fixed polysaccharides of *Ganoderma lucidum* by using ELISA (enzyme-linked immunosorbent assay). Such study revealed that several receptors (Kupffer cell receptor, DC-SIGN, Langerin, macrophage mannose receptor, Dectin-1, TLR2, and TLR4) linked to innate immunity displayed interaction with polysaccharides from *Ganoderma lucidum*. Furthermore, their study proved use of innate immune receptor-based ELISA as a high throughput profiling approach for identification and quality control of polysaccharides with medicinal values [138]. Development of evanescent-field fluorescence-detection principle allowed studying multiple lectin carbohydrate weak interactions, having $K_d$ (dissociation constant) >10–6 M under controlled conditions, with high sensitivity and real-time detection. In another study, investigators used lectin microarray with five spots (0.25 mg/ml each) and used 0 to 3.2 mg/ml concentration of Cy3 tagged ASF (asialofetuin) probe. Washing and scanning without removing the probe was carried out in order to maintain equilibrium. Increase in signal intensity with increased concentration of probe strongly suggested that signals observed on the glass slide are due to specific lectin affinity (RCA120) to target glycans. In order to study real time observation of interaction of two different lectins, RCA120 and ECA having same carbohydrate specificities for terminal Lac/LacNAc, both lectins were immobilized on the same glass slide added probe Cy3-ASF and scanned continuously. A time dependent increase in signal intensity for both lectins was observed with appearance of saturation within 180 minutes. Data obtained revealed a considerable difference between association constants for ASF between RCA120 and ECA. Approach indicated here can be used to know optimal points for simultaneous analysis of lectin-glycan interactions

[139]. Another group of researchers in 2003 identified a novel anti-cellulose antibody by using a unique glycan array to determine carbohydrate-binding antibodies from a collection of IgG antibodies of a healthy person. They used a glycan based on mono and oligosaccharides immobilized through covalent linkage. Detected antibody showed binding to β4-linked carbohydrates with specificity to glucopyranose as compared to galactopyranose. Results obtained in this study emphasized the role of this glycan array in identification of carbohydrate specific antibodies and proteins which can be important biomarkers for personalized medicine [140].

Manimala and his groups in 2007 observed that cross reactivity of antibodies with other carbohydrates or glycans can lead to false or inaccurate results in biopsy studies as saccharides with side reactions can be wrongly taken as candidate antigen. The group made this claim while studying the antibody preferences to 27 carbohydrate antigens such as Lewis antigens (121SLE, 15C02ZC-18C, 2-25LE, BR55, CA199.02, 7LE, PR.5C5, T218, K21 A70-C/C8, T174, F3, FR4A5), autoantigens A, B, H (B389, 92FR-A2, 81FR2.2, B376, B460, B480, B393, CLCP-19B) and antigens present on tumor cells (5B5, B1.1, B389, 1A4) using glycoarray immobilized with 80 different carbohydrates and glycoproteins, [141]. Researchers have also developed approaches to monitor antibody production against glycan epitopes after cancer vaccination or disease progression. Breast cells show the abundant presence of tumor-associated carbohydrate antigens like Globo H. A group of investigators used three monoclonal antibodies (Mbr1, anti-SSEA-3, and VK-9) to calculate their dissociation constant value by using a self-designed Globo H and its analogous glycan microarray in order to find their binding preferences. Same glycoarray screened anticancer antibodies from blood of normal and breast tumor patients. Breast cancer patient showed higher production of antibodies against Globo H antigen than normal person's blood. In comparison to ELISA this method was found to be efficient and delicate with very less requirement of the materials [142].

Recently Tseng and coworkers have reported development of glycan microarray by using glass slides covered with aluminum oxide (ACG) [143]. These arrays have better spatial arrangement and uniformity as compared to normally used arrays made from glass slides activated with *N*-hydroxy succinimide. To test for this, an array of α-5-pentylphosphonic acid-based mannose derivative as standard was prepared. Fluorescent signal generated from sugar interaction with concanavalin A (ConA), labeled with the fluorescent molecule A488, was used to optimize array. Atomic force microscopy (AFM), ellipsometry, and scanning electron microscopy (SEM) were used to characterize ACG array. Array was also tested for its homogeneity, sugar density, and arrangement using AFM, confocal microscopy and a GenePix scanner [143].

### 1.3.5   GLYCOMIMETICS

Glycomimetics are basically the chemical species that mimic the biological nature of carbohydrates. As discussed earlier, carbohydrates have an important role in fundamental cellular activities and any imbalance can lead to various carbohydrate related disorders. In fact, carbohydrates are susceptible to damage under acidic and basic conditions, degraded by carbohydrate processing enzymes and chemically unstable. All these limitations paved the way for the development of molecules which mimic the carbohydrates and have helped to understand role of carbohydrates and other intermediates in biological processes. In glycomimetics, while developing a mimetic molecule, common structural changes to carbohydrate skeleton are replacement of the ring oxygen or replacement of the glycosidic oxygen, with nitrogen, sulfur or carbon. These substitutions usually help in providing metabolic as well as biological stability to the compounds being designed or produced that mimic transition-state structures from glycohydrolase reactions. Some approaches have also been made to replace entire sialic acid moieties with simple charged functionally active molecules to develop useful biological probes. The role of glycomimetics in carbohydrate therapeutics and vaccine development has been already discussed in this chapter. However, a comprehensive account of carbohydrate mimetic approach with few examples from recent past are presented here.

### 1.3.6   GLYCOMIMETICS IN THERAPEUTICS

Heparanase, an endoglycosidase which cleaves heparan sulfate (HS) and controls reorganization of extracellular matrix has been found to be often over-expressed in tumor cells and can be a potent anticancer drug target. However, drug development has been hampered due to high cost, low efficiency in terms of target binding and complicated manufacturing process. Zubkova and coworkers synthesized a novel class of highly targeted single entity dendrimer glycomimetics molecules covered with simple saccharides like glucose and fructose. Many of these mimetic molecules showed a low $IC_{50}$ (measure of inhibitory concentration required) with a potential to inhibit heparin as compared to other drugs in clinical progress. They were able to provide a novel class of antitumor compounds with diverse therapeutic utility that could take advantage of targeted inhibition of heparin protein in viral pathogenesis, diabetic nephropathy, fibrosis, and inflammatory disorders. Further these compounds have been found to be free of any cytotoxicity and anticoagulant activity [144]. CLEC10A ($Ca^{2+}$-dependent) is an endocytic receptor for $N$-acetylgalactosamine (GalNAc) on antigen-presenting cells. CLEC10A has a crucial part in maturation of dendritic cells (DCs) and eliciting an immune reaction. A group of researchers undertook research in this aspect for answering whether peptidomimetics (peptide that attaches to CLEC10A at GalNAc) can play a role in producing an immune response against ovarian cancers. For this, two tetravalent peptides namely svL4 and sv6D were synthesized from tri-lysine core of GalNAc-specific lectin of a phage. Evaluation of binding specificity of both the peptides to CLEC10A was done under *in-vitro* and *in-silico* conditions. Endotoxin free peptide solutions were injected into mice sub-cutaneously to increase population of immune cells. After injection of peptide in mice cytokine level and reaction of peritoneal immune cells was used to determine biological activity of peptides. It was observed that peptides bind recombinant CLEC10A with high affinity. Monotherapy of synthesized sv6D peptide proved effective in crushing ascites in a cancerous murine ovarian model. Same peptide in combination with Paclitaxel, a chemotherapeutic drug, displayed no discharge of cytotoxic amount of cytokines. It was concluded that peptide svD6 is a CLEC10A functional molecule which prolonged survival of ovarian cancer models with toxicity and non-antigenic profile [145].

Another study for the first time identified mannose and fucose based glycomimetic molecules which can interact with Dectin-2, a C-type lectin receptor (CLR) [146]. Scaffold of cyclohexane embedded with either $\alpha$-mannose or a $\beta$-fucose was used to design glycomimetic molecule. Scaffold was further modified by enriching it with side groups from amide. Process was carried out by designing these glycomimetic molecules in a microarray platform against C-type lectin-like DC-SIGNR, Dectin-2, Langerin, and DC-SIGN [146].

Multidrug antibiotic resistance has posed a serious challenge for treatment of bacterial infectious diseases, demanding investigations to find alternative therapies for controlling bacterial infections. One such example is of Uropathogenic *Escherichia coli* (UPEC) bacteria which uses chaperones and usher proteins for pili synthesis capped with adhesins having specificity to wider range of receptors. This enables the bacteria to reside in various host organisms and other places. Galactose (Gal) or N-acetylgalactosamine epitopes are targeted by UPEC F9 pili on inflamed bladder and kidney. A group of researchers synthesized aryl galactosides and N-acetylgalactosaminosides (29β-NAc) which obstruct growth of the F9 pilus adhesin FmlH. Furthermore, it has been seen that aryl galactosides and N-acetylgalactosaminosides attach firmly to adhesin FmlH receptors by participation of Y46 residues forming salt-bridge interaction between R142 and its carboxylate group, hydrogen bonding between by K132 and its N-acetyl group. Upon injection of mimetic compound 29β-NAc in a mouse model of urinary tract infection (UTI), urinary tract and kidney showed significant reduction in bacterial load. When 29β-NAc along with mannoside 4Z269 (for type 1 pilus adhesin FimH) was co-injected to mice urinary tract (UT), it led to enhanced removal of bacteria from the UT than any of the compounds used alone. These saccharides represent a unique approach for UPEC-mediated UTI treatment [147].

### 1.3.7   CHARACTERIZATION OF GLYCANS BY MASS SPECTROMETRY (MS)

Metabolic, structural, and functional properties of glycans make them candidate molecules in driving cellular machinery. All of these glycan properties are determined by their underlying stereochemistry. In order to elucidate the role of glycans in different biological process their structural analysis is a pre-requisite. Being well known that glycans possess highly branched, non-linear, stereochemically complex structures with different configurations and ring forms, traditional MS based tools have proved to be insufficient in distinguishing them as they exist in closely related isomeric forms. With time, progress in MS based techniques like liquid chromatography-mass spectrometry has increased our understanding of glycans and their interactions with other biomolecules.

### 1.3.8   LIQUID CHROMATOGRAPHY MASS SPECTROMETRY

For elucidating the structural details of glycoconjugates, MS is extensively considered as a method of superiority. MS has been used in combination with other techniques such as LC for characterization of carbohydrate isomers by separation strategy and study of N- and O-linked sialylation of glycoproteins. In fact, MS-based techniques have somehow reduced the ambiguity in results obtained from glycome profiling. A large number of LC conjugated MS techniques like hydrophilic liquid interaction chromatography (HILIC), reversed-phase (RP) chromatography, and porous graphitized carbon (PGC) chromatography have been utilized in structural and functional analysis of biomolecules [148–153]. Gao and coworkers used microfluidic Chip-LC-ESI (electron spray ionization)-QTOF (quad time of flight) MS and MS/MS for characterization of *N*-glycome of rat serum. *O*-acetylated sialic acid-containing 282 *N*-glycans along with some isomers were observed in their study [154]. Out of these, 27 *N*-glycans were found to be novel and also reported a new type of mixed *N*-glycan containing *N*-glycosylneuraminic acid (NeuGc) and *N*-acetylneuraminic acid (NeuAc). Upon comparison of glycomic profiles of humans, rat, and mice, it has been reported that human and rat glycans show considerable resemblance as compared to rat and mice. In fact, due to this similarity in glycomic profiling the use of rats as model organisms for studying human disorders has been demonstrated [154]. Additionally, for generation of *N*-glycome profile of mouse, PGC nano LC/MS/MS technique has been used wherein almost 300 *N*-linked glycans (NLGs) along with isomers have been identified by using 100 distinct *N*-linked mixtures. Analysis of glycome profile to detect structural details facilitated in observation of dehydration, lactylation, and O-acetylation related phenomenon. Moreover, a theoretical library to automate the characterization of glycome profile has been prepared through compilation of data from pre-existing glycome studies [155].

### 1.3.9   *ION MOBILITY AND GAS PHASE MASS SPECTROMETRY (MS)*

Due to the polar nature of carbohydrates and glyco-conjugates, their separation by different LC-MS techniques like HILIC, PGC, and RPC has been found to be very difficult. In this context, recent gas phase analysis techniques like ion mobility mass spectrometry (IM-MS) and gas phase spectroscopy have been used to overcome these limitations. Additionally, as compared to traditional LC-MS, gas phase analysis technique has been found to be a less time consuming technique. Implementation of IM-MS has improved analysis of glycan isomers and their structural characteristics. Ion mobility spectroscopy separates ions on the basis of size, shape, and charge. An electric field is used to pass ions through the drift cells containing gas wherein these ions collide with a buffer gas and get separated [156]. Within a single investigation, one can detect the separated ions, calculate their mass/charge ratio and drift time of ions by coupling IMS to MS. Additional information about compound identity and their stereochemistry can be obtained by calculating collision cross section (CSS) on the basis of the time taken by a particular ion to pass through drift cell [157–160]. May et al. [161] have recently indicated that all biomolecules can be separated by interpreting data obtained from trend lines in CSS plot against mass/charge ratio, e.g., glycans have shorter drift time in IM cells as compared to peptides and lipids of same mass/charge ratio [161].

Achievements of IM-MS have opened up some new fields of the metabolome research. Sarbu and coworkers have recently used ion MS to study human brain gangliosides wherein they were able to detect a pattern of highly diverse ceramide chains and different glycoforms in the gangliosides of fetal brain. In fact, they were able to characterize a modified unique and strong biomarker namely tetrasialylated O-GalNAc associated with development of brain. Additionally, for the first time reported presence of gangliosides with abundance of sialic acid residues ranging from mono to octasialylated gangliosides in human brain [162]. In contrast to IM spectroscopy, gas phase spectroscopy provides information on secondary structure of molecules along with their size and shape. Use of Infrared radiation in spectroscopy, i.e., IR gas phase spectroscopy, can provide information on structural properties of peptides such as hydrogen bonding, stereochemistry, etc., [163, 164]. IR spectroscopy of carbohydrates allows selection of ions on the basis of mass/charge ratio which further enables characterization of each species within a complex mixture [165]. It has been observed that during ion mobility spectroscopy, poorly resolved spectra are obtained for analytes due to flexibility in shape and thermal activation. To overcome this limitation, cold ion spectroscopy has been used which works on the principle of vibrational energy differences among various isomers of analyte (glycans). Using cold ion spectroscopy, highly resolved absorption features or fingerprints of oligosaccharides have been revealed. These absorption features can be utilized to trace minimal structural differences for diagnostic purposes in various diseases [166].

## 1.4   CONCLUSION AND FUTURE PERSPECTIVES

Advancements in glycobiology have reformed the concept that only nucleic acids contain biological information and control cellular activities through expression in the form of proteins and other ways in and outside cells. In fact, rapid progress in glycobiology has demonstrated that glycans are not only for the structure and energy demands of a cell but it is now possible to think beyond doubt that "sugar code" of a cell possesses very complex language coded information. In addition to their role in various cellular and physiological processes, glycans play a crucial role in glyco-sylation-associated disorders like congenital glycosylation disorders. More importantly, discovery of various cancer biomarkers such as TACAs has made it possible to use carbohydrates as therapeutic agents. Development from pure carbohydrate vaccine to mono- and multiconjugated vaccines, using proteins and nanoparticles as carriers, has attained significant success in certain human clinical trials. Availability of synthetic glycans in ample amount due to one-pot synthesis and other chemical methods, as compared to the complex process of isolation from natural resources, has further complemented the progress of glyco-profiling and the subsequent structural analysis of complex and species specific glycome. Modified MS based approaches have played an important role in characterization of closely related isomers and structural variations associated with them in certain disease conditions.

Although a large number of trial based clinical studies have been carried out, but they are still unavailable for routine use due to high cost, infrastruc-tural limitations, and lack of complete disease profiles. More investigations should be carried out to develop glycan therapies and interdisciplinary approach to decipher the disease causes and subsequent development of drugs. However, along with the development of these highly sophisticated techniques, it becomes mandatory to design simple methods in glycobiology so as to reach for some practical solution. Future investigations should also focus on the role of glycans in the formation and organization of cellular architecture and their role in cell signaling networks. More importantly, diversity of glycome and their binding partners under cellular systems should be explored for identification of possible potent biomarkers for dreadful diseases like cancer and diabetes.

## KEYWORDS

- **atomic force microscopy**
- **conjugate vaccines**
- **glycan arrays**
- **glycan mass spectrometry**
- **glycomimetics**
- **therapeutics**

## REFERENCES

1. Apweiler, R., Hermjakob, H., & Sharon, N., (1999). On the frequency of protein glycosylation, as deduced from analysis of the Swiss-Prot database. *Biochim. Biophys. Acta, 473*, 4–8.
2. Lha, H., & Yamada, M., (2013). *Glycan Profiling of Adult T-Cell Leukemia (ATL) Cells with the High-Resolution Lectin Microarrays.* Intech open Publishers: UK.
3. Mauthana, S. M., Campbell, C., & Gildersleeve, J. C., (2012). Modifications of glycans: Biological significance and therapeutic opportunities. *ACS Chem. Biol., 7*, 31–43.
4. Marth, J. D., & Grewal, P. K., (2008). Mammalian glycosylation in immunity. *Nat. Rev. Immunol., 8*, 874–887.
5. Varki, A., (2008). Sialic acids in human health and disease. *Trends Mol. Med., 14*, 351–360.
6. Ohtsubo, K., & Marth, J. D., (2006). Glycosylation in cellular mechanisms of health and disease. *Cell, 126*, 855–867.
7. Jaeken, J., & Matthijs, G., (2007). Congenital disorders of glycosylation: A rapidly expanding disease family. *Annu. Rev. Genom. Hum. Genet., 8*, 261–278.
8. Grunewald, S., Matthijs, G., & Jaeken, J., (2009). Encyclopedia of neuroscience. *Glycosylation and its Disorders: General Overview.* Elsevier: UK.
9. Henrissat, B., Surolia, A., & Stanley, P. A., (2017). Genomic view of glycobiology. *Essentials of Glycobiology* (3rd edn.). CSHL.
10. Nairn, A. V., & Moremen, K. W., (2009). In: *Handbook of Glycomics.* Academic Press: San Diego.
11. Hollenberg, M. D., Fishman, P. H., Bennett, V., & Cuatrecasas, P., (1974). Cholera toxin and cell growth: Role of membrane gangliosides. *Proc. Natl. Acad. Sci., 71*, 4224–4228.
12. Varki, A., & Gagneux, P., (2012). Multifarious roles of sialic acids in immunity. *Ann. N.Y. Acad. Sci., 1253*, 16–36.
13. Bishop, J. R., & Gagneux, P., (2007). Evolution of carbohydrate antigens-microbial forces shaping host glycomes. *Glycobiology, 5*, 23R–34R.
14. Gabius, H. J., (2009). *The Sugar Code: Fundamentals of Glycoscience.* Wiley-Blackwell: New Jersey.
15. Messner, P., & Schaffer, C., (2003). Prokaryotic glycoproteins. *Prog. Chem. Org. Nat. Prod., 85*, 51–124.
16. Werz, D. B., Ranzinger, R., Herget, S., Adibekian, A., Vonder, L. C. W., & Seeberger, P. H., (2007). Exploring the structural diversity of mammalian carbohydrates (Glycospace) by statistical data bank analysis. *ACS Chem. Biol., 2*, 685–691.
17. Ritchie, G. E., Moffatt, B. E., Morgan, B. P., Morgan, B. P., Dwek, R. A., & Rudd, P. M., (2002). Glycosylation and the complement system. *Chem. Rev., 102*, 305–319.
18. Dove, A., (2001). The bittersweet promise of glycobiology. *Nat. Biotechnol., 19*, 913–917.
19. Ohtsubo, K., & Marth, J. D., (2006). Glycosylation in cellular mechanisms of health and disease. *Cell, 126*, 855–867.
20. Magnani, J. L., & Ernst, B., (2009). Glycomimetic drugs: A new source of therapeutic opportunities. *Discov. Med., 8*, 247–252.
21. Osborn, H. M. I., Evans, P. G., Gemmell, N., & Osborne, S. D., (2004). Carbohydrate-based therapeutics. *J. Pharm. Pharmacol., 56*, 691–702.

22. Boltje, T. J., Buskas, T. G., & Boons, G., (2009). Opportunities and challenges in synthetic oligosaccharide and glycoconjugate research. *Nat. Chem., 1*, 611–622.

23. Davies, B. G., (2000). Recent developments in oligosaccharide synthesis. *J. Chem. Soc. Perkin Trans., 14*, 2137–2160.

24. Wang, C. C., Lee, J. C., Luo, S. Y., Suvarn, S., Huang, Y. W., & Chang, K. L., (2007). Regioselective one-pot protection of carbohydrates. *Nature, 446*, 896–899.

25. Simanek, E. E., Mc Garvey, J., Jablonowski, J. A., & Wong, C. H., (1998). Selectin-carbohydrate interactions: From natural ligands to designed mimics. *Chem. Rev., 98*, 833–862.

26. Wun, T., Styles, L., Decastrol, R., Telen, M. J., Kuypers, F., & Cheung, A., (2014). Phase-1 study of the E-selectin inhibitor GMI 1070 in patients with sickle cell anemia. *PLoS One, 9*, e111690.

27. Chang, J., Patton, J. T., Sarkar, A., Ernst, B., Magnani, J. L., & Frenette, P. S., (2010). GMI-1070, a novel pan-selectin antagonist, reverses acute vascular occlusions in sickle cell mice. *Blood, 116*, 1779–1786.

28. Chien, S., Haq, S. U., Pawlus, M., Moon, R. T., Estey, E. H., Frederick, R., et al., (2013). Adhesion of acute myeloid leukemia blasts to E-selectin in the vascular niche enhances their survival by mechanisms such as Wnt activation. *Blood, 122*, 61.

29. Kitov, P. I., Sadowska, J. M., Mulvey, G., Armstrong, G. D., Ling, H., & Pannu, N. S., (2000). Shiga- like toxins is neutralized by tailored multivalent carbohydrate ligands. *Nature, 403*, 669–672.

30. Hanessian, S., (1983). Total synthesis of natural products: The "Chiron" approach. Pergamon Oxford. *J. Chem. Edu., 62*, A190.

31. Hollingsworth, R. I., & Wang, G., (2000). Toward a carbohydrate-based chemistry: Progress in the development of general-purpose chiral synthons from carbohydrates. *Chem. Rev., 100*, 4267.

32. Hirschmann, R. F., Nicolaou, K. C., Angeles, A. R., Chen, J. S., & Smithlll, A. B., (2009). The beta-D Glucose scaffold as a gamma-turn mimetic. *Acc. Chem. Res., 42*, 1511–1520.

33. Jr, L. O., & Wade, J. D., (2014). Current challenges in peptide-based drug discovery. *Front. Chem., 62*, 1–4.

34. Olson, G. L., Bolin, D. R., Bonner, M. P., Bos, M., Cook, C. M., Fry, D. C., et al., (1993). Concept and progress in the development of peptide mimetics. *J. Med. Chem., 21*, 3039–3049.

35. Farmer, P. S., (1980). *Bridging the Gap between Bioactive Peptides and Nonpeptides: Some Perspectives in Design* (Vol. 6, pp. 119–143). Academic Press: NY.

36. Le Diguarher, T., Boudon, A., Elwell, C., Paterson, D. E., & Billington, D. C., (1996). Synthesis of potential peptidomimetics based on highly substituted glucose and allose scaffolds. *Biorg. Med. Chem. Lett., 6*, 1983–1988.

37. Edwards, A. A., Ichihara, O., Murfin, S., Wilkes, R., Whittaker, M., Watkin, D. J., et al., (2004). Tetrahydrofuran-based amino acids as library scaffolds. *J. Comb. Chem., 6*, 230–238.

38. Heidelberger, M., & Avery, O. T., (1923). The soluble specific substance of *Pneumococcus. J. Exp. Med., 38*, 73–79.

39. Bernardes, G. J., Castagner, B., & Seeberger, P. H., (2009). Combined approaches to the synthesis and study of glycoproteins. *ACS Chem. Biol., 4*, 703–713.

40. Mond, J. J., Lees, A., & Snapper, C. M., (1995). T cell-independent antigens type-2. *Annu. Rev. Immunol., 13,* 655–692.

41. Kalka-Moll, W. M., Tzianabos, A. O., Bryant, P. W., Niemeyer, M., Ploegh, H. L., & Kasper, D. L., (2002). Zwitterionic polysaccharides stimulate T-cells by MHC class II-dependent interactions. *J. Immunol., 169,* 6149–6153.

42. Mishra, S., Upadhayay, K., Mishra, K. B., Tripathi, R. P., & Tiwari, V. K., (2016). Carbohydrate-based therapeutics: A frontier in drug discovery and development. *Studies in Natural Product Chemistry, 49,* 307–361.

43. Seeberger, P. H., & Werz, D. B., (2007). Synthesis and medical applications of oligosaccharides. *Nature, 446,* 1046–1051.

44. Fernandez-Santana, V., Cardoso, F., Rodriguez, A., Carmenate, T., Pena, L., Valdes, Y., et al., (2004). Antigenicity and immunogenicity of a synthetic oligosaccharide-protein conjugate vaccine against *Haemophilus influenzae* type B. *Infect. Immun., 72,* 7115–7123.

45. Verez-Bencomo, V., Fernandez-Santana, V., Hardy, E., Toledo, M. E., Rodriguez, M. C., Heynngnezz, L., et al., (2004). A synthetic conjugate polysaccharide vaccine against *Haemophilus influenzae* type B. *Science, 305,* 522–525.

46. Peltola, H., (2000). Worldwide *Haemophilus influenzae* type B disease at the beginning of the 21st century: Global analysis of the disease burden 25 years after the use of the polysaccharide vaccine and a decade after the advent of conjugates. *Clin. Microbiol. Rev., 13,* 302–317.

47. Robbins, J. B., Austrian, R., Lee, C. J., Rastogi, S. C., Schiffman, G., Henrichsen, J., et al., (1983). Considerations for formulating the second-generation pneumococcal capsular polysaccharide vaccine with emphasis on the cross-reactive types within groups. *J. Infect. Dis., 148,* 1136–1159.

48. Henrichsen, J., (1995). Six newly recognized types of *Streptococcus pneumoniae. J. Clin. Microbiol., 33,* 2759–2762.

49. Siegel, R. L., Miller, K. D., & Jemal, A., (2016). Cancer statistics. *CA Cancer J. Clin., 66,* 7–30.

50. Shashidharamurthy, R., Bozeman, E. N., Patel, J., Kaur, R., Meganathan, J., & Selvaraj, P., (2012). Immunotherapeutic strategies for cancer treatment: A novel protein transfer approach for cancer vaccine development. *Med. Res. Rev., 32,* 1197–1119.

51. Ribatti, D., (2017). The concept of immune surveillance against tumors: The first theories. *Oncotarget, 8,* 7175–7180.

52. Dube, D. H., & Bertozzi, C. R., (2005). Glycans in cancer and inflammation-potential for therapeutics and diagnostics. *Nat. Rev. Drug. Discov., 4,* 477–488.

53. Becker, T., Dziadek, S., Wittrock, S., & Kunz, H., (2006). Synthetic glycopeptides from the mucin family as potential tools in cancer immunotherapy. *Curr. Cancer Drug Tar., 6,* 491–517.

54. Lakshminarayanan, V., Thompson, P., Wolfert, M. A., Buskas, T., Bradley, J. M., Pathangey, L. B., et al., (2012). Immune recognition of tumor-associated mucin *MUC1* is achieved by a fully synthetic aberrantly glycosylated MUC1 tripartite vaccine. *Proc. Natl. Acad. Sci., 109,* 261–266.

55. Brinas, R. P., Sundgren, A., Sahoo, P., Morey, S., Rittenhouse-Olson, K., Wilding, G. E., et al., (2012). Design and synthesis of multifunctional gold nanoparticles bearing tumor-associated glycopeptides antigens as potential cancer vaccines. *Bioconjugate Chem., 23,* 1513–1523.

56. Wilson, R. M., & Danishefsky, S. J., (2013). A vision for vaccines built from fully synthetic tumor-associated antigens: From the laboratory to the clinic. *J. Am. Chem. Soc., 135,* 14462–14472.

57. Yin, Z., Aragones, M. C., Chowdhury, S., Bentley, P., Kaczanowska, K., Ben, M. L., et al., (2013). Boosting immunity to small tumor-associated carbohydrates with bacteriophage Q-capsids. *ACS Chem. Biol., 8,* 1253–1262.

58. Liu, C. C., & Ye, X. S., (2012). Carbohydrate-based cancer vaccines: Target cancer with sugar bullets. *Glycoconj. J., 9,* 259–271.

59. Fukuda, M., (1996). Possible roles of tumor-associated carbohydrate antigens. *Cancer Res., 56,* 2237–4224.

60. Hakomori, S. I., & Zhang, Y., (1997). Glycosphingolipid antigens and cancer therapy. *Chem. Biol., 4,* 97–104.

61. Werther, J. L., Tatematsu, M., Klein, R., et al., (1996). Sialosyl-Tn, antigen as a marker of gastric cancer progression: An international study. *Int. J. Cancer., 69,* 193–199.

62. Helling, F., Shang, A., Calves, M., Zhang, S., Ren, S., Yu, R. K., et al., (1994). GD3 vaccines for melanoma: Superior immunogenicity of keyhole limpet hemocyanin conjugates vaccines. *Cancer Res., 54,* 197–203.

63. Plante, O. J., Palmacci, E. R., & Seeberger, P. H., (2001). Automated solid-phase synthesis of oligosaccharides. *Science, 291,* 1523–1527.

64. Zhang, Z., Ollmann, I. R., Ye, X. S., Wischnat, R., Baaso, T., & Wong, C. H., (1999). Programmable one-Pot oligosaccharide synthesis. *J. Am. Chem. Soc., 121,* 734–753.

65. Sears, P., & Wong, C. H., (2001). Toward automated synthesis of oligosaccharides and glycoproteins. *Science, 291,* 2344–2350.

66. Huang, X., Huang, L., Wang, H., & Ye, X. S., (2004). Iterative one-pot synthesis of oligosaccharides. *Ange. Chem. Int. Ed., 43,* 5221–5224.

67. Wu, Y., Xiong, D. C., Chen, S. C., Wang, Y. S., & Ye, X. S., (2017). Total synthesis of mycobacterial arabino-galactan containing 92 monosaccharide units. *Nat. Commun., 8,* 14851.

68. Monsan, P., Remaudsimeon, M., & Andre, I., (2010). Transglucosidases as efficient tools for oligosaccharide and glucoconjugate synthesis. *Curr. Opin. Microbiol., 13,* 293–300.

69. Danishefsky, S. J., & Allen, J. R., (2000). From the laboratory to the clinic: A retrospective on fully synthetic carbohydrate-based anticancer vaccines. *Angew. Chem. Int. Ed., 39,* 836–863.

70. Germain, R. N., & Margulies, D. H., (1993). The biochemistry and cell biology of antigen processing and presentation. *Annu. Rev. Immunol., 11,* 403–450.

71. Haurum, J. S., Arsequell, G., Lellouch, A. C., Wong, S. Y., Dwek, R. A., McMichael, A., et al., (1994). Recognition of carbohydrate by major histocompatibility complex class I-restricted, glycopeptide- specific cytotoxic T-lymphocytes. *J. Exp. Med., 180,* 739–744.

72. Abdel-Motal, U. M., Berg, L., Rosen, A., Bengtsson, M., Thorpe, C. J., Kihlberg, J., et al., (1996). Immunization with glycosylated Kb-binding peptides generates carbohydrate-specific, unrestricted cytotoxic T-cells. *Eur. J. Immunol., 26,* 544–551.

73. Speir, J. A., Abdel-Motal, U. M., Jondal, M., & Wilson, I. A., (1999). Crystal structure of an MHC class I presented glycopeptides that generates carbohydrate-specific CTL. *Immunity, 10,* 51–56.

74. Dudler, T., Altmann, F., Carballido, J. M., & Blaser, K., (1995). Carbohydrate-dependent, HLA class II-restricted, human T-cell response to the bee venom allergen phospholipase A2 in allergic patients. *Eur. J. Immunol., 25*, 538–542.

75. Dengjel, J., Rammensee, H. G., & Stevanovic, S., (2005). Glycan side chains on naturally presented MHC class II ligands. *J. Mass Spectrom., 40*, 100–104.

76. Kennedy, R., & Celis, E., (2008). Multiple roles for CD4+ T-cells in anti-tumor immune responses. *Immunol. Rev., 222*, 129–144.

77. Liu, C. C., & Ye, X. S., (2012). Carbohydrate-based cancer vaccines: Target cancer with sugar bullets. *Glycoconjugate J., 9*, 259–271.

78. Cyster, J. G., (1999). Chemokines and cell migration in secondary lymphoid organs. *Science, 286*, 2098–2100.

79. Wimmers, F., De Hass, N., Scholzen, A., Schreibelt, G., Simonetti, E., Elveld, M. J., et al., (2017). Monitoring of dynamic changes in keyhole limpet hemocyanin (KLH)-specific B-cells in KLH-vaccinated cancer patients. *Sci. Rep., 7*, 43486.

80. Kagan, E., Ragupathi, G., Yi, S. S., Reis, C. A., Gildersleeve, J., Kahne, D., et al., (2005). Comparison of antigen constructs and carrier molecules for augmenting the immunogenicity of the monosaccharide epithelial cancer antigen Tn. *Cancer Immunol. Immunother., 54*, 424–430.

81. Feng, D., Shaikh, A. S., & Wang, F., (2016). Recent advance in tumor-associated carbohydrate antigens (TACAs)-based antitumor vaccines. *ACS Chem. Biol., 11*, 850–863.

82. Gilewski, T., Ragupathi, G., Bhuta, S., Williams, L. J., Musselli, C., Zhang, X. F., et al., (2001). Immunization of metastatic breast cancer patients with a fully synthetic Globo H conjugate: A phase I trial. *Proc. Natl. Acad. Sci., 98*, 3270–3275.

83. Huang, Y. L., Hung, J. T., Cheung, S. K., Lee, H. Y., Chu, K. C., Li, S. T., et al., (2013). Carbohydrate- based vaccines with a glycolipid adjuvant for breast cancer. *Proc. Natl. Acad. Sci., 110*, 2517–2522.

84. Livingston, P. O., Adluri, S., Helling, F., Lee, H. Y., Chu, K. C., Li, S. T., et al., (1994). Phase-1 trial of immunological adjuvant QS-21 with a GM2 ganglioside-keyhole limpet hemocyanin conjugate vaccine in patients with malignant melanoma. *Vaccine, 12*, 1275–1280.

85. Bundle, D. R., (2011). *Vaccine Design: Innovative Approaches and Novel Strategies.* Academic Press: San Diego.

86. Buskas, T., Li, Y., & Boons, G. J., (2004). The immunogenicity of the tumor-associated antigen Lewis Y may be suppressed by a bifunctional cross-linker required for coupling to a carrier protein. *Chem. Eur. J., 10*, 3517–3524.

87. Kuduk, S. D., Schwarz, J. B., Chen, X. T., Glunz, P. W., Sames, D., Raghupathi, G., et al., (1998). Synthetic and immunological studies on clustered modes of mucin-related Tn and TF O-linked antigens: The preparation of a glycopeptide-based vaccine for clinical trials against prostate cancer. *J. Am. Chem. Soc., 120*, 12474–12485.

88. Adluri, S., Helling, F., Ogata, S., Itzkowitz, S. H., Zhang, S., Lloyd, K. O., et al., (1995). Immunogenicity of synthetic TF-KLH (keyhole limpet hemocyanin) and sTn-KLH conjugates in colorectal carcinoma patients. *Cancer Immunol. Immunother., 41*, 185–192.

89. Fung, P. Y., Madej, M., Koganty, R. R., & Longenecker, B. M., (1990). Active specific immunotherapy of a murine mammary adeno-carcinoma using a synthetic tumor-associated glycoconjugates. *Cancer Res., 50*, 4308–4314.

90. Helling, F., Zhang, S., Shang, A., Adluri, S., Calves, M., Koganty, R., et al., (1995). GM2-KLH conjugate vaccine: Increased immunogenicity in melanoma patients after administration with immunological adjuvant QS-21. *Cancer Res., 55,* 2783–2788.

91. Gilewski, T., Ragupathi, G., Bhuta, S., Williams, L. J., Musseli, C., Zhang, X. F., et al., (2001). Immunization of metastatic breast cancer patients with a fully synthetic Globo H conjugate: A phase I trial. *Proc. Natl. Acad. Sci., 98,* 3270–3275.

92. Ragupathi, G., Slovin, S. F., Adluri, S., et al., (1999). A fully synthetic Globo H carbohydrate vaccine induces a focused humoral response in prostate cancer patients: A proof of principle. *Angew. Chem. Int. Ed., 38,* 563–566.

93. Slovin, S. F., Ragupathi, G., Adluri, S., Sames, D., Kim, I. J., Kim, H. M., et al., (1999). Carbohydrate vaccines in cancer: Immunogenicity of a fully synthetic Globo H hexasaccharide conjugates in man. *Proc. Natl. Acad. Sci., 96,* 5710–5715.

94. Kudryashov, V., Kim, H. M., Ragupathi, G., Danishefsky, S. J., Livingston, P. O., & Lloyd, K. O., (1998). Immunogenicity of synthetic conjugates of Lewis Y oligosaccharide with proteins in mice: Towards the design of anticancer vaccines. *Cancer Immunol. Immunother., 45,* 281–286.

95. Sabbatini, P. J., Kudryashov, V., Ragupathi, G., Danishefaky, S. J., Livingston, P. O., Bornmann, W., et al., (2000). Immunization of ovarian cancer patients with a synthetic Lewis Y-protein conjugate vaccine: A phase 1 trial. *Int. J. Cancer, 87,* 79–85.

96. Miles, D., Roche, H., Martin, M., Perren, T. J., Cameron, D. A., Glaspy, J., et al., (2011). Phase III multicenter clinical trial of the sialyl-TN (STn)-keyhole limpet hemocyanin (KLH) vaccine for metastatic breast cancer. *Oncologist, 16,* 1092–1100.

97. Slovin, S. F., Keding, S. J., & Ragupathi, G., (2005). Carbohydrate vaccines as immunotherapy for cancer. *Immunol. Cell Biol., 83,* 418–428.

98. Slovin, S. F., Ragupathi, G., Musselli, C., Fernandez, C., Diani, M., Verbel, D., et al., (2005). Thomsen-Friedenreich (TF) antigen as a target for prostate cancer vaccine: Clinical trial results with TF cluster (c)-KLH plus QS21 conjugate vaccine in patients with biochemically relapsed prostate cancer. *Cancer Immunol. Immunother., 54,* 694–702.

99. Denis, J., Majeau, N., Acosta, R. E., Savard, C., Bedard, M. C., Simard, S., et al., (2007). Immunogenicity of papaya mosaic virus-like particles fused to a hepatitis C virus epitope: Evidence for the critical function of multimerization. *Virology, 363,* 59–68.

100. Jegerlehner, A., Storni, T., Lipowsky, G., Schmid, M., Pumpens, P., & Bachmann, M. F., (2002). Regulation of IgG antibody responses by epitope density and CD21-mediated costimulation. *Eur. J. Immunol., 32,* 3305–3314.

101. Bachmann, M. F., & Zinkernagel, R. M., (1997). Neutralizing antiviral B-cell responses. *Annu. Rev. Immunol., 15,* 235–270.

102. Bachmann, M. F., Hengartner, H., & Zinkernagel, R. M., (1995). T helper cell-independent neutralizing B cell response against vesicular stomatitis virus: Role of antigen patterns in B cell induction? *Eur. J. Immunol., 25,* 3445–3451.

103. Bachmann, M. F., Rohrer, U. H., Kundig, T. M., Burki, K., Hengartner, H., & Zinkernagel, R. M., (1993). The influence of antigen organization on B-cell responsiveness. *Science, 262,* 1448–1451.

104. Sun, Z. Y., Chen, P. G., Liu, Y. F., Shi, L., Zhang, B. D., Wu, J. J., et al., (2017). Self-assembled nano-immunostimulant for synergistic immune activation. *Chembiochem, 18,* 1721–1729.

105. Krishnamachari, Y., Geary, S. M., Lemke, C. D., & Salem, A. K., (2011). Nanoparticle delivery systems in cancer vaccines. *Pharm. Res., 28*, 215–236.

106. Hong, S. Y., Tobias, G., Al Jamal, K. T., Ballesteros, B., Ali-Baucetta, H., Lozano-Perez, S., et al., (2010). Filled and glycosylated carbon nano-tubes for *in vivo* radioemitter localization and imaging. *Nat. Mater., 9*, 485–490.

107. Van, G. F., Hadida, S., Grootenhuis, P. D., Burton, B., Cao, D., Neuberger, T., et al., (2009). Rescue of CF airway epithelial cell function *in vitro* by a CFTR potentiator, VX-770. *Proc. Natl. Acad. Sci., 106*, 18825–18830.

108. Raymond, P., Sundgren, B. A., Sahoo, P., Morey, S., Rittenhouse-Olson, K., Greg, E., et al., (2012). Design and synthesis of multifunctional gold nanoparticles bearing tumor-associated glycopeptide antigens as potential cancer vaccines. *Bioconjug. Chem., 23*, 1513–1523.

109. Biswas, S., Medina, S. H., Joseph, J., & Barchi, Jr., (2015). Synthesis and cell-selective antitumor properties of amino acid conjugated tumor-associated carbohydrate antigen-coated gold nanoparticles. *Carbohydr. Res., 405*, 93–101.

110. Moll, W. M. K., Tzianabos, A. O., Bryant, P. W., Niemeyer, M., Ploegh, H. L., & Kasper, D. L., (2002). Zwitterionic polysaccharides stimulate T-cells by MHC class II-dependent interactions. *J. Immunol., 169*, 6149–6153.

111. Pantosti, A., Tzianabos, A. O., Onderdonk, A. B., & Kasper, D. L., (1991). Immunochemical characterization of two surface polysaccharides of Bacteroides fragilis. *Infect. Immun., 59*, 2075–2082.

112. Wang, Y., Kalka, M. W. M., Roehr, M. H., & Kasper, D. L., (2000). Structural basis of the abscess- modulating polysaccharide A2 from Bacteroides fragilis. *Proc. Natl. Acad. Sci., 97*, 13478–13483.

113. Stingele, F., Corthesy, B., Kusy, N., Porcelli, S. A., Kasper, D. L., & Tzianabos, A. O., (2004). Zwitterionic polysaccharides stimulate T-cells with no preferential Vβ usage and promote energy, resulting in protection against experimental abscess formation. *Immunol., 172*, 1483–1490.

114. Tzianabos, A. O., Wang, J. Y., & Lee, J. C., (2001). Structural rationale for the modulation of abscess formation by *Staphylococcus aureus* capsular polysaccharides. *Proc. Natl. Acad. Sci., 98*, 9365–9937.

115. Tzianabos, A. O., Onderdonk, A. B., Rosner, B., Cisneros, R. L., & Kasper, D. L., (1993). Structural features of polysaccharides that induce intra-abdominal abscesses. *Science, 262*, 416–419.

116. Nishat, S., & Andreana, P. R., (2016). Entirely carbohydrate-based vaccines: An emerging field for specific and selective immune responses. *Vaccine, 34*, 19–53.

117. De Silva, R. A., Wang, Q., Chidley, T., Appulage, D. K., & Andreana, P. R., (2009). Immunological response from an entirely carbohydrate antigen: Design of synthetic vaccines based on Tn-PS A1 conjugates. *J. Am. Chem. Soc., 131*, 9622–9623.

118. Bourgault, J. P., Trabbic, K. R., Shi, M., & Andreana, P. R., (2014). Synthesis of the tumor associative α-aminooxy disaccharide of the TF antigen and its conjugation to a polysaccharide immune stimulant. *Org. Biomol. Chem., 12*, 1699–1702.

119. Trabbic, K. R., De Silva, R. A., & Andreana, P. R., (2014). Elucidating structural features of an entirely carbohydrate cancer vaccine construct employing circular dichroism and fluorescent labeling. *Med. Chem. Comm., 5*, 1143–1149.

120. Sharmeen, N., & Andreana, P. R., (2016). Entirely carbohydrate-based vaccines: An emerging field for specific and selective immune responses. *Vaccine, 34*, 19–53.

121. Nores, G. A., Dohi, T., Taniguchi, M., & Hakomori, S., (1987). Density-dependent recognition of cell surface GM3 by a certain anti-melanoma antibody, and GM3 lactone as a possible immunogen. *J. Immunol., 139*, 3171–3176.

122. Qiu, L., Gong, X., Wang, Q., Li, J., Hu, H., Wu, Q., et al., (2012). Combining synthetic carbohydrate vaccines with cancer cell glycoengineering for effective cancer immunotherapy. *Cancer Immunol., 61*, 2045–2054.

123. Zou, W., Borrelli, S., Gilbert, M., Liu, T., Pon, R. A., & Jennings, H. J., (2004). Bioengineering of surface GD3 ganglioside for immune targeting human melanoma cells. *J. Biol. Chem., 279*, 25390–25399.

124. Pan, Y., Chefalo, P., Nagy, N., Harding, C., & Guo, Z., (2005). Synthesis and immunological properties of N-modified GM3 antigens as therapeutic cancer vaccines. *J. Med. Chem., 48*, 875–883.

125. Paulson, J. C., Blixt, O., & Collins, B. E., (2006). Sweet spots in functional glycomics. *Nat. Chem. Biol., 2*, 238–248.

126. Liang, P. H., Wu, C. Y., Greenberg, W. A., & Wong, C. H., (2008). Glycan arrays: Biological and medical applications. *Curr. Opin. Chem. Biol., 12*, 86–92.

127. Smith, D. F., Song, X., & Cummings, R. D., (2010). Use of glycan microarrays to explore specificity of glycan-binding proteins. *Meth. Enzymol., 480*, 417–444.

128. Fukui, S., Feizi, T., Galustian, C., Lawson, A. M., & Chai, W., (2002). Oligosaccharide microarrays for high throughput detection and specificity assignments of carbohydrate-protein interactions. *Nat. Biotechnol., 20*, 1011–1017.

129. Wang, D., Liu, S., Trummer, B. J., Deng, C., & Wang, A., (2002). Carbohydrate microarrays for the recognition of cross-reactive molecular markers of microbes and host cells. *Nat. Biotechnol., 20*, 275–281.

130. Bryan, M. C., Plettenburg, O., Sears, P., Rabuka, D., Wacowich-Sgarbi, S., & Wong, C. H., (2002). Saccharide display on microtiter plates. *Chem. Biol., 9*, 713–720.

131. Fazio, F., Bryan, M. C., Blixt, O., Paulson, J. C., & Wong, C. H., (2002). Synthesis of sugar arrays in microtiter plate. *J. Am. Chem. Soc., 124*, 14397–14402.

132. MacBeath, G., & Schreiber, S. L., (2000). Printing proteins as microarrays for high-throughput function determination. *Science, 289*, 1760–1763.

133. Haslam, S. M., North, S. J., & Dell, A., (2006). Mass spectrometric analysis of N and O-glycosylation of tissues and cells. *Curr. Opin. Struct. Biol., 16*, 584–591.

134. Park, S., & Shin, I., (2002). Fabrication of carbohydrate chips for studying proteins carbohydrate interactions. *Angew. Chem. Int. Ed. Engl., 41*, 3180–3182.

135. Houseman, B. T., & Mrksich, M., (2002). Carbohydrate arrays for the evaluation of protein binding and enzymatic modifications. *Chem. Biol., 9*, 443–454.

136. Zhi, Z. L., Laurent, N., Powel, A. K., Karamanska, R., Fais, M., Voglmeir, J., et al., (2008). A versatile gold surface approach for fabrication and interrogation of glycoarrays. *Chembiochem., 9*, 1568–1575.

137. Karamanska, R., Clarke, J., Blixt, O., MacRae, J. I., Zhang, J. Q., Crocker, P. R., et al., (2008). Surface plasmon resonance imaging for real-time, label-free analysis of protein interactions with carbohydrate microarrays. *Glycoconj. J., 25*, 69–74.

138. Hsu, T. L., Cheng, S. C., Yang, W. B., Chin, S. W., Chen, B. H., Huang, M. T., et al., (2009). Profiling carbohydrate-receptor interaction with recombinant innate immunity receptor-Fc fusion proteins. *J. Biol. Chem., 284*, 34479–34489.

139. Kuno, A., Uchiyama, N., Koseki-Kuno, S., Ebe, Y., Takashima, S., Yamada, M., & Hirabayashi, J., (2005). Evanescent-field fluorescence-assisted lectin microarray: A new strategy for glycan profiling. *Nat. Methods, 2,* 851–856.

140. Schwarz, M., Spector, L., Gargir, A., Shtevi, A., Gortler, M., & Rom, T., (2003). A new kind of carbohydrate array, its use for profiling antiglycan antibodies, and the discovery of a novel human cellulose-binding antibody. *Glycobiology, 11,* 749–754.

141. Manimala, J. C., Roach, T. A., Li, Z., & Gildersleeve, J. C., (2007). High-throughput carbohydrate microarray profiling of 27 antibodies demonstrates widespread specificity problems. *Glycobiology, 17,* 17–23.

142. Wang, C. C., Huang, Y. L., Ren, C. T., Lin, C. W., Hung, J. T., Yu, J. C., et al., (2008). Glycan microarray of Globo H and related structures for quantitative analysis of breast cancer. *Proc. Natl. Acad. Sci. U.S.A, 105,* 11661–11666.

143. Tseng, S. Y., Cho, W. H., Su, J., Chang, S. H., Chiang, D., & Wu, C. Y., (2016). Preparation of aluminum oxide coated glass slides for glycan microarrays. *ACS Omega, 1,* 773–783.

144. Zubkova, O. V., Ahmed, Y. A., Guimond, S. E., Noble, S. L., Miller, J. H., Smith, R. A. A., et al., (2018). Dendrimer heparan sulfate glycomimetics: Potent heparanase inhibitors for anticancer therapy. *ACS Chem. Biol., 12,* 3236–3242.

145. Eggink, L. L., Roby, K. F., Cote, R., & Hoober, J. K., (2018). An innovative immunotherapeutic strategy for ovarian cancer: CLEC10A and glycomimetic peptides. *J. Immunother. Cancer, 6,* 28.

146. Medve, L., Achilli, S., Serna, S., Zuccotto, F., Varga, N., Thepaut, M., et al., (2018). On-chip screening of a glycomimetic library with C-type lectins reveals structural features responsible for preferential binding of dectin-2 over DC-SIGN/R and langerin. *Chem. Eur. J., 24,* 14448–14460.

147. Kalasa, V., Hibbinga, M. E., Maddiralac, A. R., Chuganic, R., Pinknera, J. S., Mydock-McGranec, L. K., et al., (2018). Structure-based discovery of glycomimetic *FmlH* ligands as inhibitors of bacterial adhesion during urinary tract infection. *Proc. Nati. Acad. Sci. U.S.A, 15,* E2819–2828.

148. Dell, A., & Morris, H. R., (2001). Glycoprotein structure determination by mass spectrometry. *Science, 291,* 2351–2356.

149. Mechref, Y., & Novotny, M. V., (2002). Structural investigations of glycoconjugates at high sensitivity. *Chem. Rev., 102,* 321–369.

150. Novotny, M. V., & Mechref, Y., (2005). Combining lectin micro-columns with high-resolution separation techniques for enrichment of glycoproteins and glycopeptides. *J. Sep. Sci., 28,* 1956–1968.

151. Alpert, A. J., (1990). Hydrophilic-interaction chromatography for the separation of peptides, nucleic acids and other polar compounds. *J. Chromatogr., 499,* 177–196.

152. Anumula, K. R., (2006). Advances in fluorescence derivatization methods for high-performance liquid chromatographic analysis of glycoprotein carbohydrates. *Anal. Biochem., 350,* 1–23.

153. Wuhrer, M., De Boer, A. R., & Deelder, A. M., (2009). Structural glycomics using hydrophilic interaction chromatography (HILIC) with mass spectrometry. *Mass Spectrom. Rev., 28,* 192–206.

154. Gao, W. N., Yau, L. F., Liu, L., Zeng, X., Chen, D. C., Jiang, M., et al., (2015). Microfluidic chip-LC/MS-based glycomic analysis revealed distinct N-glycan profile of rat serum. *Sci. Rep., 5,* 844.

155. Hua, S., Jeong, H. N., Dimapascos, L. M., Kang, I., Han, C., Choi, J. S., et al., (2013). Isomer-specific LC/MS and LC/MS/MS profiling of the mouse serum N-glycome revealing a number of novel sialylated N-glycans. *Anal. Chem., 85,* 4636–4643.

156. Palmisano, G., Larsen, M. R., Packer, H. N., & Andersen, M. T., (2013). Structural analysis of glycoprotein sialylation - part II: LC-MS based detection. *RSC Adv.,* **3,** 22706–22726.

157. May, J. C., & McLean, J. A., (2015). Ion mobility-mass spectrometry: Time-dispersive instrumentation. *Anal. Chem., 87,* 1422–1436.

158. Bohrer, B. C., Merenbloom, S. I., Koeniger, S. L., Hilderbrand, A. E., & Clemmer, D. E., (2008). Biomolecule analysis by ion mobility spectrometry. *Annu. Rev. Anal. Chem., 1,* 293–327.

159. Kanu, A. B., Dwivedi, P., Tam, M., Matz, L., & Hill, H. H., (2008). Ion mobility-mass spectrometry. *J. Mass Spectrom., 43,* 1–22.

160. Jurneczko, E., & Barran, P. E., (2011). How useful is ion mobility mass spectrometry for structural biology? The relationship between protein crystal structures and their collision cross sections in the gas phase. *Analyst, 136,* 20–28.

161. May, J. C., Goodwin, C. R., Lareau, N. M., Leaptrot, K. L., Morris, C. B., Kurulugama, R. T., et al., (2014). Conformational ordering of biomolecules in the gas phase: Nitrogen collision cross sections measured on a prototype high resolution drift tube ion mobility-mass spectrometer. *Anal. Chem., 86,* 2107–2116.

162. Sarbu, M., Robu, A. C., Ghiulai, R. M., Vukelic, Z., Clemmer, D. E., & Zamfir, A. D., (2016). Electrospray ionization ion mobility mass spectrometry of human brain gangliosides. *Anal. Chem., 10,* 5166–5178.

163. Seo, J., Hoffmann, W., Warnke, S., Bowers, M. T., Pagel, K., & Von, H. G., (2016). Retention of native protein structures in the absence of solvent: A coupled ion mobility and spectroscopic study. *Angew. Chem. Int. Ed. Engl., 55,* 14173–14176.

164. Gonzalez, F. A. I., Mucha, E., Ahn, D. S., Gewinner, S., Schollkopf, W., Pagel, K., et al., (2016). Charge-induced unzipping of isolated proteins to a defined secondary structure. *Angew. Chem. Int. Ed. Engl., 55,* 3295–3299.

165. Oomens, J., Sartakov, B. G., Meijer, G., & Von, H. G., (2006). Gas-phase infrared multiple photon dissociation spectroscopy of mass-selected molecular ions. *Int. J. Mass. Spectrom., 254,* 1–19.

166. Mucha, E., Gonzalez, F. A. L., Marianski, M., Thomas, D. A., Hoffmann, W., & Struwe, W. B., (2017). Glycan fingerprinting via cold-ion infrared spectroscopy. *Angew. Chem. Int. Ed. Engl., 56,* 11248–11251.

# CHAPTER 2

# Defects in the Human Glycome: Congenital and Non-Congenital Disorders

MUMTAZ ANWAR

*Department of Pharmacology, University of Illinois at Chicago, Chicago – 60612, Illinois, USA*

## ABSTRACT

Glycome is defined as a huge gamut of glycan structures wherein the constituent monosaccharide residues are assembled to form glycans through the process of glycosylation-an enzyme-directed and a highly regulated process. More importantly, the structure and function of a particular glycosylated protein are described by both its glycan as well as its polypeptide part. Various genetic diseases have been identified in the past decade that predominantly alter glycan synthesis and their structure, which in turn has been shown to affect the functioning of almost all organ systems. Among these, congenital disorders of glycosylation (CDG) are an increasing class of multi-systemic diseases/disorders, with severe clinical implications. In addition to this, non-congenital disorders arise due to defects in glycogen synthesis or its breakdown that, in turn, lead to the development of glycogen storage diseases (GSDs). Most of these disorders occur due to improper functioning/defects in metabolic enzymes. This chapter will provide an update about the major advances in these glycan disorders and also highlight the future directions in this regard.

## 2.1 INTRODUCTION

The complete repertoire of glycans that are present in an organism/cell/tissue comprises glycome, which is studied under the field of glycomics [1]. These glycans represent one of the primary cellular constituents necessary for the cellular/organism life as the entire glycan deficiency, both at the genetic and embryological level, has been shown to be fatal [2]. Glycans, being the most abundant and diverse biopolymer class, are itself made up of saccharides typically attached to a variety of molecules, including proteins, via an enzyme-regulated process of glycosylation to augment their function.

Complex machinery of enzymes, transcription factors, and other proteins are involved in glycan synthesis. Both genetic and environmental factors do influence glycan biosynthesis. Glycans have been shown to play a major role in all fundamental functions of organisms, including the immune system, cell recognition, etc. Given the role of glycans in major biological processes, defects in glycan synthesis have been recognized as direct causal agents of an increasingly diverse range of diseases. In fact, many of the glycans have been considered to act as markers of these diseases as well as promising therapeutic targets. Keeping in view the advances in the glycomics field and the involvement of glycans in a number of diseases, this chapter will focus on some important congenital and non-congenital glycan disorders. Future insights in this direction have also been highlighted.

## 2.2   BIOSYNTHETIC PATHWAYS FOR GLYCOSYLATION

### 2.2.1   N AND O-LINKED PROTEIN GLYCOSYLATION

Initiation of an amide linkage occurs on target proteins with the beginning of N-glycosylation between asparagine and N-acetylglucosamine (GlcNAc). Two GlcNAc, three glucose, and nine mannose units are built by a chain of glycosyltransferases on a carrier lipid that is contained in an oligosaccharide (LLO) with lipid-linked 14-sugar precursor [3]. The glycans are then transferred to nascent proteins with a convenient asparagine in an appropriate atmosphere.

Golgi-endoplasmic reticulum (ER) channel is the best channel through which nearly all N-glycosylated proteins travel, and this alteration can influence or determine the protein stability, folding, localization, trafficking, and oligomerization with special inference to cell signaling and cellular interaction. This alteration is significantly lethal in yeast to mammalian species, which is manifested by the absence of all N-glycans [4].

Van den et al. [5] reported that O-glycosylation in contradiction to N-glycosylation occurs with single monosaccharide (GalNAc) addition, and not due to transfer of sugars from precursor molecules. O-linked fucoses and glucose are found in the protein domains of epidermal growth factor on a particular consensus sequence, in peculiar O-glycosylation types that contradicts the conventional pathway/s [6]. O-linked GlcNAc is one of the mechanisms by which cytoplasmic and nuclear proteins can be glycosylated reversibly, which is analogous to phosphorylation [7, 8].

A careful enzyme-directed and synchronized glycosylation process is required for glycans to get assembled from monosaccharide residues [1]. The configuration of cellular memory is represented by the glycome, which is regulated by contemporary cellular physiology in the cell on the basis of past events [9], as the environmental and genetic components amalgamate at the extent of glycan biosynthesis [10].

## 2.3   CONGENITAL DISORDERS OF GLYCOSYLATION (CDG)

The congenital disorders of glycosylation (CDGs) were previously known as carbohydrate-deficient glycoprotein syndromes and are swiftly growing into the heterogeneous class of diseases that begin with the alteration in the formation and synthesis of protein and lipid glycosylation [173]. It is a multi-systemic condition affecting different pathways of glycosylation [11]. CDG-I and CDG-II are the two groups in which CDG patients have been divided based on the nature of the defective gene (Table 2.1). Gene mutations that correspond to vesicular trafficking from the ER and then to the ER-Golgi intermediate compartment (ERGIC), homeostasis, and or stability of the Golgi apparatus (GA) and ER have been found to be related to CDG [173]. The CDGs are one of the broad classes of diseases/conditions produced due to either the faulty biosynthesis of glycans or their degradation, respectively [12].

**TABLE 2.1**   Congenital Disorders of Glycosylation

| Gene | Enzyme | Notes | References |
|------|--------|-------|------------|
| **Asparagine (N)-Linked Glycosylation** | | | |
| PMM2 | Monosaccharide activation | Phosphomannomutase deficiency (CDG-Ia) | [13, 14] |
| MP1 | Monosaccharide activation | Phosphomannose isomerase deficiency (CDG-Ib) | [15] |
| DPAGT1 | Glycosyltransferase | N-Acetyl glucosaminyltransferase I deficiency (CDG-Ij) | [16] |

**TABLE 2.1**   *(Continued)*

| Gene | Enzyme | Notes | References |
|------|--------|-------|------------|
| HMT1 | Glycosyltransferase | Mannosyltransferase I deficiency (CDG-Ik) | [4] |
| ALG2 | Glycosyltransferase | Mannosyltransferase II deficiency (CDG-Ii) | [17] |
| RFT-1 | Substrate localization | RFT1 deficiency (CDG-In) | [4] |
| NOT56L | Glycosyltransferase | Mannosyltransferase VI deficiency (CDG-Id) | [18] |
| DIBD1 | Glycosyltransferase | Mannosyltransferase VII/IX deficiency (CDG-IL) | [19] |
| ALG12 | Glycosyltransferase | Mannosyltransferase VIII deficiency (CDG-Ig) | [20] |
| ALG6 | Glycosyltransferase | Glucozyltransferase I deficiency (CDG-Ic) | [21] |
| ALG8 | Glycosyltransferase | Glucozyltransferase II deficiency (CDG-Ih) | [22] |
| N33/TUSC3 | Glycosyltransferase | Oligosaccharyltransferase deficiency | [4] |
| IAP | Glycosyltransferase | Oligosaccharyltransferase deficiency | [4] |
| GLS1 | Glycosidase | Glucosidase I deficiency (CDG-IIb) | [4] |
| MGAT2 | Glycosyltransferase | N-Acetyl glucosaminyltransferase II deficiency (CDG-IIa) | [4] |
| GNPTAB | Glycosyltransferase | I-Cell disease/mucolipidosis II and III | [23] |
| **O-Linked Glycosylation** | | | |
| GALNT3 | Glycosyltransferase | Familial tumoral calcinosis | [24] |
| COSMC | Chaperone | Tn syndrome | [25] |
| POMT1 | Glycosyltransferase | WWS, type II lissencephaly, LGMD2K | [26] |
| POMT2 | Glycosyltransferase | WWS, type II lissencephaly | [27] |
| POMGnT1 | Glycosyltransferase | MEB, type II lissencephaly | [28] |
| FKTN | Unknown | Fukuyama CMD, WWS, LGMD2M | [29] |
| FKRP | Unknown | MDC1C, WWS, MEB, LGMD2I | [30] |
| LARGE | Putative glycosyltransferase | MDC1D, WWS | [31] |

**TABLE 2.1** *(Continued)*

| Gene | Enzyme | Notes | References |
|------|--------|-------|-----------|
| GYLTL1B | Putative glycosyltransferase | Non detected | [4] |
| LFNG | Glycosyltransferase | Spondylocostal dysostosis type 3 | [32] |
| B3GALTL | Glycosyltransferase | Peter's plus syndrome | [33] |
| **Lipid Glycosylation** | | | |
| SIAT9 | Glycosyltransferase | Amish infantile-onset epilepsy | [34] |
| PIGA | Glycosyltransferase | Marchiafava-Micheli syndrome/ PNH | [35] |
| PIGM | Glycosyltransferase | A hepatic and portal veins thrombosis with an epilepsy syndrome | [36] |
| **Glycosaminoglycan Biosynthesis** | | | |
| B4GALT7 | Glycosyltransferase | Progeria variant of Ehlers-Danlos syndrome | [37, 38] |
| EXT1 | Glycosyltransferase | Hereditary multiple exostoses | [39] |
| EXT2 | Glycosyltransferase | Hereditary multiple exostoses | [39] |
| **Two or More glycosylation pathways** | | | |
| DPM1 | Monosaccharide activation | Dol-P-Man synthase deficiency (CDG-Ie) | [40] |
| MPDU1 | Substrate utilization | MPDU1 deficiency (CDG-If) | [41] |
| DK1 | Lipid anchor activation | DK1 deficiency (CDG-Im) | [42] |
| B4GALT1 | Glycosyltransferase | B4GALT1 deficiency (CDG-IId) | [43] |
| SLC35A1 | Nucleotide sugar transport | SLC35A1 deficiency (CDG-IIf) | [44] |
| SLC35C1 | Nucleotide sugar transport | SLC35C1 deficiency (CDG-IIc) | [45] |
| COG1 | Golgi function | COG1 deficiency (CDG-IIg) | [46] |
| COG7 | Golgi function | COG7 deficiency (CDG-IIe) | [47] |
| COG8 | Golgi function | COG8 deficiency (CDG-IIh) | [48] |
| ATP6V0A2 | Golgi function | Autosomal recessive cutis laxa type II | [49] |
| DTDST | Ion transporter | Diastrophic dysplasia, achondrogenesis type IB | [50] |
| ATPSK2 | Substrate activation | Spondyloepimetaphyzeal dysplasia | [51] |
| GNE | Monosaccharide biosynthesis | Hereditary inclusion body myopathy-II (IBM2). Adult-onset myopathy | [52] |

### 2.3.1   TYPE I CDG

#### 2.3.1.1   CDG-IA (PHOSPHOMANNOMUTASE (PMM) DEFICIENCY)

An autosomal recessive disease known as CDG Ia, which is characterized by PMM2 gene alterations and a deficit in phosphomannomutase (PMM) enzyme action that is located on chromosome 16p13 [53, 54]. The onset of this type of disease occurs in early infancy with typical clinical features (fat pads and inverted nipples) besides failure to thrive, muscular hypotonia, strabismus, and raised transaminases and mostly involves the central nervous system (CNS); thus, it was recognized as a multi-system disorder [53, 55].

The disease mortality rate in childhood is 25% approximately, thus leading to acute conditions like sepsis and or organ collapse. A variable degree of mental retardation occurs at a later age of life along with retinitis pigmentosa and cerebellar dysfunction, leading to impairment of the neurologic system. Seizures and strokes have also been reported in some children. The disease in adults is mainly characterized by peripheral neuropathy, stable mental retardation, and nonprogressive ataxia. Most of the patients are bounded by wheelchair. Hypergonadotropic hypogonadism is also being presented by adult female patients [56].

#### 2.3.1.2   CDG-IB (PHOSPHOMANNOSE ISOMERASE [MPI DEFICIENCY])

CDG Ib is a metabolic disorder in which there is a deficiency of phosphomannose isomerase [57]. This disease has a different clinical presentation than that of CDG-Ia and other CDGs. Coagulopathy without conspicuous neurologic manifestations, enteropathy, malabsorption, and hepatic fibrosis [58, 59], are the major clinical features being presented by these patients and among others are hypoglycemia [55] and vomiting [58]. Oral mannose therapy is the only successful treatment for these patients after early diagnosis.

#### 2.3.1.3   CDG-IC (ALG 6 DEFICIENCY)

ALG6, a congenital disorder of glycosylation, is an inherited condition that affects many parts of the body [60]. Neurologic sickness is predominantly caused by a shortage of ALG6, which is minimal than that in the case of

CDG-Ia. Polyneuropathy, cerebellar hypoplasia, inverted nipples, and fat pads are some of the features of CDG-Ia that are missing in this disorder [61].

The signs and symptoms of this condition develop in individuals with ALG-CDG during infancy. The individuals may grow at the calculated rate (faltering weight) and also have difficulty in gaining weight. Infants affected with this disease often have developmental delay and weak muscle tone (hypotonia) [62]. Problems with coordination and balance (ataxia), seizures, or stroke-like episodes that include lack of energy (lethargy) and temporary paralysis along with blood clotting disorders may develop in people with ALG6-CDG [63].

### 2.3.1.4   CDG-ID (ALG 3 DEFICIENCY)

Mannosyltransferase VI inadequacy leads to the development of this disorder. Eye abnormalities, optic atrophy, intractable seizures, profound psychomotor retardation, hypsarrhythmia, and postnatal microcephaly are among the neurological impairments that CDG-Id individuals suffer with. The first was the child patient who reported to be the CDG "type IV positive in 1995 [64].

### 2.3.1.5   CDG-IE (MANNOSE SYNTHASE 1 INSUFFICIENCY) DPM1

Mannosyltransferase 1 insufficiency or fault in Mannose synthase I catalytic subunit leads to this condition. Mannose synthase I enzyme complex is formed by three different subunits, viz.; DPM1, DPM2, and DPM3. The catalytic subunit of DPM1 gets stabilized due to the association of the DPM3 subunit of the C-terminal domain, and the DPM2 subunit stabilizes this interaction [65].

GPI-anchored proteins, N-glycan en-bloc sugars/oligosaccharides, and some distinct Man-carrying glycoconjugates are produced by mannose elongation reactions, which require Dol-P-Man. Similar to CDG-Id patients, CDG-Ie patients also endure acute neurological problems that comprise intractable seizures, hypotonia, psychomotor retardation, cortical blindness, weight faltering, and eye deformity [66].

## 2.3.1.6    CDG-IF (MPDU1-CDG)

CDG-If, a recently discovered glycosylation disorder, with serious malfunctions like seizures, retardation at psychomotor level, weight faltering, erythroderma along with scaling, dry skin, and damaged vision are some of the clinical problems these patients suffer with. It is the first disorder that affects the use, rather than the biosynthesis, of donor substrates for lipid-linked oligosaccharides (LLO).

CDG-If results due to a deficiency of the gene MPDU1 (hamster Lec35, also a human homolog), and this mayhem is the first one that influences the usage of patron substrates for sugars/oligosaccharides that are lipid-linked, rather than its biosynthesis [41]. An amino acid of MPDU1, which is semi-conserved, carries a point mutation that is homozygous (221T->C, L74S) in these patients.

## 2.3.1.7    CDG-IG (ALG12-CDG)

Facial dystrophism, psychomotor retardation, and hypotonia are some of the clinical attributes related to CDG-Ig, and this disorder occurs due to the defect in the enzyme mannosyltransferase VIII. Microcephaly, feeding problems and frequent respiratory tract infections, and convulsions, also occur in some patients [67, 68].

## 2.3.1.8    CDG-IH (ALG8-CDG)

Deficiency in glucosyltransferase II results in CDg-Ih, which has been reported in only five children so far. The symptoms of CDG-Ih are hypotonia, moderate hepatomegaly, anemia, lung hypoplasia, and thrombocytopenia. There are 15 exons in the ALG8 gene, and two isoforms viz.; a and b of the enzyme are formed by the alternative mRNA splicing of these exons (526 and 467 amino acids respectively) and is located on chromosome 11q14.1 [69, 70].

## 2.3.1.9    CDG-II (ALG2-CDG)

This disorder is the consequence of a deficiency in the mannosyltransferase II enzyme. Only one case with this kind of disorder (CDG-Ii) has been identified so far, which revealed hypsarrhythmia, severe psychomotor retardation, irregular nystagmus (involuntary eye movements), and spasms. ALG2 gene is composed of only three exons [71, 72].

### 2.3.1.10    CDG-Ij (DPAGT1-CDG)

Defects in acetylglucosamine phosphotransferase enzyme lead to CDG-Ij and have been observed in only one patient so far. The patient had severe psychomotor retardation, suffered from intractable spasms, was also hypotonic [16]. Micrognathia with prominent dysmorphy, microcephaly, and esotropia may also occur (inward movement of both the eyes) [73].

### 2.3.1.11    CDG-Ik (ALG1-CDG)

This disease results due to the deficiency of the enzyme mannosyltransferase I. It is also known as β-1,4-mannosyltransferase. Psychomotor retardation, seizures, and early death were observed in infants that were found to be CDG-Ik positive [74]. Dysmorphia, liver dysfunction, microcephaly, acute infections, hypotonia, and defects in coagulation are the most common and unpredictable symptoms that were observed in these patients [75].

### 2.3.1.12    MAGT1-CDG

Deficiency of magnesium transporter 1 leads to MAGT1-CDG. The enzyme complex N-oligosaccharyltransferase (OST) transfers LLO (Glc3Man-9GlcNAc2-Dol) to the asparagine residue after the interaction of magnesium transporter 1 protein in nascent polypeptides. This reaction is the central step in the case of N-linked glycosylation of proteins [169].

The interaction of MAGT1 with OST specifies the role of MAGT1 in protein N-glycosylation that accounts for alterations in the MAGT1 gene that confirm its correspondence with congenital glycosylation disorders. Mental retardation X-linked type 95, MXR95 are disorders, which are associated with the alterations in the MAGT1 gene [76].

## 2.3.1.13   TUSC3-CDG

Tumor suppressor candidate 3 gene deficiency leads to TUSC3-CDG disorder. N-oligosaccharyltransferase complex (OST) interaction with TUSC3 encoded protein, transfers LLO structure (Glc3Man9GlcNAc2-Dol) to the asparagine residue in a nascent polypeptide chain [77]. This is the central reaction in the N-linked glycosylation of proteins. Autosomal recessive non-syndromic mental retardation-7 (MRT7) is related to the alterations in the TUCS3 gene. The patients who were identified to suffer acute non-syndromic mental retardation harbor mutations in this gene only.

## 2.3.2   TYPE II CDG FORMS

### 2.3.2.1   CDG-IIA TYPE DISORDER (DEFICIENCY OF N-ACETYL-GLUCOSAMINYLTRANSFERASE II)

MGAT2-CDG and/or CDG-IIa is the sequel of the 2-β-N-acetylglucosaminyl transferase enzyme deficiency. An interchangeable mono-antennary type glycan, i.e., acetyl-lactosamine, was established by glycan structural studies in sufferers with dyserythropoietic congenital type II anemia, which in turn confirmed the deficiency of this enzyme [78].

Complex sialylated structures are lost significantly in this defect, and solely mono-sialylated structures are formed upon a shift from N-linked residues to an N-block sugars/oligosaccharide. Craniofacial dysmorphia, Hypotonia, acute psychomotor retardation, frequent infections, stereotypic hand movement, and coarse facies are the symptoms of CDG-IIa [79]. The MGAT2 gene is an intronless gene that is positioned to the 14q21 chromosome that codes for 447 amino acid protein.

### 2.3.2.2   CDG-IIB (GLUCOSIDASE I DEFICIENCY)

Mannosyl-oligosaccharide glucosidase deficiency leads to MOGS; also called glucosidase 1 or GCS1-CDG. Hepatomegaly, seizures, feeding difficulty, hypotonia, failure to thrive, and peculiar dysmorphia was found in a patient identified with CDG-IIb, and the patient did not survive [80].

Tetrasaccharide Glc($\alpha$1-2),($\alpha$1-3),($\alpha$1-3)-Man accumulation has been observed in the urine of these patients [81].

### 2.3.2.3 CDG-IIC (DEFICIENCY OF TRANSPORTER GDP-FUCOSE)

LAD II syndrome (Leukocyte adhesion deficiency) is also called CDG-IIc (SLC35C1-CDG). Defects in leukocyte function lead to primary immuno-deficiency syndromes, of which LAD II is one of the classes of disorders. Symptoms of LAD II are characterized by defective neutrophil chemotaxis, persistent leukocytosis, recurrent infections, unique facial features; mental retardation, and severe growth are the characteristic symptoms of LAD II patients [82]. The defective neutrophil function seen in LAD II patients leads to recurrent infections. In innate immunity responses, mostly neutrophils are involved in bacterial infection. At the surface of the endothelium, neutrophils must adhere to perform their job at the site of inflammation in host protection mechanisms, the event mediated by cell surface adhesion molecules.

To arbitrate the inceptive process of adherence of neutrophils to the endothelium, the animal lectins are necessary, especially E, L, and P selectins of the selectin family. The deficiency of CD18, a leukocyte integrin subunit $\beta$2 found on the surface of monocytes and neutrophils, leads to the allied disorder known as LAD I. The location of the SLC35C1 gene corresponds to the 11p11.2 chromosome. This comprises of five exons that together create mRNA splicing, and these spliced isoforms codes for two distinct isoforms of proteins. The isoform a and b are composed of 364 and 351 amino acids, respectively [83, 84].

### 2.3.2.4 CDG-IID (B4GALT1-CDG)

The polypeptide 1,$\beta$-1,4-galactosyltransferase enzyme deficiency results in CDG-IId. In the human genome, there are seven B4GALT genes. All are membrane-bound type II glycoproteins that encode enzymes which exhibit utter selectivity as a donor substrate for the UDP-galactose that is shifted to either Glc or xylulose (Xyl), or GlcNAc. CDG-IId has been observed in one patient only [85] with spontaneous bleeding, mild psychomotor retardation, myopathy, and hydrocephalus. Two mutations were found in the affected individual in the B4GALT1 gene (9p13) [86].

### 2.3.2.5   CDG-IIE (COG7-CDG)

An oligomeric component of Golgi complex 7 deficiency results in CDG-IIe (COG7-CDG) disorder. The protein encoded by this gene (COG7) is one among eight oligomeric Golgi complexes (COG) conserved subunits [87]. Since this defect properly distorts trafficking along the Golgi and its sequels are very much apparent in the processes of both O- and N-linked pathways of glycosylation. Hypotonia, progressive severe microcephaly, growth retardation, pseudo-obstruction of the gastrointestinal tract, adducted thumbs, failure to thrive, wrinkled skin, cardiac anomalies, and extreme hyperthermia episodes are among the features presented by these patients [88].

### 2.3.2.6   CDG-IIF (SLC35A1-CDG)

The absence of nucleotide sugar transporter (NST) results in CDG-IIf, as this transporter/enzyme is responsible for the transport of sialic acid conjugated cytidine monophosphate CMP into the ER/Golgi lumen from the cytosol. NST is a solute carrier family member of transporters. Transporter deficiency results in neutropenia and marked thrombocytopenia along with periodic infections. The cyclical infections are ascribed by imperceptible sialyl-Lewis on the neutrophil cell surface that restrains them at sites of inflammation from sticking to the endothelium [89].

Cutaneous hemorrhages and spontaneous extensive bleeding of the right eye posterior chamber were among the clinical findings presented at the age of four months. Recurring infections and pronounced thrombocytopenia were seen following acute hemorrhages. Testing revealed neutropenia, macro-thrombocytopenia, and immunodeficiency. The patient succumbed to complications from bone-marrow transplantation at the age of 37 months [90].

### 2.3.2.7   CDG-IIG (COG1-CDG)

CDG-IIg is a kind of disorder that results from the deficiency of a component of a gene that encodes an oligomeric subunit of Golgi complex 1. Jaeken and Matthijs [86] demonstrated that the protein encoded by the COG1 gene is one among the preserved oligomeric subunits of the Golgi (COG) complex.

Golgi trafficking is properly disrupted by the defect in this gene, which is evident by the effective O- and N-linked glycosylation pathways. This disorder has been described in only a few patients so far, but the patients were presented with generalized hypotonia, growth retardation with rhizomelic short stature, failure to thrive, mild psychomotor retardation, and hepatosplenomegaly in infancy. Cerebrocostomandibular (CCMS)-like syndrome has also been described in one of the patients [46].

### 2.3.3   PROTEIN O-GLYCOSYLATION DEFECTS

#### 2.3.3.1   MUSCULAR DYSTROPHY-DYSTROGLYCANOPATHIES (MDDG)

Various known disorders that occur due to error in the O-mannosylation of the α-dystroglycan protein of the dystrophin-glycoprotein complex (DGC) are known and have been termed as muscular dystrophy-dystroglycanopathies (MDDG) [91]. The diseases or defects of muscle-eye-brain diseases (less severe forms) and Walker Warburg (most severe forms) syndrome are among the characterized class of disorders/defects that often result due to defects in O-mannosylation [92, 93]. POMT1, POMT2, or POMGNT1 deficiencies result in these disorders. The genes POMT1 and 2 encode two different isoforms of the enzyme O-mannosyltransferase respectively, while POMGNT1 participates in O-mannosyl glycosylation and is specific for alpha linked terminal mannose.

#### 2.3.3.2   B4GALT7-CDG

The enzyme β1,4-galactosyltransferase I (B4GALT1) that usually elongates glycoprotein glycans is a Golgi inhabitant enzyme [94]. This enzyme is often called galactosyltransferase I, which is encoded by the B4GALT7 gene in humans [37, 38] and is numbered as one among the seven enzymes of β-1,4-galactosyltransferase (β4GalT). These are also categorized as membrane-bound type II glycoproteins [95]. The Ehlers-Danlos syndrome (EDS), a progeroid variant, also known as accelerated aging and multiple cartilaginous exostoses (MCE), both occur due to the defect in O-xylosylation [179].

The glucuronyltransferase and N-acetylglucosaminyltransferase, which are encoded by either EXT1 or EXT2 genes, respectively, results in different forms of MCE [39]. The defect in protein O-glycosylation leads to disorder EDS spondylodysplastic type 1 EDSSPD1, which is the result of B4GALT7 gene deficiency that encodes β-1,4-galactosyltransferase 7 enzymes.

### 2.3.4    *DIAGNOSIS OF CDGS*

The CDGs are currently diagnosed by analyzing the glycoforms of serum transferrin (transferrin isoforms analysis). DNA testing is usually done after this test comes positive for that particular patient. DNA analysis/testing is helpful in identifying the mutations in that particular gene, which leads to the development of the particular disorder.

## 2.4    GLYCOGEN STORAGE DISEASES (GSDS)

The enzymes of carbohydrate metabolism and synthesis, which are being affected by various genetic variants, have been identified, and the glycogen storage diseases (GSDs) occur due to these variants. The incidence of these in live births ranges from 1 in 20,000 to 1 in 43,000 [96]. The complete list of GSDs are provided in Table 2.2 with detailed discussion as under:

**TABLE 2.2**    Glycogen Storage Diseases

| Type | Notes | Development/ Prognosis | References |
|---|---|---|---|
| GSD Type 0 | Glycogen Synthase | Growth failure in some cases | [96] |
| GSD Type I (von Gierke's disease) | Glycogen-6-phosphatase | Growth failure | [97] |
| GSD Type II (Pompe's disease) | Acid alpha-glucosidase | Death at the age of 2 years | [177] |
| GSD Type III (Cori's disease or Forbes' disease) | Glycogen debranching enzyme | | [98] |
| GSD Type IV (Andersen disease) | Glycogen branching enzyme | Death at the age of 5 years | [99] |
| GSD Type V (McArdle disease) | Muscle glycogen phosphorylase | | [100] |
| GSD Type VI (Hers' disease) | Liver glycogen phosphorylase | Growth retardation | [101] |
| GSD Type VII (Tarui's disease) | Muscle phosphofructokinase | Growth retardation | [102] |

**TABLE 2.2**    *(Continued)*

| Type | Notes | Development/ Prognosis | References |
|---|---|---|---|
| GSD Type IX | Phosphorylase kinase | Delayed motor development | [103] |
| GSD Type XI (Fanconi-Bickel syndrome) | Glucose transporter, GLUT2 | | [104] |
| GSD Type XII (Red cell aldolase deficiency) | Aldolase A | | [105] |
| GSD Type XIII | β-enolase | Myalgias with increasing intensity | [106] |
| GSD XV | Glycogenin-1 | | [107] |
| E3 ubiquitin ligase Deficiency | E3 ubiquitin ligase | | [96] |

## 2.4.1   MCARDLE DISEASE

Brian McArdle, in 1951 [108], was the first to describe this disorder in a considerate that during the ischemic workout was totally unable to make lactate [108]. Gene alterations in glycogen phosphorylase, muscle associated PYGM gene, leads to the development of this disorder [109]. Various PYGM mutations have been related to this disorder. A lot of clinical variabilities have also been observed in this disease in which some patients have proximal muscle weakness [100]; others report widespread muscle weakness. In addition to this, a rapidly progressive neonatal fatal form is also described [110]. The exact epidemiological data is not available, but it is believed that the prevalence rate ranges from 1 in 100,000 to 1 in 167,000 [111]. Myalgia, rapid fatigue, and cramping are among the symptoms associated with exercise. The main finding in sufferers of this particular disorder is the prompt refinement of symptoms due to rest, called the "second-wind phenomenon." The diagnosis at the clinical level demands a high grade of suspicion in mild to moderately affected patients, when workout intolerance is the only symptom, especially in older patients. Recognition of PYGM gene pathogenic allelic variants confirms the diagnosis, as this is the only gene related to McArdle disorder, and it encodes for muscle phosphorylase protein [111].

### 2.4.2   CORI-FORBES DISEASE

Recessive autosomal alteration in the gene AGL, results in type III GSD, which is also referred to as Cori-Forbes disorder. This results in a deficit of the enzyme amylo, α-1,6-glucosidase, 4α-glucanotransferase (glycogen debranching enzyme). There is less than 1 in 100,000 prevalence rate of this disorder. It has four subtypes viz.; IIIa, IIIb, IIIc, and IIId. Most of the patients had IIIa type disorder, which constitutes the maximum number of GSD III. These patients had the absence of this enzyme in the liver and muscle. However, people with type IIIb comprise only 15% of patients with this disorder, and they lack the enzyme only in the liver [112]. While as in the case of IIIa type, late-onset cardiomyopathy results due to the deficiency of this enzyme in cardiac muscles [98, 113].

### 2.4.3   TARUI DISEASE

Tarui disease occurs due to the deficiency of phosphofructokinase and was the first disorder identified to affect glycolysis directly. This disorder is named after Seiichiro Tarui, a Japanese physician, who in 1965, was the first to identify the deficiency of phosphofructokinase leading to GSD VII (Tarui disease) [102]. Alterations in the PFKM gene lead to this condition, and this gene encodes the PFK muscle forms. There are three different subunits of the PFK gene, and each subunit is encoded by a different gene, so there is a possibility of different gene mutations like PFKP (platelet type) and PFKL (liver type) [114, 115]. Muscle contracture, exercise intolerance, rhabdomyolysis, and pain are among the symptoms of Tarui disease, which are in common with other glycolytic disorders.

Approximately 100 cases have been described so far according to published reports, with this rare condition known as Tarui disease. The consanguineous inheritance and or Ashkenazi Jewish genesis individuals are having a high prevalence rate of this disorder [116]. An animal model like dogs would represent the best future study models, as the disease is more common in them [117].

### 2.4.4   B-ENOLASE DEFICIENCY

The conversion of 2-phosphoglycerate to phosphoenolpyruvate occurs due to the presence of enzyme β-enolase, and the deficiency of this enzyme leads to the rare disorder, GSD XIII. Comi et al. in 2001 [106] reported this for the first time. Alterations in the ENO3 gene correspond to the registered cases of this disorder till date [118]. Because of the presence of various subunits (α, β, and γ) in β-enolase, there is a possibility of other gene mutations as well [119].

The deficiency of β-Enolase is less severe as compared to McArdle disease, and myalgia, exercise intolerance, post-workout muscle fatigue, myoglobinuria, contracture, and a high proportion of creatine are among the major manifestations observed in the patients who suffer with this disorder [106].

### 2.4.5   PHOSPHORYLASE SS KINASE DEFICIENCY

The deficiency of this particular enzyme was first outlined in the 1960s [120], and the gene alterations in PHKA1, A2, and B or G2 recessive autosomal or X-linked inheritance were found to be responsible for the deficiency of this enzyme. The deficiency of Phosphorylase ß kinase leads to disorder GSD IX and has a prevalence rate of about 1 in 100000 [121]. The deficit of this enzyme in the liver or muscle is the characteristic feature. The active form of phosphorylase-a is formed from the non-functioning form of specific phosphorylase (phosphorylase-b) due to the normal functioning of this enzyme. Different GSD types (IXa, IXb, IXc, and IXd), are the result of the deficiency of the enzyme phosphorylase b kinase, thus considering the fact that this enzyme consists of a tetramer of four different protein subunits (α, β, γ, and δ), each encoded by a distinct gene [122].

### 2.4.6   PHOSPHOGLUCOMUTASE-1 DEFICIENCY

The crucial transitional molecule during the processes that precede to N-glycosylation of proteins and glucose homeostasis is glucose-1-phosphate. Glucose-6-phosphate and Glucose-1-phosphate interconversion is catalyzed by Phosphoglucomutase 1 [123]. Deficiency of Phosphoglucomutase-1 (PGM1) has been reported as a source of rhabdomyolysis and workout intolerance in adults [124]. Dysmorphic features, short stature, accessory thumbs, and cleft palate are among the clinical features of this disorder. Increased proportion of creatine kinase, cardiomyopathy, myopathy, and diseases of the liver are included as other features of this disease [98].

### 2.4.7  POMPS DISEASE

α-glucosidase acid (GAA), the main enzyme that hydrolyzes glycogen inside the lysosomes into glucose, and loss of action/activity lead to the development of Infantile-onset type Pompe disease, which is a recessive autosomal glycogen storage defect. Glycogen gets accumulated within the lysosomes in the absence of GAA activity, and this condition is related to the foremost pathological manifestations in cardiac and skeletal muscle [177].

Three categories/classes of Pomps defect/disease have been described. Infantile onset classic, Non-classic, and Late-Onset. The classic form of this disorder starts within a few months of birth. Hepatomegaly, muscle hypotonia, muscle myopathy, and heart defects are experienced by these types of patients. These patients have a problem in gaining weight and also have breathing problems. Death may result, if it remains untreated [125].

The non-classic form generally begins when the person is one year old. Progressive muscle weakness and delayed motor skills are the characteristic features of this form of the disease. Cardiomegaly occurs in these individuals, but maybe protected from heart failure. The muscle weakness causes extreme breathing problems [126].

The late-onset form of the disease begins in the later childhood stages, or adulthood. This type of disease is not as severe as that of the other two forms. The disruptive weakness of muscles is mostly experienced by these patients, especially in the trunk and legs. Respiratory failure may occur in the late stages of the disease [180].

## 2.5  LYSOSOMAL STORAGE DISEASES AND MODELING LYSOSOMAL STORAGE DISORDERS (LSDS) WITH PATIENT-DERIVED IPSCS

The inherited metabolic disorders that occur due to lysosomal dysfunction lead to the development of lysosomal storage diseases. These disorders occur due to the deficiency or consequence of an enzyme necessary for the metabolism of glycoproteins and lipids, together known as mucopolysaccharides [127]. Details of lysosomal storage diseases are summarized in Table 2.3 [128].

Lysosomal storage disorders (LSDs), including Pompe disease [129, 130], Hurler syndrome [131], Type C Niemann-Pick [132], Gaucher disease (GD) [133–135], and Fabry disease [136] are another domains of iPSC disease modeling technology. The alterations in proteins facilitate the degradation of specific glycan products that result in these storage disorders. Deprivation of these proteins culminates in an expansion of glycan pieces and leads to significant physiological and developmental effects [137]. LSDs exhibit various cell-type-specific phenotypes, and these can best be elucidated by differentiating iPSCs towards the most afflicted cell-type (Table 2.3).

**TABLE 2.3**  Lysosomal Storage Diseases

| Disease Type | Enzyme Deficiency | Notes | References |
|---|---|---|---|
| Alpha-Mannosidosis | Alpha-D-mannosidosis | Splenomegaly, Intellectual Disability, Hearing Loss, Skeletal abnormalities | [138, 139] |
| Aspartylglucosaminuria | Aspartylglucosaminidase | Seizures, Respiratory Infections, Scoliosis, Progressive change in Facial features | [140, 141] |
| Batten Disease | Mutations in the CLN3 gene | Clumsiness, or stumbling, repetitive speech or echolalia | [128] |
| Beta-Mannosidosis | Beta-mannosidase | Hearing loss, angiokeratomas | [142] |
| Cystinosis | Mutation in the gene CTNS | Photophobia, Kidney Failure, Muscle deterioration, Blindness, Inability to swallow, Diabetes | [143] |
| Danon Disease | Defect in LAMP2 | Wolff-Parkinson-White syndrome, Cardiomyopathy, etc. | [144] |
| Fabry Disease | Alpha-galactosidase A deficiency | Acroparesthesia, kidney insufficiency and kidney failure, restrictive cardiomyopathy | [145] |
| Farber Disease | Ceramidase deficiency | Kidneys, liver, and heart may also be affected, swollen lymph nodes | [146] |
| Fucosidosis | Deficiency of alpha-L-fucosidase | Enlarged Spleen, Liver/or heart, Seizures, Intellectual Disability | [147] |

**TABLE 2.3**   *(Continued)*

| Disease Type | Enzyme Deficiency | Notes | References |
|---|---|---|---|
| Galactosialidosis | Deficiency of beta-galactosidase (GLB1) and neuraminidase 1 | Hepatosplenomegaly, cardiomegaly, inguinal hernia | [148] |
| Gaucher disease | Glucocerebrosidase | Fatigue, bruising, anemia, low blood platelet count | [149] |
| Gangliosidosis<br>• M1-Gangliosidosis | Beta-galactosidase deficiency | Lose motor skills like sitting, turning over, and crawling, seizures, vision, and hearing loss | [150] |
| • M2-Gangliosidosis AB Variant | Beta-hexosaminidase deficiency | | [151] |
| Krabbe Disease | Galactosylceramidase (GALC) deficiency | Seizures, Limb stiffness, Vomiting, Feeding difficulties, and slowing of mental and motor development. | [152] |
| Metachromatic Leukodystrophy | Arylsulfatase A deficiency | Muscle rigidity, Convulsions, Paralysis, and dementia | [153] |
| Mucopolysaccharidoses Disorders<br>• MPS 1-Hurler syndrome | α-L-iduronidase deficiency | Micrognathia, Intellectual disability, corneal clouding, cardiomyopathy, hepatosplenomegaly, retinal degeneration, | [128]<br><br>[154] |
| • MPS II-Hunter syndrome | Iduronate sulfatase deficiency | Intellectual disability, X-linked recessive inheritance | [155] |
| • MPS III-Sanfilippo<br>• MPS III A | Heparan sulfamidase | Severe hyperactivity, developmental delay, spasticity, death by the second decade, motor dysfunction, | [128]<br>[156, 157] |
| • MPS III B | N-acetylglucosaminidase | | |
| • MPS III C | Heparan-α-glucosaminide N-acetyltransferase | | |
| • MPS III D | N-acetylglucosamine 6-sulfatase | | |

**TABLE 2.3** *(Continued)*

| Disease Type | Enzyme Deficiency | Notes | References |
|---|---|---|---|
| • MPS IVA-Morquio | Galactose-6-sulfate sulfatase | Motor dysfunction, severe skeletal dysplasia, short stature | [158] |
| • MPS IVB-Morquio | β-galactosidase | | |
| • MPS IX-Hyaluronidase Deficiency | Hyaluronidase | Episodes of painful swelling of the masses, mild facial changes, short stature, normal joint movement, normal intelligence | [128] |
| • MPS VI-Maroteaux-Lamy | N-acetylgalactosamine-4-sulfatase | Motor dysfunction, severe skeletal dysplasia, heart defects, short stature, | [159] |
| • MPS VII-Sly syndrome | β-glucuronidase | Corneal clouding, short stature, hepatomegaly, developmental delay, skeletal dysplasia, | [160] |
| Mucolipidosis type I (Sialidosis) | Neuraminidase deficiency | Visual problems, ataxia, and myoclonic epilepsy | [161] |
| Mucolipidosis type II (I-Cell disease) | GlcNAc-1-phosphotransferase | Dislocated hips, kyphosis, unusually shaped long bones, clubfeet. | [128] |
| Mucolipidosis type III (Pseudo-Hurler polydystrophy) | N-acetylglucosamine-1-phosphotransferase | Osteoporosis, multiple skeletal abnormalities, short stature | [162] |
| Mucolipidosis type IV | Mutations in the MCOLN1 gene | Anemia, Hypotonia, Psychomotor retardation, Achlorhydria, | [163] |
| Multiple Sulfatase Deficiency | Sulfatase | Abnormalities of the skeleton, deafness, ichthyosis, coarsened facial features, enlarged liver, and spleen. | [128] |

**TABLE 2.3**    *(Continued)*

| Disease Type | Enzyme Deficiency | Notes | References |
|---|---|---|---|
| Niemann-Pick types A, B, C | Sphingomyelinase | Splenomegaly, thrombocytopenia, hepatosplenomegaly | [128] |
| Pycnodysostosis | Defect in Cathepsin K | Abnormally short fingers, coarse facial features, small jaw, scoliosis, wrinkled skin | [128] |
| Sandhoff Disease | Beta-hexosaminidase deficiency | Respiratory problems and infections, muscle/motor weakness, blindness, inability to react to stimulants, sharp reaction to loud noises, deafness, seizures | [164] |
| Schindler Disease | Alpha-NAGA (alpha-N-acetylgalactosaminidase) | Seizures, Telangiectasia, involuntary rapid eye movements, angiokeratomas, vision loss | [165] |
| Salla Disease/Sialic Acid Storage Disease | Mutation in the SLC17A5 gene | Cognitive impairment, nystagmus, as well as hypotonia, reduced muscle tone. | [166] |
| Tay-Sachs | Hexosaminidase A | Intellectual disability, vision, and hearing loss, seizures, and paralysis. | [167] |
| Wolman Disease | Lysosomal acid lipase (LAL) | Swelling of the abdomen, liver dysfunction, Anemia | [168] |

## 2.6   CONCLUSION

From the last one decade, the subject of CDG has been expeditiously expanding, the increasing awareness and understanding of the clinical variability of CDG, glycogen storage disorders, and lysosomal disorders, and its use in genetic technology has a tremendous impact on the diagnostics of the human glycosylation disorders. The most important part is in diagnosing the disease states in children in their infancy, which if corrected makes their life easier and, in adults as well. For example, by using serum transferrin IEF as a diagnostic tool as this was the first testing done for detecting the presence of any kind of CDGs. Glycans play important roles in tumor cell dissociation, cancer cell signaling, invasion, angiogenesis, cell-matrix interactions, immune modulation, and metastasis. It is rather certain that various new disorders of glycosylation are awaiting recognition as it is believed that a lot of genes are involved in the synthesis of oligosaccharides and hence their function. Moreover, the establishment of cell and animal models offers the promise to test the therapeutic approaches which will be beneficial in grasping the pathogenesis of these disorders.

## KEYWORDS

- **central nervous system**
- **dystrophin-glycoprotein complex**
- **endoplasmic reticulum**
- **glycogen storage diseases**
- **Golgi apparatus**
- **lipid-linked oligosaccharide**

## REFERENCES

1. Lauc, G., Pezer, M., Rudan, I., & Campbell, H., (2016). Mechanisms of disease: The human N-glycome. *Biochim. Biophys. Acta, 1860*(8), 1574–1582.
2. Marek, K. W., Vijay, I. K., & Marth, J. D., (1999). A recessive deletion in the GlcNAc-1-phosphotransferase gene results in peri-implantation embryonic lethality. *Glycobiology, 9*(11), 1263–1271.
3. Kornfeld, R., & Kornfeld, S., (1985). Assembly of asparagine-linked oligosaccharides. *Annu. Rev. Biochem., 54*, 631–664.
4. Freeze, H. H., (2006). Genetic defects in the human glycome. *Nat. Rev. Genet., 7*(7), 537–551.
5. Van, D. S. P., Rudd, P. M., Dwek, R. A., & Opdenakker, G., (1998). Concepts and principles of O-linked glycosylation. *Crit. Rev. Biochem. Mol. Biol., 33*(3), 151–208.

6. Harris, R. J., & Spellman, M. W., (1993). O-linked fucose and other post-translational modifications unique to EGF modules. *Glycobiology, 3*(3), 219–224.
7. Michaëlsson, E., Malmström, V., Reis, S., Engström, A., Burkhardt, H., & Holmdahl, R., (1994). T-cell recognition of carbohydrates on type II collagen. *J. Exp. Med., 180*(2), 745–749.
8. Hart, G. W., (1997). Dynamic O-linked glycosylation of nuclear and cytoskeletal proteins. *Annu. Rev. Biochem., 66*, 315–335.
9. Lauc, G., Vojta, A., & Zoldoš, V., (2014). Epigenetic regulation of glycosylation is the quantum mechanics of biology. *Biochim. Biophys. Acta, 1840*(1), 65–70.
10. Knezevic, A., Gornik, O., Polasek, O., Pucic, M., Redzic, I., Novokmet, M., et al., (2010). Effects of aging, body mass index, plasma lipid profiles, and smoking on human plasma N-glycans. *Glycobiology, 20*(8), 959–969.
11. Hülsmeier, A. J., Tobler, M., Burda, P., & Hennet, T., (2016). Glycosylation site occupancy in health, congenital disorder of glycosylation and fatty liver disease. *Sci. Rep., 6*, 33927.
12. Berger, R. P., Dookwah, M., Steet, R., & Dalton, S., (2016). Glycosylation and stem cells: Regulatory roles and application of iPSCs in the study of glycosylation-related disorders. *Bioessays, 38*(12), 1255–1265.
13. Neumann, L. M., Von, M. A., Kunze, J., Blankenstein, O., & Marquardt, T., (2003). Congenital disorder of glycosylation type Ia in a macrosomic 16-month-old boy with an atypical phenotype and homozygosity of the N216I mutation. *Eur. J. Pediatr., 162*(10), 710–713.
14. Bjursell, C., Wahlström, J., Berg, K., Stibler, H., Kristiansson, B., Matthijs, G., et al., (1998). Detailed mapping of the phosphomannomutase 2 (PMM2) genes and mutation detection enable improved analysis for Scandinavian CDG type I families. *Eur. J. Hum. Genet., 6*(6), 603–611.
15. Schollen, E., Dorland, L., De Koning, T. J., Van, D. O. P., Huijmans, J. G., & Marquardt, T., (2000). Genomic organization of the human phosphomannose isomerase (MPI) gene and mutation analysis in patients with congenital disorders of glycosylation type Ib (CDG-Ib). *Hum. Mutat., 16*(3), 247–252.
16. Wu, X., Rush, J. S., Karaoglu, D., Krasnewich, D., Lubinsky, M. S., Waechter, C. J., et al., (2003). Deficiency of UDP-GlcNAc: Dolichol phosphate N-acetylglucosamine-1 phosphate transferase (DPAGT1) causes a novel congenital disorder of glycosylation type Ij. *Hum. Mutat., 22*(2), 144–150.
17. Thiel, C., Schwarz, M., Peng, J., Grzmil, M., Hasilik, M., Braulke, T., et al., (2003). A new type of congenital disorders of glycosylation (CDG-Ii) provides new insights into the early steps of dolichol-linked oligosaccharide biosynthesis. *J. Biol. Chem., 278*(25), 22498–22505.
18. Denecke, J., Kranz, C., Kemming, D., Koch, H. G., & Marquardt, T., (2004). An activated 5' cryptic splice site in the human ALG3 gene generates a premature termination codon insensitive to nonsense-mediated mRNA decay in a new case of congenital disorder of glycosylation type Id (CDG-Id). *Hum. Mutat., 23*(5), 477–486.
19. Weinstein, M., Schollen, E., Matthijs, G., Neupert, C., Hennet, T., Grubenmann, C. E., et al., (2005). CDG-IL: An infant with a novel mutation in the ALG9 gene and additional phenotypic features. *Am. J. Med. Genet. A., 136*(2), 194–197.

20. Eklund, E. A., Newell, J. W., Sun, L., Seo, N. S., Alper, G., & Willert, J., (2005). Molecular and clinical description of the first US patients with congenital disorder of glycosylation Ig. *Mol. Genet. Metab., 84*(1), 25–31.

21. Imbach, T., Grünewald, S., Schenk, B., Burda, P., Schollen, E., Wevers, R. A., et al., (2000). Multi-allelic origin of congenital disorder of glycosylation (CDG)-Ic. *Hum. Genet., 106*(5), 538–545.

22. Schollen, E., Frank, C. G., Keldermans, L., Reyntjens, R., Grubenmann, C. E., Clayton, P. T., et al., (2004). Clinical and molecular features of three patients with congenital disorders of glycosylation type Ih (CDG-Ih) (ALG8 deficiency). *J. Med. Genet., 41*(7), 550–556.

23. Coutinho, M. F., Encarnação, M., Laranjeira, F., Lacerda, L., Prata, M. J., & Alves, S., (2016). Solving a case of allelic dropout in the GNPTAB gene: Implications in the molecular diagnosis of mucolipidosis type III alpha/beta. *J. Pediatr. Endocrinol. Metab., 29*(10), 1225–1228.

24. Rafaelsen, S., Johansson, S., Ræder, H., & Bjerknes, R., (2014). Long-term clinical outcome and phenotypic variability in hyperphosphatemic familial tumoral calcinosis and hyperphosphatemic hyperostosis syndrome caused by a novel GALNT3 mutation, case report and review of the literature. *BMC Genetics, 15*, 98.

25. Wang, Y., Ju, T., Ding, X., Xia, B., Wang, W., Xia, L., et al., (2010). Cosmic is an essential chaperone for correct protein O-glycosylation. *Proc. Natl. Acad Sci. The U.S.A, 107*(20), 9228–9233.

26. Judas, M., Sedmak, G., Rados, M., Sarnavka, V., Fumić, K., Willer, T., et al., (2009). POMT1-associated walker-Warburg syndrome: A disorder of dendritic development of neocortical neurons. *Neuropediatrics, 40*(1), 6–14.

27. Martinez, H. R., Craigen, W. J., Ummat, M., Adesina, A. M., Lotze, T. E., & Jefferies, J. L., (2014). Novel cardiovascular findings in association with a POMT2 mutation: Three siblings with α-dystroglycanopathy. *Eur. J. Hum. Genet., 22*(4), 486–491.

28. Bouchet, C., Gonzales, M., Vuillaumier-Barrot, S., Devisme, L., Lebizec, C., Alanio, E., et al., (2007). Molecular heterogeneity in fetal forms of type II lissencephaly. *Hum. Mutat., 28*(10), 1020–1027.

29. Saito, Y., Yamamoto, T., Mizuguchi, M., Kobayashi, M., Saito, K., Ohno, K., et al., (2006). Altered glycosylation of alpha-dystroglycan in neurons of Fukuyama congenital muscular dystrophy brains. *Brain Res., 1075*(1), 223–228.

30. Beltran-Valero, D. B. D., Voit, T., Longman, C., Steinbrecher, A., Straub, V., Yuva, Y., et al., (2004). Mutations in the FKRP gene can cause muscle-eye-brain disease and Walker-Warburg syndrome. *J. Med. Genet., 41*(5), e61.

31. Longman, C., Brockington, M., Torelli, S., Jimenez-Mallebrera, C., Kennedy, C., Khalil, N., et al., (2003). Mutations in the human LARGE gene cause MDC1D, a novel form of congenital muscular dystrophy with severe mental retardation and abnormal glycosylation of alpha-dystroglycan. *Hum. Mol. Genet., 12*(21), 2853–2861.

32. Dunwoodie, S. L., (2009). Mutation of the fucose-specific beta1, 3 N-acetylglucosaminyltransferase LFNG results in abnormal formation of the spine. *Biochim. Biophys. Acta, 1792*(2), 100–111.

33. Weh, E., Reis, L. M., Tyler, R. C., Bick, D., Rhead, W. J., Wallace, S., et al., (2014). Novel B3GALTL mutations in classic Peters plus syndrome and lack of mutations in a large cohort of patients with similar phenotypes. *Clin. Genet., 86*(2), 142–148.

34. Simpson, M. A., Cross, H., Proukakis, C., Priestman, D. A., Neville, D. C., Reinkensmeier, G., et al., (2004). Infantile-onset symptomatic epilepsy syndrome caused by a homozygous loss-of-function mutation of GM3 synthase. *Nat. Genet., 36*(11), 1225–1229.

35. Mortazavi, Y., Merk, B., McIntosh, J., Marsh, J. C., Schrezenmeier, H., Rutherford, T. R., et al., (2003). The spectrum of PIG-A gene mutations in aplastic anemia/paroxysmal nocturnal hemoglobinuria (AA/PNH): A high incidence of multiple mutations and evidence of a mutational hot spot. *Blood, 101*(7), 2833–2841.

36. Freeze, H. H., & Ng, B. G., (2011). Golgi glycosylation and human inherited diseases. *Cold Spring Harb. Perspect. Biol., 3*(9), a005371.

37. Okajima, T., Yoshida, K., Kondo, T., & Furukawa, K., (1999). Human homolog of *Caenorhabditis elegans* sqv-3 gene is galactosyltransferase I involved in the biosynthesis of the glycosaminoglycan-protein linkage region of proteoglycans. *J. Biol. Chem., 274*(33), 22915–22918.

38. Almeida, R., Levery, S. B., Mandel, U., Kresse, H., Schwientek, T., Bennett, E. P., et al., (1999). Cloning and expression of a proteoglycan UDP-galactose: Beta-xylose beta1, 4-galactosyltransferase I. A seventh member of the human beta4-galactosyltransferase gene family. *J. Biol. Chem., 274*(37), 26165–26171.

39. Cousminer, D. L., Arkader, A., Voight, B. F., Pacifici, M., & Grant, S. F., (2016). Assessing the general population frequency of rare coding variants in the EXT1 and EXT2 genes previously implicated in hereditary multiple exostoses. *Bone, 92*, 196–200.

40. Kim, S., Westphal, V., Srikrishna, G., Mehta, D. P., Peterson, S., Filiano, J., et al., (2000). Dolichol phosphate mannose synthase (DPM1) mutations define congenital disorder of glycosylation Ie (CDG-Ie). *J. Clin. Invest., 105*(2), 191–198.

41. Kranz, C., Denecke, J., Lehrman, M. A., Ray, S., Kienz, P., Kreissel, G., et al., (2001). A mutation in the human MPDU1 gene causes congenital disorder of glycosylation type If (CDG-If). *J. Clin. Invest., 108*(11), 1613–1619.

42. Helander, A., Stödberg, T., Jaeken, J., Matthijs, G., Eriksson, M., & Eggertsen, G., (2013). Dolichol kinase deficiency (DOLK-CDG) with a purely neurological presentation caused by a novel mutation. *Mol. Genet. Metab., 110*(3), 342–344.

43. Guillard, M., Morava, E., De Ruijter, J., Roscioli, T., Penzien, J., Van, D. H. L., et al., (2011). B4GALT1-congenital disorders of glycosylation presents as a non-neurologic glycosylation disorder with hepatointestinal involvement. *J. Pediatr., 159*(6), 1041–1043.e2.

44. Riemersma, M., Sandrock, J., Boltje, T. J., Büll, C., Heise, T., Ashikov, A., et al., (2015). Disease mutations in CMP-sialic acid transporter SLC35A1 result in abnormal α-dystroglycan O-mannosylation, independent from sialic acid. *Hum. Mol. Genet., 24*(8), 2241–2246.

45. Cagdas, D., Yilmaz, M., Kandemir, N., Tezcan, I., Etzioni, A., & Sanal, Ö., (2014). A novel mutation in leukocyte adhesion deficiency type II/CDGIIc. *J. Clin. Immunol., 34*(8), 1009–1014.

46. Foulquier, F., Vasile, E., Schollen, E., Callewaert, N., Raemaekers, T., Quelhas, D., et al., (2006). Conserved oligomeric Golgi complex subunit 1 deficiency reveals a previously uncharacterized congenital disorder of glycosylation type II. *Proc. Natl. Acad. Sci. The U.S.A., 103*(10), 3764–3769.

47. Spaapen, L. J., Bakker, J. A., Van, D. M. S. B., Sijstermans, H. J., Steet, R. A., Wevers, R. A., et al., (2005). Clinical and biochemical presentation of siblings with COG-7

deficiency, a lethal multiple O- and N-glycosylation disorder. *J. Inherit. Metab. Dis.,* *28*(5), 707–714.

48. Foulquier, F., Ungar, D., Reynders, E., Zeevaert, R., Mills, P., García-Silva, M. T., et al., (2007). A new inborn error of glycosylation due to a Cog8 deficiency reveals a critical role for the Cog1-Cog8 interaction in COG complex formation. *Hum. Mol. Genet.,* *16*(7), 717–730.

49. Kornak, U., Reynders, E., Dimopoulou, A., Van, R. J., Fischer, B., Rajab, A., et al., (2008). Impaired glycosylation and cutis laxa caused by mutations in the vesicular H+-ATPase subunit ATP6V0A2. *Nat. Genet., 40*(1), 32–34.

50. Dwyer, E., Hyland, J., Modaff, P., & Pauli, R. M., (2010). Genotype-phenotype correlation in DTDST dysplasias: Atelosteogenesis type II and diastrophic dysplasia variant in one family. *Am. J. Med. Genet. A., 152A*(12), 3043–3050.

51. Faiyaz-Ul, H. M., King, L. M., Krakow, D., Cantor, R. M., Rusiniak, M. E., Swank, R. T., et al., (1998). Mutations in orthologous genes in human spondyloepimetaphyseal dysplasia and the brachymorphic mouse. *Nat. Genet., 20*(2), 157–162.

52. Nishino, I., Carrillo-Carrasco, N., & Argov, Z., (2015). GNE myopathy: Current update and future therapy. *J. Neurol. Neurosurg. Psychiatry, 86*(4), 385–392.

53. Jaeken, J., (2011). Congenital disorders of glycosylation (CDG): It's (nearly) all in it! *J. Inherit. Metab. Dis., 34*(4), 853–858.

54. Yuste-Checa, P., Gámez, A., Brasil, S., Desviat, L. R., Ugarte, M., Pérez-Cerdá, C., et al., (2015). The effects of PMM2-CDG-causing mutations on the folding, activity, and stability of the PMM2 protein. *Hum. Mutat., 36*(9), 851–860.

55. Grunewald, S., Matthijs, G., & Jaeken, J., (2002). Congenital disorders of glycosylation: A review. *Pediatr. Res., 52*(5), 618–624.

56. De Zegher, F., & Jaeken, J., (1995). Endocrinology of the carbohydrate-deficient glycoprotein syndrome type-1 from birth through adolescence. *Pediatr. Res., 37*(4 Pt 1), 395–401.

57. Helander, A., Jaeken, J., Matthijs, G., & Eggertsen, G., (2014). Asymptomatic phosphomannose isomerase deficiency (MPI-CDG) initially mistaken for excessive alcohol consumption. *Clin. Chim. Acta., 431*, 15–18.

58. De Koning, T. J., Dorland, L., Van, D. O. P., Boonman, A. M., De Jong, G. J., Van, N. W. L., et al., (1998). A novel disorder of N-glycosylation due to phosphomannose isomerase deficiency. *Biochem. Biophys. Res. Commun., 245*(1), 38–42.

59. Niehues, R., Hasilik, M., Alton, G., Körner, C., Schiebe-Sukumar, M., Koch, H. G., et al., (1998). Carbohydrate-deficient glycoprotein syndrome type Ib. Phosphomannose isomerise deficiency and mannose therapy. *J. Clin. Invest., 101*(7), 1414–1420.

60. Burda, P., Borsig, L., De Rijk-Van, A. J., Wevers, R., Jaeken, J., Carchon, H., et al., (1998). A novel carbohydrate-deficient glycoprotein syndrome characterized by a deficiency in glucosylation of the dolichol-linked oligosaccharide. *J. Clin. Invest., 102*(4), 647–652.

61. Grünewald, S., Imbach, T., Huijben, K., Rubio-Gozalbo, M. E., Verrips, A., De Klerk, J. B., et al., (2000). Clinical and biochemical characteristics of congenital disorder of glycosylation type Ic, the first recognized endoplasmic reticulum defect in N-glycan synthesis. *Ann. Neurol., 47*(6), 776–781.

62. Damen, G., De Klerk, H., Huijmans, J., Den, H. J., & Sinaasappel, M., (2004). Gastrointestinal and other clinical manifestations in 17 children with congenital

disorders of glycosylation type Ia, Ib, and Ic. *J. Pediatr. Gastroenterol. Nutr., 38*(3), 282–287.

63. Westphal, V., Xiao, M., Kwok, P. Y., & Freeze, H. H., (2003). Identification of a frequent variant in ALG6, the cause of congenital disorder of glycosylation-Ic. *Hum. Mutat., 22*(5), 420–421.

64. Stibler, H., Stephani, U., & Kutsch, U., (1995). Carbohydrate-deficient glycoprotein syndrome-a fourth subtype. *Neuropediatrics, 26*(5), 235–237.

65. García-Silva, M. T., Matthijs, G., Schollen, E., Cabrera, J. C., Sanchez, D. P. J., Martí, H. M., et al., (2004). Congenital disorder of glycosylation (CDG) type Ie: A new patient. *J. Inherit. Metab. Dis., 27*(5), 591–600.

66. Yang, A. C., Ng, B. G., Moore, S. A., Rush, J., Waechter, C. J., Raymond, K. M., et al., (2013). Congenital disorder of glycosylation due to DPM1 mutations presenting with dystroglycanopathy-type congenital muscular dystrophy. *Mol. Genet. Metab., 110*(3), 345–351.

67. Di Rocco, M., Hennet, T., Grubenmann, C. E., Pagliardini, S., Allegri, A. E., Frank, C. G., et al., (2005). Congenital disorder of glycosylation (CDG) Ig: Report on a patient and review of the literature. *J. Inherit. Metab. Dis., 28*(6), 1162–1164.

68. Peric, D., Durrant-Arico, C., Delenda, C., Dupré, T., De Lonlay, P., De Baulny, H. O., et al., (2010). The compartmentalization of phosphorylated free oligosaccharides in cells from a CDG Ig patient reveals a novel ER-to-cytosol translocation process. *PLoS One, 5*(7), e11675.

69. Höck, M., Wegleiter, K., Ralser, E., Kiechl-Kohlendorfer, U., Scholl-Bürgi, S., Fauth, C., et al., (2015). ALG8-CDG: Novel patients and review of the literature. *Orphanet. J. Rare Dis., 10*, 73.

70. Sorte, H., Mørkrid, L., Rødningen, O., Kulseth, M. A., Stray-Pedersen, A., Matthijs, G., et al., (2012). Severe ALG8-CDG (CDG-Ih) associated with homozygosity for two novel missense mutations detected by exome sequencing of candidate genes. *Eur. J. Med. Genet., 55*(3), 196–202.

71. Rymen, D., Peanne, R., Millón, M. B., Race, V., Sturiale, L., Garozzo, D., et al., (2013). MAN1B1 deficiency: An unexpected CDG-II. *PLoS Genet., 9*(12), e1003989.

72. Van, S. M., Timal, S., Rymen, D., Hoischen, A., Wuhrer, M., Hipgrave-Ederveen, A., et al., (2014). Diagnostic serum glycosylation profile in patients with intellectual disability as a result of MAN1B1 deficiency. *Brain, 137*(Pt. 4), 1030–1038.

73. Würde, A. E., Reunert, J., Rust, S., Hertzberg, C., Haverkämper, S., Nürnberg, G., et al., (2012). Congenital disorder of glycosylation type Ij (CDG-Ij, DPAGT1-CDG): Extending the clinical and molecular spectrum of a rare disease. *Mol. Genet. Metab., 105*(4), 634–641.

74. Bengtson, P., Ng, B. G., Jaeken, J., Matthijs, G., Freeze, H. H., & Eklund, E. A., (2016). Serum transferrin carrying the xeno-tetrasaccharide NeuAc-Gal-GlcNAc2 is a biomarker of ALG1-CDG. *J. Inherit. Metab. Dis., 39*(1), 107–114.

75. Dupré, T., Vuillaumier-Barrot, S., Chantret, I., Sadou, Y. H., Le Bizec, C., Afenjar, A., et al., (2010). Guanosine diphosphate-mannose: GlcNAc2-PP-dolichol mannosyltransferase deficiency (congenital disorders of glycosylation type Ik): Five new patients and seven novel mutations. *J. Med. Genet., 47*(11), 729–735.

76. Goytain, A., & Quamme, G. A., (2005). Identification and characterization of a novel mammalian Mg$^{2+}$ transporter with channel-like properties. *BMC Genomics, 6*, 48.

77. Anne, W., Gloria, V., Elisabeth, L., & Stefan, B., (2015). Tumor suppressor candidate 3 gene deletion correlates with mental retardation in a child. *Appl. Med. Res., 1*(1), 35–36.

78. Jaeken, J., (2012). MGAT2-CDG (CDG-IIa) and dysmorphism. *Am. J. Med. Genet. A., 158A*(11), 2974–2976.

79. Fukuda, M. N., Dell, A., & Scartezzini, P., (1987). Primary defect of congenital dyserythropoietic anemia type II. Failure in glycosylation of erythrocyte lactosaminoglycan proteins caused by lowered N-acetylglucosaminyltransferase II. *J. Biol. Chem., 262*(15), 7195–7206.

80. De Praeter, C. M., Gerwig, G. J., Bause, E., Nuytinck, L. K., Vliegenthart, J. F., Breuer, W., et al., (2000). A novel disorder caused by defective biosynthesis of N-linked oligosaccharides due to glucosidase I deficiency. *Am. J. Hum. Genet., 66*(6), 1744–1756.

81. Hong, Y., Sundaram, S., Shin, D. J., & Stanley, P., (2004). The Lec23 Chinese hamster ovary mutant is a sensitive host for detecting mutations in alpha-glucosidase I that give rise to congenital disorder of glycosylation IIb (CDG IIb). *J. Biol. Chem., 279*(48), 49894–49901.

82. Varki, A., Cummings, R. D., Esko, J. D., et al., (2009). *Essentials of Glycobiology* (2nd edn.). Cold Spring Harbor (NY): Cold Spring Harbor Laboratory Press.

83. Sturla, L., Rampal, R., Haltiwanger, R. S., Fruscione, F., Etzioni, A., & Tonetti, M., (2003). Differential terminal fucosylation of N-linked glycans versus protein O-fucosylation in leukocyte adhesion deficiency type II (CDG IIc). *J. Biol. Chem., 278*(29), 26727–26733.

84. Carolyn, R. B., Hudson, H. F., Ajit, V., & Jeffrey, D. E., (2009). *Glycans in Biotechnology and the Pharmaceutical Industry.* Cold Spring Harbor (NY): Cold Spring Harbor Laboratory Press.

85. Freeze, H. H., (2007). Congenital disorders of glycosylation: CDG-I, CDG-II, and beyond. *Curr. Mol. Med., 7*, 389–396.

86. Jaeken, J., & Matthijs, G., (2007). Congenital disorders of glycosylation: A rapidly expanding disease family. *Annu. Rev. Genomics Hum. Genet., 8*, 261–278.

87. Wu, X., Steet, R. A., Bohorov, O., Bakker, J., Newell, J., Krieger, M., et al., (2004). Mutation of the COG complex subunit gene COG7 causes a lethal congenital disorder. *Nat. Med., 10*(5), 518–523.

88. Morava, E., Zeevaert, R., Korsch, E., Huijben, K., Wopereis, S., Matthijs, G., et al., (2007). A common mutation in the COG7 gene with a consistent phenotype including microcephaly, adducted thumbs, growth retardation, VSD, and episodes of hyperthermia. *Eur. J. Hum. Genet., 15*(6), 638–645.

89. Martinez-Duncker, I., Dupré, T., Piller, V., Piller, F., Candelier, J. J., Trichet, C., et al., (2005). Genetic complementation reveals a novel human congenital disorder of glycosylation of type, I. I., due to inactivation of the Golgi CMP-sialic acid transporter. *Blood, 105*(7), 2671–2676.

90. Salinas-Marín, R., Mollicone, R., & Martínez-Duncker, I., (2016). A functional splice variant of the human Golgi CMP-sialic acid transporter. *Glycoconj. J., 33*(6), 897–906.

91. Brancaccio, A., (2005). Alpha-dystroglycan, the usual suspect? *Neuromuscul. Disord., 15*(12), 825–828.

92. Valk, M. J., Loer, S. A., & Schober, P., (2015). Perioperative considerations in Walker-Warburg syndrome. *Clin. Case Rep., 3*(9), 744–748.

93. Falsaperla, R., Praticò, A. D., Ruggieri, M., Parano, E., Rizzo, R., & Corsello, G., (2016). Congenital muscular dystrophy: From muscle to brain. *Ital. J. Pediatr., 42*(1), 78.

94. Geisler, C., Mabashi-Asazuma, H., Kuo, C. W., Khoo, K. H., & Jarvis, D. L., (2015). Engineering β1, 4-galactosyltransferase I to reduce secretion and enhance N-glycanelongation in insect cells. *J. Biotechnol., 193*, 52–65.

95. Jiang, J., Kanabar, V., Padilla, B., Man, F., Pitchford, S. C., Page, C. P., & Wagner, G. K., (2016). Uncharged nucleoside inhibitors of β-1, 4-galactosyltransferase with activity in cells. *Chem. Commun (Camb)., 52*(20), 3955–3958.

96. Ozen, H., (2007). Glycogen storage diseases: New perspectives. *World J. Gastroenterol., 13*(18), 2541–2553.

97. Moses, S. W., (2002). Historical highlights and unsolved problems in glycogen storage disease type-1. *Eur. J. Pediatr., 161*(1), S2–9.

98. Godfrey, R., & Quinlivan, R., (2016). Skeletal muscle disorders of glycogenolysis and glycolysis. *Nat. Rev. Neurol., 12*(7), 393–402.

99. Lee, K., Ernst, T., Løhaugen, G., Zhang, X., & Chang, L., (2017). Neural correlates of adaptive working memory training in a glycogen storage disease type-IV patient. *Ann. Clin. Transl. Neurol., 4*(3), 217–222.

100. Wolfe, G. I., Baker, N. S., Haller, R. G., Burns, D. K., & Barohn, R. J., (2000). McArdle's disease presenting with asymmetric, late-onset arm weakness. *Muscle Nerve, 23*(4), 641–645.

101. Roscher, A., Patel, J., Hewson, S., Nagy, L., Feigenbaum, A., Kronick, J., et al., (2014). The natural history of glycogen storage disease types VI and IX: Long-term outcome from the largest metabolic center in Canada. *Mol. Genet. Metab., 113*(3), 171–176.

102. Tarui, S., Okuno, G., Ikura, Y., Tanaka, T., Suda, M., & Nishikawa, M., (1965). Phosphofructokinase deficiency in skeletal muscle. a new type of glycogenosis. *Biochem. Biophys. Res. Commun., 19*, 517–523.

103. Hodax, J. K., Uysal, S., Quintos, J. B., & Phornphutkul, C., (2017). Glycogen storage disease type IX and growth hormone deficiency presenting as severe ketotic hypoglycemia. *J. Pediatr. Endocrinol. Metab., 30*(2), 247–251.

104. Thorens, B., (2015). GLUT2, glucose sensing and glucose homeostasis. *Diabetologia, 58*(2), 221–232.

105. Mamoune, A., Bahuau, M., Hamel, Y., Serre, V., Pelosi, M., Habarou, F., et al., (2014). A thermolabile aldolase A mutant causes fever-induced recurrent rhabdomyolysis without hemolytic anemia. *PLoS Genet., 10*(11), e1004711.

106. Comi, G. P., Fortunato, F., Lucchiari, S., Bordoni, A., Prelle, A., Jann, S., Keller, A., et al., (2001). Beta-enolase deficiency, a new metabolic myopathy of distal glycolysis. *Ann. Neurol., 50*(2), 202–207.

107. Krag, T. O., Ruiz-Ruiz, C., & Vissing, J., (2017). Glycogen synthesis in glycogenin 1-deficient patients: A role for glycogenin 2 in muscle. *J. Clin. Endocrinol. Metab., 102*(8), 2690–2700.

108. Mcardle, B., (1951). Myopathy due to a defect in muscle glycogen breakdown. *Clin. Sci., 10*(1), 13–35.

109. Tsujino, S., Shanske, S., & DiMauro, S., (1993). Molecular genetic heterogeneity of myophosphorylase deficiency (McArdle's disease). *N. Engl. J. Med., 329*(4), 241–245.

110. DiMauro, S., & Hartlage, P. L., (1978). Fatal infantile form of muscle phosphorylase deficiency. *Neurology, 28*(11), 1124–1129.

111. De Castro, M., Johnston, J., & Biesecker, L., (2015). Determining the prevalence of McArdle disease from gene frequency by analysis of next-generation sequencing data. *Genet. Med., 17*(12), 1002–1006.

112. Shen, J., Bao, Y., Liu, H. M., Lee, P., Leonard, J. V., & Chen, Y. T., (1996). Mutations in exon 3 of the glycogen debranching enzyme gene are associated with glycogen storage disease type III that is differentially expressed in liver and muscle. *J. Clin. Invest., 98*(2), 352–357.

113. Endo, Y., Horinishi, A., Vorgerd, M., Aoyama, Y., Ebara, T., Murase, T., et al., (2006). Molecular analysis of the AGL gene: Heterogeneity of mutations in patients with glycogen storage disease type III from Germany, Canada, Afghanistan, Iran, and Turkey. *J. Hum. Genet., 51*(11), 958–963.

114. Vora, S., Seaman, C., Durham, S., & Piomelli, S., (1980). Isozymes of human phosphofructokinase: Identification and subunit structural characterization of a new system. *Proc. Natl. Acad. Sci. U.S.A., 77*(1), 62–66.

115. Toscano, A., & Musumeci, O., (2007). Tarui disease and distal glycogenoses: Clinical and genetic update. *Acta Myol., 26*(2), 105–107.

116. Sherman, J. B., Raben, N., Nicastri, C., Argov, Z., Nakajima, H., Adams, E. M., et al., (1994). Common mutations in the phosphofructokinase-M gene in Ashkenazi Jewish patients with glycogenesis VII- and their population frequency. *Am. J. Hum. Genet., 55*(2), 305–313.

117. Smith, B. F., Stedman, H., Rajpurohit, Y., Henthorn, P. S., Wolfe, J. H., Patterson, D. F., & Giger, U., (1996). Molecular basis of canine muscle type phosphofructokinase deficiency. *J. Biol. Chem., 271*(33), 20070–20074.

118. Musumeci, O., Brady, S., Rodolico, C., Ciranni, A., Montagnese, F., Aguennouz, M., et al., (2014). Recurrent rhabdomyolysis due to muscle β-enolase deficiency: Very rare or underestimated? *J. Neurol., 261*(12), 2424–2428.

119. Chen, S. H., & Giblett, E. R., (1976). Enolase: Human tissue distribution and evidence for three different loci. *Ann. Hum. Genet., 39*(3), 277–280.

120. Hug, G., Schubert, W. K., & Chuck, G., (1966). Phosphorylase kinase of the liver: Deficiency in a girl with increased hepatic glycogen. *Science, 153*(3743), 1534, 1535.

121. Maichele, A. J., Burwinkel, B., Maire, I., Søvik, O., & Kilimann, M. W., (1996). Mutations in the testis/liver isoform of the phosphorylase kinase gamma subunit (PHKG2) cause autosomal liver glycogenosis in the GSD rat and in humans. *Nat. Genet., 14*(3), 337–340.

122. Beauchamp, N. J., Dalton, A., Ramaswami, U., Niinikoski, H., Mention, K., Kenny, P., et al., (2007). Glycogen storage disease type IX: High variability in clinical phenotype. *Mol. Genet. Metab., 92*(1/2), 88–99.

123. Tegtmeyer, L. C., Rust, S., Van, S. M., Ng, B. G., Losfeld, M. E., Timal, S., et al., (2014). Multiple phenotypes in phosphoglucomutase 1 deficiency. *N. Engl. J. Med., 370*(6), 533–542.

124. Stojkovic, T., Vissing, J., Petit, F., Piraud, M., Orngreen, M. C., & Andersen, G., (2009). Muscle glycogenosis due to phosphoglucomutase 1 deficiency. *N. Engl. J. Med., 361*(4), 425–427.

125. Bembi, B., Cerini, E., Danesino, C., Donati, M. A., Gasperini, S., Morandi, L., et al., (2008). Diagnosis of glycogenosis type II. *Neurology, 71*(2/23), S4–11.

126. Taurisano, R., D'Amico, A., Colafati, G. S., Pichiecchio, A., Catteruccia, M., Bertini, E., et al., (2015). Long-term follow-up of two siblings with a non-classic infantile variant form of pompe disease. *J. Neuromuscul. Dis., 2*(s1), S70–S71.

127. Fagan, N., Alexander, A., Irani, N., Saade, C., & Naffaa, L., (2016). Magnetic resonance imaging findings of central nervous system in lysosomal storage diseases: A pictorial review. *J. Med. Imaging Radiat. Oncol., 61*(3), 344–352.

128. Fuller, M., Meikle, P. J., & Hopwood, J. J., (2006). Epidemiology of lysosomal storage diseases: An overview. *Fabry Disease: Perspectives from 5 Years of FOS.* Oxford: Oxford Pharma Genesis, Chapter 2.

129. Higuchi, T., Kawagoe, S., Otsu, M., Shimada, Y., Kobayashi, H., Hirayama, R., et al., (2014). The generation of induced pluripotent stem cells (iPSCs) from patients with infantile and late-onset types of Pompe disease and the effects of treatment with acid-α-glucosidase in Pompe's iPSCs. *Mol. Genet. Metab., 112*(1), 44–48.

130. Raval, K. K., Tao, R., White, B. E., De Lange, W. J., Koonce, C. H., Yu, J., et al., (2015). Pompe disease results in a Golgi-based glycosylation deficit in human induced pluripotent stem cell-derived cardiomyocytes. *J. Biol. Chem., 290*(5), 3121–3136.

131. Tolar, J., Park, I. H., Xia, L., Lees, C. J., Peacock, B., Webber, B., et al., (2011). Hematopoietic differentiation of induced pluripotent stem cells from patients with mucopolysaccharidosis type I (Hurler syndrome). *Blood, 117*(3), 839–847.

132. Maetzel, D., Sarkar, S., Wang, H., Abi-Mosleh, L., Xu, P., Cheng, A. W., et al., (2014). Genetic and chemical correction of cholesterol accumulation and impaired autophagy in hepatic and neural cells derived from Niemann-pick type C patient-specific iPS cells. *Stem Cell Reports, 2*(6), 866–880.

133. Awad, O., Sarkar, C., Panicker, L. M., Miller, D., Zeng, X., Sgambato, J. A., et al., (2015). Altered TFEB-mediated lysosomal biogenesis in gaucher disease iPSC-derived neuronal cells. *Hum. Mol. Genet., 24*(20), 5775–5788.

134. Aflaki, E., Stubblefield, B. K., Maniwang, E., Lopez, G., Moaven, N., & Goldin, E., (2014). Macrophage models of gaucher disease for evaluating disease pathogenesis and candidate drugs. *Sci. Transl. Med., 6*(240), 240ra73.

135. Panicker, L. M., Miller, D., Park, T. S., Patel, B., Azevedo, J. L., Awad, O., et al., (2012). Induced pluripotent stem cell model recapitulates pathologic hallmarks of gaucher disease. *Proc. Natl. Acad. Sci. U.S.A, 109*(44), 18054–18059.

136. Kawagoe, S., Higuchi, T., Otaka, M., Shimada, Y., Kobayashi, H., Ida, H., et al., (2013). Morphological features of iPS cells generated from Fabry disease skin fibroblasts using Sendai virus vector (SeVdp). *Mol. Genet. Metab., 109*(4), 386–389.

137. Grabowski, G. A., (2012). Gaucher disease and other storage disorders. *Hematology Am. Soc. Hematol. Educ. Program,* 13–18.

138. Paciotti, S., Codini, M., Tasegian, A., Ceccarini, M. R., Cataldi, S., & Arcuri, C., (2017). Lysosomal alpha-mannosidase and alpha-mannosidosis. *Front Biosci. (Landmark Ed), 22,* 157–167.

139. Beck, M., Olsen, K. J., Wraith, J. E., Zeman, J., Michalski, J. C., & Saftig, P., (2013). Natural history of alpha mannosidosis a longitudinal study. *Orphanet. J. Rare Dis., 8,* 88.

140. Aronson, N. N. Jr., (1999). Aspartylglycosaminuria: Biochemistry and molecular biology. *Biochim. Biophys. Acta, 1455*(2/3), 139–154.

141. Arvio, M. A., Peippo, M. M., Arvio, P. J., & Kääriäinen, H. A., (2004). Dysmorphic facial features in aspartylglucosaminuria patients and carriers. *Clin. Dysmorphol., 13*(1), 11–15.

142. Zhu, M., Lovell, K. L., Patterson, J. S., Saunders, T. L., Hughes, E. D., & Friderici, K. H., (2006). Beta-mannosidosis mice: A model for the human lysosomal storage disease. *Hum. Mol. Genet., 15*(3), 493–500.

143. Nesterova, G., & Gahl, W. A., (2013). Cystinosis: The evolution of a treatable disease. *Pediatr. Nephrol., 28*(1), 51–59.

144. D'souza, R. S., Levandowski, C., Slavov, D., Graw, S. L., Allen, L. A., Adler, E., et al., (2014). Danon disease: Clinical features, evaluation, and management. *Circ. Heart Fail, 7*(5), 843–849.

145. Deegan, P. B., Baehner, A. F., Barba, R. M. A., Hughes, D. A., Kampmann, C., & Beck, M., (2006). European FOS investigators. Natural history of Fabry disease in females in the Fabry outcome survey. *J. Med. Genet., 43*(4), 347–352.

146. Alves, M. Q., Le Trionnaire, E., Ribeiro, I., Carpentier, S., Harzer, K., Levade, T., et al., (2013). Molecular basis of acid ceramidase deficiency in a neonatal form of Farber disease: Identification of the first large deletion in ASAH1 gene. *Mol. Genet. Metab., 109*(3), 276–281.

147. Kanitakis, J., Allombert, C., Doebelin, B., Deroo-Berger, M. C., Grande, S., Blanc, S., & Claudy, A., (2005). Fucosidosis with angiokeratoma. Immunohistochemical and electron microscopic study of a new case and literature review. *J. Cutan. Pathol., 32*(7), 506–511.

148. Matsumoto, N., Gondo, K., Kukita, J., Higaki, K., Paragison, R. C., & Nanba, E., (2008). A case of galactosialidosis with a homozygous Q49R point mutation. *Brain Dev., 30*(9), 595–598.

149. Mignot, C., Gelot, A., & De Villemeur, T. B., (2013). Gaucher disease. *Handb. Clin. Neurol., 113*, 1709–1715.

150. Hofer, D., Paul, K., Fantur, K., Beck, M., Roubergue, A., Vellodi, A., Poorthuis, B. J., Michelakakis, H., Plecko, B., & Paschke, E., (2010). Phenotype determining alleles in GM1 gangliosidosis patients bearing novel GLB1 mutations. *Clin. Genet., 78*(3), 236–246.

151. Mahuran, D. J., (1999). Biochemical consequences of mutations causing the GM2 gangliosidoses. *Biochim. Biophys. Acta, 1455*(2/3), 105–138.

152. Sakai, N., (2009). Pathogenesis of leukodystrophy for Krabbe disease: Molecular mechanism and clinical treatment. *Brain Dev., 31*(7), 485–487.

153. Al-Hassnan, Z. N., Al Dhalaan, H., Patay, Z., Faqeih, E., Al-Owain, M., Al-Duraihem, A., & Faiyaz-Ul-Haque, M., (2009). Sphingolipid activator protein B deficiency: Report of 9 Saudi patients and review of the literature. *J. Child Neurol., 24*(12), 1513–1519.

154. Campos, D., & Monaga, M., (2012). Mucopolysaccharidosis type I: Current knowledge on its pathophysiological mechanisms. *Metab. Brain Dis., 27*(2), 121–129.

155. Clarke, L. A., (2008). The mucopolysaccharidoses: A success of molecular medicine. *Expert Rev. Mol. Med., 10*, e1.

156. Bodamer, O. A., Giugliani, R., & Wood, T., (2014). The laboratory diagnosis of mucopolysaccharidosis III (Sanfilippo syndrome): A changing landscape. *Mol. Genet. Metab., 113*(1/2), 34–41.

157. Gilkes, J. A., & Heldermon, C. D., (2014). Mucopolysaccharidosis III (Sanfilippo syndrome)-disease presentation and experimental therapies. Pediatr, Endocrinol, Rev., 12(1), 133–140.

158. Montaño, A. M., Tomatsu, S., Gottesman, G. S., Smith, M., & Orii, T., (2007). International Morquio a registry: Clinical manifestation and natural course of Morquio a disease. *J. Inherit. Metab. Dis., 30*(2), 165–174.

159. Garrido, E., Cormand, B., Hopwood, J. J., Chabás, A., Grinberg, D., & Vilageliu, L., (2008). Maroteaux-Lamy syndrome: Functional characterization of pathogenic mutations and polymorphisms in the arylsulfatase B-gene. *Mol. Genet. Metab., 94*(3), 305–312.

160. Metcalf, J. A., Zhang, Y., Hilton, M. J., Long, F., & Ponder, K. P., (2009). Mechanism of shortened bones in mucopolysaccharidosis VII. *Mol. Genet. Metab., 97*(3), 202–211.

161. Lai, S. C., Chen, R. S., Wu, C. Y. H., Chang, H. C., Kao, L. Y., Huang, Y. Z., et al., (2009). A longitudinal study of Taiwanese sialidosis type 1: An insight into the concept of cherry-red spot myoclonus syndrome. *Eur. J. Neurol., 16*(8), 912–919.

162. Cathey, S. S., Kudo, M., Tiede, S., Raas-Rothschild, A., Braulke, T., Beck, M., et al., (2008). Molecular order in mucolipidosis II and III nomenclature. *Am. J. Med. Genet. A., 146A*(4), 512, 513.

163. Dong, X. P., Cheng, X., Mills, E., Delling, M., Wang, F., Kurz, T., et al., (2008). The type IV mucolipidosis-associated protein TRPML1 is an endolysosomal iron release channel. *Nature, 455*(7215), 992–996.

164. Hendriksz, C. J., Corry, P. C., Wraith, J. E., Besley, G. T., Cooper, A., & Ferrie, C. D., (2004). Juvenile Sandhoff disease-nine new cases and a review of the literature. *J. Inherit. Metab. Dis., 27*(2), 241–249.

165. Clark, N. E., & Garman, S. C., (2009). The 1.9 a structure of human alpha-N-acetylgalactosaminidase: The molecular basis of Schindler and Kanzaki diseases. *J. Mol. Biol., 393*(2), 435–447.

166. Aula, N., Kopra, O., Jalanko, A., & Peltonen, L., (2004). Stalin expression in the CNS implicates extralysosomal function in neurons. *Neurobiol. Dis., 15*(2), 251–261.

167. Maegawa, G. H., Stockley, T., Tropak, M., Banwell, B., Blaser, S., Kok, F., et al., (2006). The natural history of juvenile or subacute GM2 gangliosidosis: 21 new cases and literature review of 134 previously reported. *Pediatrics, 118*(5), e1550–1562.

168. Zhang, B., & Porto, A. F., (2013). Cholesteryl ester storage disease: Protean presentations of lysosomal acid lipase deficiency. *J. Pediatr. Gastroenterol. Nutr., 56*(6), 682–685.

169. Brigida, I., Chiriaco, M., Di Cesare, S., Cittaro, D., Di Matteo, G., Giannelli, S., et al., (2017). Large deletion of MAGT1 gene in a patient with classic Kaposi sarcoma, CD4 lymphopenia, and EBV infection. *J. Clin. Immunol., 37*(1), 32–35.

170. Cunningham, S., Gerlach, J. Q., Kane, M., & Joshi, L., (2010). Glyco-biosensors: Recent advances and applications for the detection of free and bound carbohydrates. *Analyst, 135*, 2471–2480.

171. Dwek, R. A., (1996). Glycobiology: Toward understanding the function of sugars. *Chem. Rev., 96*, 683–720.

172. Kuzmanov, U., Kosanam, H., & Diamandis, E. P., (2013). The sweet and sour of serological glycoprotein tumor biomarker quantification. *BMC Med., 11*, 31.

173. Matalonga, L., Bravo, M., Serra-Peinado, C., García-Pelegrí, E., Ugarteburu, O., Vidal, S., et al., (2017). Mutations in TRAPPC11 are associated with a congenital disorder of glycosylation. *Hum. Mutat., 38*(2), 148–151.

174. Moremen, K. W., Tiemeyer, M., & Nairn, A. V., (2012). Vertebrate protein glycosylation: diversity, synthesis and function. *Nat. Rev. Mol. Cell Biol., 13*(7), 448–462.

175. Ohtsubo, K., & Marth, J. D., (2006). Glycosylation in cellular mechanisms of health and disease. *Cell, 126*(5), 855–867.

176. Reinders, J., & Sickmann, A., (2007). Modificomics: Posttranslational modifications beyond protein phosphorylation and glycosylation. *Biomol. Eng., 24*, 169–177.

177. Schänzer, A., Kaiser, A. K., Mühlfeld, C., Kulessa, M., Paulus, W., Von, P. H., et al., (2017). Quantification of muscle pathology in infantile Pompe disease. *Neuromuscul. Disord., 27*(2), 141–152.

178. Spiro, R. G., (2002). Protein glycosylation: Nature, distribution, enzymatic formation, and disease implications of glycopeptide bonds. *Glycobiology, 12*, 43R–56R.

179. Stern, C. M., Pepin, M. J., Stoler, J. M., Kramer, D. E., Spencer, S. A., & Stein, C. J., (2017). Musculoskeletal conditions in a pediatric population with Ehlers-Danlos syndrome. *J. Pediatr., 181*, 261–266.

180. Taisne, N., Desnuelle, C., Juntas, M. R., Ferrer, M. X., Sacconi, S., Duval, F., et al., (2017). Bent spine syndrome as the initial symptom of late-onset Pompe disease. *Muscle Nerve, 56*(1), 167–170.

# CHAPTER 3

# Glycome in Immunological Processes: Current Scenario and Future Prospects

ZEBA MUEED,[1] MANALI SINGH,[1] ABHAY MISHRA,[2] and
NITESH KUMAR PODDAR[3*]

[1]Department of Biotechnology, Invertis Institute of Engineering and
Technology, Invertis University, Bareilly, Uttar Pradesh – 243123, India

[2]Department of Biotechnology, Khandelwal College of Management
Science and Technology, NH 74, Bareilly, Uttar Pradesh – 243122, India

[3]Department of Biosciences, Manipal University Jaipur,
Jaipur-Ajmer Express Highway, Dehmi Kalan, Near GVK Toll Plaza,
Jaipur, Rajasthan – 303007, India

*Corresponding author. E-mail: niteshkumar.poddar@jaipur.manipal.edu

## ABSTRACT

Glycans, being a fundamental part of cell components, are involved in diverse metabolic functions such as cell signaling, cell recognition, and immunological processes. With respect to immunological processes, glycans have been implicated in both innate as well as adaptive immune systems through their involvement in immune cell differentiation, B-, and T-cell activation, protein folding, and antibody functioning, complement activation, antigen processing and presentation, and apoptosis. Aberrant glycosylation during these critical processes may lead to impaired protein expression and altered immune signaling pathways. Due to their involvement in immunological processes, glycans have been found to be associated with numerous immunological diseases known to be grouped as congenital disorders of glycosylation (CDG). In addition to this, the role of glycosylation is also important in host-pathogen interaction, wherein the interaction is mediated by the glycosylated surface receptors on the host cell and the glycan moieties on microbes. Keeping this in view, this chapter was designed to provide a deep insight into the mechanism involved behind altered glycosylation in an effort to sort out ways for the prevention of immunological disorders. Future perspectives with respect to carbohydrate vaccines and other medications such as engineered glycans and glycosidase inhibitors that may provide a valuable therapeutic approach are also included. In addition to this, the chapter also discusses the development of carbohydrate vaccines in conjugation with a suitable protein carrier.

## 3.1  INTRODUCTION

Glycans, more commonly known as saccharides or sugars, are one of the fundamental cellular components of the cell, being involved in an array of cellular activities varying from cell recognition to performing unique roles in the immune system (Figure 3.1). The carbohydrate entities are covalently linked via glycosidic linkages to proteins and lipids referred to as glycoproteins or glycolipids, respectively [1]. The cellular enzymatic process that produces the glycosidic linkages of saccharides to other saccharides, lipids, or proteins is referred to as glycosylation. These multiple repertoires of cellular glycans, either free or attached to proteins and lipids, are collectively called glycome glycosylation, being most prominent in endoplasmic reticulum (ER) lumen and Golgi apparatus (GA), is the most common posttranslational modification required for protein activation and function. Glycans serve a variety of structural and functional roles, few of them include discriminating immune self-cells from non-self-cells, receptor binding, and activation; marking the cells for apoptosis, cell signaling, and cell-cell interaction [2].

Among all these, Glycosylation has been shown to be a complex and highly regulated process with respect to the immune system. In fact, it has been shown that several specific immunologic pathways have demonstrated their dependence on appropriate glycosylation. In the context of innate immunity, glycans have been found to mediate host-pathogen interaction, activation of complement pathways, and inflammatory response, where their function mainly involves the process of mobilizing cells to sites where rolling on the vascular endothelium has occurred [3] (Figure 3.1). In adaptive immunity, glycans have ordered multiple aspects in the functioning of B-cells as well as of T-cells such as immune cell differentiation, antibody functioning, antigen processing, and presentation, etc., [2] (Figure 3.1). This chapter mainly focuses on the role of glycosylation in diverse immunological processes with respect to both innate and adaptive immunity and the components involved therein. The chapter further provides insight as to how aberrant glycosylation may alter the normal functioning of the immune system and contribute to innumerable diseases such as rheumatoid arthritis (RA), SLE, cancer, etc. Engineered glycans, glycan mimics, and carbohydrate vaccines have proved to be a good therapeutic approach to treat altered glycosylation. Moreover, inhibitors of glycan-binding receptors, as well as glycosidase inhibitors, are being considered as a potential target for treating cancer.

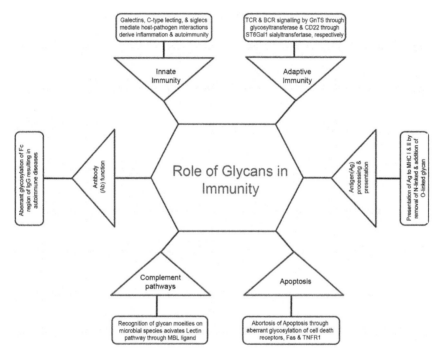

**FIGURE 3.1** Role of glycans in immunological processes such as antigen processing, antibody function, innate immunity, adaptive immunity, complement pathways, and apoptosis of cells.

## 3.2 GLYCOSYLATION: A PROCESS OF ATTACHING GLYCANS

Glycosylation is a process which involves the addition of glycosyl group (glycan) to some other functional group (protein or lipid or any other organic molecule) commonly known as glycosyl acceptor, resulting in the production of one of the fundamental biopolymer found in a cell along with DNA, RNA, and proteins. The process of glycosylation involves the combinatorial effect of an array of enzymes, ultimately producing a molecule in a fully activated form [4]. It is an important event during co-translational or posttranslational modification, after which a molecule becomes committed to its specific function. Glycosylation is an enzyme directed and a site-specific process. The two most important subsets of enzymes involved in this process are glycosyltransferases and glycosidases. Apart from the ER lumen and Golgi, glycosylation is also reported in the cytoplasm and nucleus. The process of glycosylation has been observed in both prokaryotes and eukaryotes with the help of analytical techniques and genome sequencing. Glycosylation has links in human diseases such as during inflammation, in cellular stress, or even in cancer, which eventually leads to alteration of expressed glycome [4]. Glycosylation is an important epigenetic modification of the protein, which affects their localization, expression, and function of different proteins, which are needed for the immune system to function normally. Glycosylation of proteins is basically of two types, N-linked glycosylation, where carbohydrate is attached through N-acetylglucosamine and the amino acid involved here is asparagine, and O-linked glycosylation, where attachment between carbohydrate and protein is through N-acetylgalactosamine and the amino acid involved is serine or threonine. Glycoproteins have a significant role in cell-cell recognition and interactions. In general, any modification in glycosylation of proteins takes place by enzymes known as glucosyltransferases along the ER/Golgi pathway. Any changes in glycosylation can lead to congenital disorders of the immune system. For the movement of leukocytes to the site of the invasion, cascades of events occur like capturing of leukocytes, which are free-flowing inside the blood vessels where they roll on the walls of the blood vessels. Both capture and rolling mechanisms of the leukocyte is mediated with the help of receptor and ligands of the selectin family, which binds to carbohydrate motifs of the glycans. The selectins are of different types, like P-selectins, E-selectin, and L-selectin. The P-selectins are found inside the endothelial cells of blood vessels and in granules of platelets. E-selectins are not stored inside the cells like P-selectins; rather, they are synthesized during the transcription and translational events. At the same time, L-selectins are expressed on the surface of leukocytes, which can be removed later when not required. This process of rolling is mediated by the glycan ligand sLex (a common outer structural element on both N- and O-glycans is the sub-terminal disaccharide N-acetyllactosamine (LacNAc)). The outer LacNAc residues are frequently capped by sialic acid residues. The sialyl-Lewis X (SLe$^x$) motif facilitates cell-cell interactions by providing an important ligand for selectins, mediating processes such as intercellular adhesion. The absence of sLex, as seen in patients with SLC35C1 mutations, results in impairment of this process and leukocyte adhesion deficiency type 2. It has also been postulated that the virus envelop utilizes the host glycosylation mechanism leading to the production of glycoproteins, which is involved in invasion [5].

### 3.2.1  HISTORY OF GLYCAN STRUCTURES IN MEDICINE

Glycans have immense therapeutic benefits and thus are under great consideration. However, with the invention of the genetic code and novel DNA-based therapeutic techniques, glycans, and lipids at that time could not gain so much importance as treatments with DNA and proteins became readily accessible for the scientific society. Still, the therapeutic relevance of glycan and lipids cannot be ignored [6]. Lipids and glycans have played a vital role in obesity and type II diabetes. Alpha-glucosidase inhibitors (AGIs), as if acarbose, miglitol, and voguelibose are most commonly used for treating type 2 diabetes patients. AGIs lower the effect on postprandial blood glucose and insulin levels by delaying the absorption of carbohydrates from the small intestine [7]. Thus, the emergence of glycans provided some of the first major breakthroughs in modern medicine in the scientific society. It was in 1900, Karl Landsteiner reported three types of blood groups, i.e., A, B, and O, which govern the matching of the blood groups among the donor and recipient and pioneered successful blood transfusion in 1907 [8]. For this discovery, he was awarded Nobel Prize in Medicine in 1930. Still, much progress for identifying the chemical entities of various blood types and their distinct functions are largely unknown and uncharacterized, until the identification of different blood group types on the basis of glycans became fruitful in the 1950s.

The main component of the blood group system is H-antigen, in which fucose is the main oligosaccharide upon which addition of N-acetylgalactosamine (GalNAc) or galactose (Gal) occurs to form the A and B antigens, respectively [9–11]. By using the selective alkylation technique, the whole structure was then elucidated in the 1960s, along with enzymatic and acid/base hydrolysis to distinguish the monosaccharide linkages and components [12]. The treatment of thrombosis in humans after the invention of polysaccharide heparin has made a great and lasting impact on the medical field.

The discovery of heparin in 1916 [13] and advancement in its isolation from various animal sources in the 1930s [14] made it relevant for clinical trials in various therapeutic purposes. Heparin is basically a glycosaminoglycan (GAG) consisting of sulfated glucosamine and iduronic acid [15]. The repeating unit (except for keratan) consists of an amino sugar (N-acetylglucosamine or N-acetylgalactosamine) along with an uronic sugar (glucuronic acid or iduronic acid) or galactose. Shedding of a "soluble-specific substance" by pneumococcus reacted with antisera that were obtained from people infected with the same pathogen was already reported [16]. Later on, after 5 years, Avery and *Heidelberger* reported that this compound was a type-specific polysaccharide-based soluble material [17] which could be used as a significant component for vaccine production against pneumococcus [18]. This polysaccharide-based therapeutic product has a varied clinical significance and is utilized in the vaccine Pneumovax (PPV23), which contains 23 purified capsular polysaccharides from *Streptococcus pneumonia* [19]. Whereas, some polysaccharides isolated from other pathogens were able to alone provide effective antibody response against vaccines. These investigations led to the conclusion that carbohydrate entities could make effective vaccines. Small glycan molecules known as aminoglycosides are a class of amine synthesized by gram-positive bacteria, micromonospora, and streptomyces. Streptomycin, the first aminoglycoside, was investigated in 1943 and had great medicinal importance and served as the first antibiotic for treating tuberculosis [20]. Gentamicin, kanamycin, and neomycin are the other members of this widely used class of antibiotics. Many of them function as inhibitors of protein synthesis [21]. The invention of positron emission tomography (PET), for the first time in the 1950s, led to the synthesis and usage of 2-fluorodeoxy-D-glucose (FDG) for studies related to oncology, identification of causal genes and their effects [22]. Interestingly, 18-FDG is readily taken up by cells with a higher metabolic rate than other cells, and thereby leading to the elucidation of tumors and imaging of the brain [23]. Alternative methods would take several decades for the direct imaging of glycans and other biomolecules extensively [24, 25].

## 3.3   PROTEIN GLYCOSYLATION

Glycosylation is the most varied, complex yet common process during post or co-translational modification of proteins to make them functional and active, which is eventually needed to carry out a sundry of biological processes of the cell such as protein folding, stability, activity, etc., [25]. The importance of glycosylation is highlighted by the fact that almost 50% of the proteome is governed by glycoproteins [26]. Thereby, appropriate glycosylation during posttranslational modification is critical to the biological activity of a protein. Glycosylation site has been shown to mainly depend on the type of amino acid residues and their position in a protein. Most commonly, it has been seen that during glycosylation, the sugar chain is linked either to the nitrogen atom of asparagine or to the oxygen atom of serine or threonine by covalent bonds.

Apart from these conventional glycosylation types, some other glycosylation linkages involve other amino acids as well, such as tryptophan or tyrosine in C-glycosylation, cysteine, or methionine having sulfur atom in S-linked glycosylation, and C-mannosylation involving tryptophan. These reactions, each being carried out by different enzymes, are tissue-specific and are found to be a major regulator of numerous physiological processes [27]. Glycosylation begins as soon as the protein synthesis starts, and ER acts as a control check, recognizing proteins on the basis of their glycosylation profile. Glycans serve as an address guiding a protein to its destination, which may be any other organelle or back to ER. Protein displaying aberrant glycosylation is ubiquitinated or degraded [28]. Thus it can be implicated that proper glycosylation of protein is critical for a diverse range of biological processes ranging from protein folding and dynamics to cell growth and differentiation. Any perturbation in the glycosylation cascade is likely to lead to a variety of disorders like diabetes, neurodegenerative disorders, etc. [29].

### 3.3.1  EFFECT OF GLYCOSYLATION ON PROTEIN FOLDING AND STABILITY

Glycosylation involving covalent binding of glycan chains to proteins leads plays a vital role in governing the protein folding and stability. Glycosylation mainly works by preventing the formation of secondary structures and also prevents unfolding by elevating the vibrational and conformational entropy of the proteins. In addition to this, attachment of glycans to proteins also does influence the protein/enzyme kinetics by changing the rate-limiting step of protein folding reaction. Recently it has been shown that the presence of glycans increases the thermodynamic stability of proteins by stabilizing secondary structures during folding [30] due to the fact that as compared to non-glycosylated proteins, glycosylated proteins have less proportion of exposed non-polar groups [31]. A study carried out on glycosylated and non-glycosylated forms of soybean agglutinin has revealed less stability of non-glycosylated form than glycosylated one at both normal as well as higher temperatures [32]. The reason mainly attributed for this was the presence of non-covalent bonds between glycans and proteins. Additionally, results from molecular dynamics simulation studies have shown that decline in free energy and an increment in vibrational energy was responsible for glycan-induced stabilization of proteins at a normal temperature [33]. Another study has shown that the addition of glycans on the Fc region of IgG greatly impacts its receptor binding activity and other biological effector functions [34]. In addition to this, it has also been seen that the removal of glycans from proteins results in their loss of activity as well as undergo misfolding and then rapid degradation [35].

It has been shown that both N- and O-glycans contribute equally in maintaining protein stability. This is achieved by the addition of hydrophilic groups, which helps in stabilizing hydrogen bonds by inhibiting the effect of solvent, which might affect hydrophobicity of certain residues around the glycosylation site [36]. Moreover, the addition of sugars on proteins dwindles its dynamics by favoring diverse interactions between glycans and proteins such as hydrogen bonds, van der Waals, and electrostatic energy, in a way enhancing protein folding [37].

## 3.4   TYPES OF GLYCOSYLATION

Glycosidic linkages between proteins and glycans have been found to involve a number of saccharides such as glucose, mannose, galactose, fucose, N-acetylglucosamine (GlcNAc), N-GalNAc, sialic acid, etc., [38, 39]. Glycosidases and Glycosyltransferase are two main enzymes involved in this process. In addition to several other types of glycosylation, the two most common and of utmost importance are N-linked glycosylation and O-linked glycosylation [40]. The following section will discuss about these two most important glycosylations.

### 3.4.1   *N-LINKED GLYCOSYLATION*

During N-linked glycosylation, a glycan chain is attached to the nitrogen atom of asparagine at specific sequence N-X-S/T, where X can be any amino acid except proline. It has been shown that that sequence containing proline and any other charged amino acid were comparatively less glycosylated than other amino acids [41]. However, the presence of an aromatic amino acid favors glycosylation mainly by enhancing the enzymatic activity of oligosaccharyltransferase [42]. Unlike O-linked glycans, N-linked glycans (NLGs) are large, consisting of 14 sugar residues, with several branches having sialic acid at their end. NLGs have generally been found to possess a core of three mannose and two GlcNAc, built on a core of pentasaccharide, Man3GlcNAc2, which is then attached to the protein. This pentasaccharide core is then processed and modified in GA, leading to three main types of NLGs, namely high-mannose, hybrid, and complex. Initial processing of glycans involves the addition of 5–9 unsubstituted mannoses to core GlcNAc2 at the terminals. In addition to unsubstituted mannose, hybrid type NLGs are characterized by the branches of substituted residues through GlcNAc. As compared to high mannose and hybrid NLGs, complex NLGs contain mannose residue only in their core structure. The characteristic feature of complex glycan is the addition of GlcNAc residues at both the α-3 and α-6 mannose positions as well the presence of sialic acid at the terminals [43].

Apart from playing a crucial role in protein folding, stability, and dynamics, NLGs also work as molecular insulators by inhibiting the formation of interactions among molecules in an overcrowded environment. Thus it can be inferred that any variation in the sequence of glycosylation site directly alters the pattern of modification and thus contributing to altering the structure and function of the protein [44]. N-linked glycosylation of proteins occurs as a co-translation event on the cytoplasmic face of ER, with a dolichol phosphate already attached to a protein. This is then followed by the addition of mannose (Man) and N-acetyl-D-glucosamine molecules, where GDP-Man and UDP-GlcNAc are acting as glycan donors who then lead to the formation of Man5GlcNAc2-P-Dol. This precursor with its attached glycan is then moved to the ER lumen where further enzymatic processes such as removal of mannose by glycosidase and addition of some other groups by glycosyltransferases in ER as well as in GA. The finalized oligosaccharide is then transferred from dolichol (Dol) to asparagine residue of protein with enzymatic action of oligosaccharyltransferase (OST), which is then committed to performing its destined function, which may be related to protein folding, stability, dynamics or cell growth and differentiation (Figure 3.2(A)).

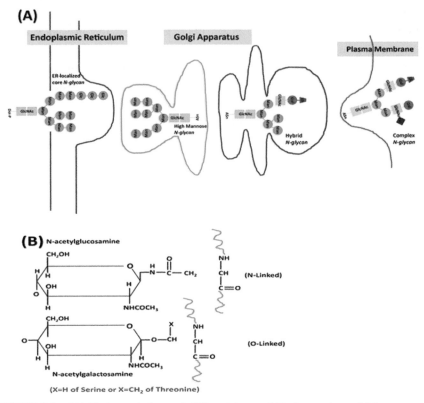

**FIGURE 3.2** (A) N-glycosylation and different types of N-glycans (core, high mannose, hybrid, and complex N-glycan) and their synthetic pathway from ER to Golgi apparatus; (B) Core structure of N-linked and O-linked glycans.

### 3.4.2 O-LINKED GLYCOSYLATION

O-glycosylation is also an essential posttranslational modification as it is known to regulate many physiological functions of proteins namely cell cycle, cell signaling and protein-protein interactions [45, 46]. Aberrations in O-glycosylation are linked with several human diseases and disease risk factors [47]. Formation of O-linked glycosylation occurs in ER and GA by the addition of short sugar residues to the -OH group of serine or threonine in a sequential manner with the help of different glycosyltransferase enzymes [48] (Figure 3.2(B)). A variety of glycans are attached to protein based on its structure and localization. Miscellaneous forms of O-linked glycosylation are known to occur, such as O-mannosylation, which involves the addition of mannose from Dol to -OH group of serine and threonine amino acid in protein, which then directs the protein for degradation [49]. Glycosylation of modified amino acids such as hydroxylysine and hydroxyproline involves the addition of Gal (galactose) or Glc (glucose) on collagen proteins in Golgi complex, whereas nuclear and cytoplasmic proteins, critical for cell signaling, also undergo O-linked glycosylation [50].

### 3.4.2.1   ROLE OF O-GLYCOSYLATION IN HUMAN INTESTINAL SYMBIONTS

Intestinal microbial populations serve several essential functions for the benefit of hosts, which aid in human metabolism, growth, and development and the body's defense mechanism and thereby boosting the host immunity. They also serve as the energy source in the form of short-chain fatty acids to the host [51] and aid in the recycling of bile acids [52]. Bacteroides have also been reported in stimulating intestinal angiogenesis during the course of development and induction of local and systemic immune functions [51]. The microbial population of the intestine is also part of inflammatory bowel syndromes, allergic disorders like asthma, colon cancer, and in spreading antibiotic resistance [52, 53].

Bacteroides are both categorized as harmful and beneficial. Their multiple unique characteristics of inhabiting and adaptation help them to grow well in the intestine. The microbial species colonize in high densities persistently in the human intestine [54]. The microbial ecosystem comprising of Bacteroides, namely *B. caccae, B. ovatus, B. thetaiotaomicron, B. uniformis,* and *B. vulgatus* (10%–20%) are abundantly found in the human colon [55]. Bacteroides can adapt themselves to several habitats and environmental niche as they have O-glycosylation system unlike eukaryotes which have N- and O-glycosylation systems with which modification of many cellular proteins can be done but protein glycosylation is very much rare in bacteria. These bacteroidal species are capable of producing fucosylated glycoproteins. Bacteroidal species have also genetic systems which are mainly involved in acquiring and metabolizing carbohydrates which cannot be digested by the host [56, 57]. Their ability to rapidly change their morphology on the cell surface by producing immense number of phase-variable capsular polysaccharides is specific for the intestinal environment [58]. Bacteroides are genetically capable of metabolizing carbohydrates as plant polysaccharides and mucosal glycans which cannot be digested by the host, allowing a rapid response to shifting food supplies [56, 57]. In the intestinal epithelia of colonized mice, the expression of fucosylated glycoconjugates has been shown by one of the bacteroides [59]. By producing enzymes, bacteroides are able to harvest exogenous fucose from host mucosal glycans and their incorporation directly into their capsular glycoproteins and polysaccharides [60]. In order to competitively colonize the mammalian intestine, *B. fragilis* is capable of synthesizing fucosylated glycoproteins [58, 60].

## 3.5  GLYCANS AND THEIR ROLE IN IMMUNOLOGICAL PROCESSES

### 3.5.1  GLYCOSYLATION AND INNATE IMMUNITY

Since the origin of life, glycans have been shown to play key role in innate immunity. In this regard, impairment of N-glycan branching is likely to lead to autoimmunity. Glycan ligands, mainly lectins are endowed with the ability to suppress immune cell-activation, whereas another glycan known as hyaluronan interacts with toll-like receptors (TLRs) and thus regulating innate immunity [61].

Immune system comprises of a variety of glycan binding receptors present on immune cells that are known to bind to a variety of glycans present on protein or lipid structures. The three most commonly known are galectins, C-type lectins, and siglecs (sialic acid-binding immunoglobulin-like lectins) [62]. Among these, galectins interact with pathogen by binding to their glycan motifs comprising of disaccharide N-acetyllactosamine (Gal-β (1-4)-GlcNAc: Gal, galactose; GlcNAc, *N*-acetylglucosamine), initiating innate immune responses. Galectins are involved in obstructing graft-versus-host disease, collagen-induced arthritis, type 1 diabetes and T-cell-mediated tumor rejection [63, 64]. C-type lectin receptors (CLRs), which are membrane-bound proteins, require calcium for binding glycans. It comprises of a large number of receptors. However, the two most important ones are the mannose-specific C-type lectins possessing an EPN (Glu-Pro-Asn) amino acid motif and specifically recognize mannose- and/or fucose-terminated glycans. In contrast, the galactose-specific C-type lectins possess the QPD (Gln-Pro-Asp) sequence in the Carbohydrate Recognition Domain (CRD) and recognize galactose-terminated or GalNAc-terminated glycan structure [65]. Siglecs harbor a large number of immunoglobulin domains which are known for specifically recognizing sialic acid-containing glycans [66]. Thus, lectin-glycan binding has the capability to start, intensify or disrupt the transmembrane signal transduction and thereby affect diverse biological processes of cell such as cell proliferation, differentiation, migration, and apoptosis, and so these lectins are being targeted for therapeutic approaches. Administration of galectin-1 inhibits autoimmunity and is also seen to prevent transplant rejection whereas administration of glycomaterial has been shown to restore T-cell dependent anti-tumor immune responses [64]. Therapeutic use of Siglec is gaining increased importance for drug delivery in a cancer cell with drug being enclosed in a liposome with attached siglec ligand just as being done by antigen-presenting cells (APCs) during presentation of antigen to T-cells [67].

Presence of glycans on microbial species and other organisms has been used as a recognition factor for the activation of TLR, an important component of innate immunity, and thus glycan structures have been considered as activators for innate immune response. Innate immune cells particularly macrophages and dendritic cells (DCs) harboring CLR, act as first line of defense for pathogen clearance which may later on lead to adaptive immune responses. C-type lectins, siglecs, as well as galectins act as recognition factor for glycans present on organism [68]. After recognition, they interact with these specific glycan motifs on species such as Lewis, GlcNAc, high-mannose structures, sialic acids, and GalNAc, internalizing them and mediating their processing for further presentation by MHCI and MHC II (major histocompatibility complexes) (Figure 3.3). Galectins act as a link between innate and adaptive immunity by its remarkable ability of modifying various innate immune cells, sometimes by working as cytokines or adhesion molecules and at other times by influencing the functions of monocytes and macrophages through ERK cascade regulation. Galectins are found to be inhibitors of graft versus host diseases and also several autoimmune diseases. On the other hand, C-type lectins and siglecs are totally involved in antigen internalization, processing, and presentation by regulating signaling cascades, thus influencing differentiation of DCs and T-cell activation [69].

In addition to this, it has been observed that abnormal glycans may be recognized as non-self by the immune system and may contribute to autoimmunity. Impairment of NLGs structure may be considered as ligands for receptors of innate immunity which then induce a chronic inflammatory response activating macrophages. Hyaluronan, a disaccharide repeat, is known to mediate innate immune responses by interacting with TLRs. This glycan is synthesized as a result of any tissue injury, where it induces inflammation by interacting with TLR2 and TLR4. Thus, the biosynthesis as well as the breakdown of hyaluronan is gaining increasing importance with respect to its role innate immunity [61]. Recent research has given ample indications that lectins such as galectins, C-type lectin and siglecs are responsible for initiating innate immune responses, maintaining immune tolerance and homeostasis, and also playing a crucial role in host-pathogen interaction.

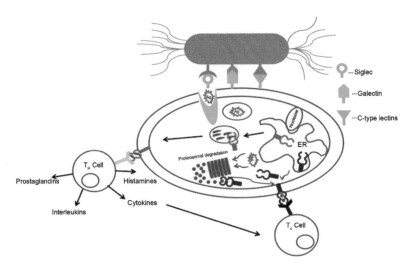

**FIGURE 3.3**   Lectin-glycan interactions between lectins (galectin, siglec, and C-type lectin) and glycan motifs on microbial surface, initiating a complex sequence of events which in turn leads to humoral and cell mediated responses.

### 3.5.1.1   GLYCANS IN HOST-PATHOGEN INTERACTION

The skin serves as the first innate immune barrier against the entry of microbes into the host organism. However, the host cell surface encompasses a wide variety of structures particularly the glycoproteins and glycolipids, which are involved in a number of key biological processes such as cell signaling, cell communication, regulation of biological and immunological pathways, etc., [70]. These sugar residues are critical for the adherence or attachment of microbial species on to the surface of host for colonization. The adherence of microbial lectins to these host cell surface receptors is a pre-requisite for the pathogenesis of tissue infection and virulence [71]. Adhesins are present on microbial cell wall membrane or fimbriae and pilli and glycoconjugates such as fibrin and fibronectin on host tissue drive this attachment process. Furthermore, several studies suggest that the occupation of lectins on the bacterial surface by exogenous sugars can prevent bacterial adherence to epithelial cells of different tissues [72]. Thus, the use of exogenous sugars with the ability to disrupt this interaction and preventing microbes to attach to these cell surface receptors can prove to be a good anti-microbial treatment [71]. Mannose and N-acetyl-D-galactosamine inhibits the adherence of *Escherichia coli* and *Pseudomonas aeruginosa* to epithelial cells [73].

### 3.5.2    GLYCOSYLATION AND ADAPTIVE IMMUNITY

Glycans have been known as key regulators in the diversity of T-cells and B-cells. Posttranslational modifications of B- and T-cells are important aspects in several immunological pathways such as receptor trafficking, association with other glycoproteins on the cell surface, signal transduction, and receptor internalization by endocytosis. This modification which occurs in GA involves the addition of N- and O-glycan chains on both B-cell receptor (BCR) and T-cell receptor (TCR) with the help of enzyme glycosyltransferases.

#### 3.5.2.1    TCR AND BCR SIGNALING

Glycosylation has been shown to play a pivotal role in T- and BCR signaling cascade pathways. For example, the GnT5 glycosyltransferase plays an important role in TCR signaling as it is involved in the formation of the β-1,6 N-glycan-branch structure on glycoproteins, such as TCR. This N-glycan branch commonly includes poly-lactosamine, an important component of this N-glycan structure works as a ligand for many lectins of the galectin family [74]. Mice model study has indicated that deficiency of GnT5 is related with decreased T-cell activation. In addition, these mice models have also displayed increased delayed-type hypersensitivity responses, autoimmune encephalomyelitis (EAE), and glomerulonephritis [75]. A number of homotypic and heterotypic TCR interactions which are essential for immunological signaling are mediated by lectins. For example, Galectin-3 is found to be chemically cross-linked with the TCR on T-cells, and any disruption of this interaction has been found to be linked with decreased TCR activation. In addition, it has been observed that treatment with galectin-3 restores normal conditions [75].

Glycoproteins that regulate the process of BCR activation include lectin CD22 (also known as siglec-2), found solely on B-cells. CD22 requires glycan ligands for binding for which it requires ST6Gal1 sialyl-transferase, whose function is to add sialic acid to galactose residue present on N-acetyl-lactosamine disaccharide [76]. In case of ST6Gal1 deficiency, CD22 cannot bind to its ligand hampering BCR activation and humoral immunity. Thus it can be deduced that glycan ligands namely, ST6Gal1 in B-cells and GnT5 in T-cells are pre-requisite for normal BCR and TCR signaling, respectively.

### 3.5.2.2    IMMUNE CELL DIFFERENTIATION

Glycoproteins such as MHC and glycan dependent mechanisms are important regulators of immune-cell differentiation. For example, sulfation of glycosaminoglycan by the sulfotransferase NDST2 ((N-deacetylase/N-sulfotransferase (heparan glucosaminyl) 2)) has been associated with the development of natural killer (NK) and T-cells [77]. O-linked glycosylation and N-linked deglycosylation are required for presentation by MHCI, and their further recognition by T-cells is then followed in a sequential cascade for neutralization of antigen [78]. Exogenous antigens also possess glycan ligands, which are critically important for presentation by MHCII and subsequent binding by TCRs. Thus altered glycosylation, due to deficiency of glycosyltransferase, is likely to hamper the formation of the TCR repertoire and thus affecting immunity [79]. Similarly, certain glycolipids are also involved in the presentation of endogenous Ag by NK cells. TCR co-receptors, CD4 (constant domain 4) on MHCII and CD8 on MHCI are the glycoproteins whose regulated sialylation confers restriction for TCR activation during thymic selection and activation. The sialylation levels are mediated by an enzyme, ST3Gal1 sialyltransferase, as deficiency of this enzyme in mice has been shown to lead to altered TCR repertoire [80].

### 3.5.3    ANTIGEN PROCESSING AND PRESENTATION

For normal immune process, varies types of processing and presentation of epitope is a pre-requisite for an efficacious Ag-Ab interaction for a normal immune response. MHC, a glycoprotein present on APCs is essential for presenting epitopes to TCR or BCR. MHCI is responsible for presenting exogenous Ag to T helper ($T_H$) cells and MHCII for presenting endog-enous Ag to $T_C$ cells (T cytotoxic) (Figure 3.3) [81]. Hanisch et al. have experimentally shown that O-linked glycosylation is more likely to influence proteolytic processing of Ag and thus their presentation by MHCI and MHCII. In fact, it has been indicated that NLGs need to be removed before cytosolic processing [82]. However, in contrast, O-linked glycans remain attached during Ag processing and finally the resulting peptide is presented by MHCI.

Several studies have shown that N- and O-linked glycosides can survive lysosomal degradation and the resultant glycopeptides can be presented in complex with MHCII [83]. A study concluded that T-cell hybridomas showed a great specificity for α-GalNAc moiety. It has also been established by previous studies that hydroxylation of galactose residue is required for T-cell activation. With the help of synthetic peptides, it was established that in collagen-induced arthritis, T-cell hybridomas specifically recognized β-D-galactopyranoside moiety of the peptide sequence CII256–270 [84, 85].

In addition, it has been observed that truncation of O-linked saccharides such as Tn (αGalNAc-Thr), STn (αNeu5Ac-(2,6)-αGalNAc-Thr) and Thomsen-Friedenreich antigen (βGal-(1,3)-αGalNAc-Thr) is responsible for epithelial carcinoma [86]. Thus, the glycosylation of antigen is an important facet when it comes to vaccine development. A major obstacle in this regard is the identification of glycosylation sites [79]. To overcome this, newly developed strategy of zinc finger nuclease-based gene targeting has been established, which involves the deficiencies of analyzing complex mixtures of glycoproteins [87].

### 3.5.4   ANTIBODY FUNCTIONING

Antibodies, produced from plasma cells as a result of B-cell activation, are essentially glycoproteins. Numerous antibody effector functions such as complement activation, phagocytosis, etc., require the binding of glycans present on the Fc region of immunoglobulin to Fc receptors [88]. Any alteration in the glycosylated part of immunoglobulins is likely to hamper its biological effector functions leading to a number of diseases like RA and IgA nephropathy. For example, people suffering from RA, terminal galactose, and sialic acid are missing from the N-glycans attached to IgG and therefore exposing N-acetylglucosamine [89] and ultimately promoting inflammatory responses [90]. Similarly, in case of IgA nephropathy, the O-glycans present on IgA are structurally disrupted exposing N-acetylglucosamine leading to inflammation of kidney.

### 3.5.5   COMPLEMENT PATHWAYS

Complement system is an integral part of innate immunity, protecting the host against a wide range of pathogenic microorganisms by mediating their destruction by using any one of the three commonly known complement pathways, i.e., classical pathway, alternate pathway and lectin pathway [91]. A number of complement proteins are involved in the complement cascades. However, all three pathways commonly involve the formation of C5 convertase which then leads to membrane attack complex (MAC) formation and ultimately cell death. The lectin pathway of complement system is activated after the recognition of glycan (mannose) moieties on microbial species by mannose binding lectin (MBL) [92]. C1q, which is an important component of complement cascade, retains its functional conformation by O-linked glycosylation which works by stabilizing the triple helical structure of C1q. The head region of C1q is highly flexible posing a hindrance while attaching to targets. This hindrance is overcome by the addition of NLGs at Asn 146, minimizing interactions among proteins between globular head domains of C1q. Almost all the complement components are glycosylated in order to be functionally active such as C3 is predominantly glycosylated with high mannose, whereas C5 is mainly sialylated [93, 94]. Properdin, an important complement component, is N-linked glycosylated at Asn 428, which is known to enhance its stability [95].

### 3.5.6   APOPTOSIS

Apoptosis, being essential process for immune response of a cell, is an important mechanism of system homeostasis, lymphoid development, and clearance of effector cells after immune activation. Glycosylation of mammals contributes to this process by generating endogenous ligand lectins, including galectins. These lectins bind to glycans that contain the disaccharide lactose and polylactosamine, present in certain cellular glycoproteins and glycolipids [96].

T cell is linked with both the expression, i.e., endogenous, and exogenous and has a role in normal functioning of galectin-1 and galectin-3. Further in thymocytes, Gelectin-1 has an interesting role in cell death or apoptosis as T-cell do not normally work for the process of the cell death but in response to galectin-1 and monitoring of galectin-3 leads to T-cell activation and apoptosis [96]. Cancer cells have been known to combat apoptosis by modifying the glycans present on cell death receptors such as CD95 (Fas) and tumor necrosis factor receptor 1 (TNFR1) [97]. This glycosylation disrupts the ligand-receptor interactions influencing the apoptotic machinery and its signaling cascade [98]. In addition, aggregation of glycosphingolipid GD3 and ceramide have also been found to be coupled with mitochondrial damage and thus programmed cell death [99].

## 3.6   CONSEQUENCES OF ALTERED GLYCOSYLATION ON IMMUNE SYSTEM

### 3.6.1   ROLE OF GLYCOSYLATED IGE IN ALLERGIC REACTIONS

Allergic reactions are primarily known to be mediated by IgE, due to its unique structure consisting of an extra constant heavy chain region which enables it to bind to Fc-ER receptors present on tissue mast cells. Cross linkage of IgE on mast cells with an allergen results in the degranulation of mast cells, releasing chemical mediators of allergic reactions or inflammation such as histamine, leukotrienes, prostaglandins, etc., [100]. These chemical substances are responsible for symptoms of allergy such as vasoconstriction, vasodilation, and respiratory symptoms including sneezing, rhinitis, asthma, and more dangerous anaphylaxis [101]. Research study has indicated that glycosylated IgE derives the allergic cascades. However, the mechanism involved behind the glycosylation of IgE still remains vague. Studies have revealed that the glycosylation occurs with the addition of N-linked oligo-mannose at asparagine-394 (N394) on constant domain 3. Any alteration in this glycosylation makes IgE incapable of binding to Fc-ER, which stops further sequence of events of the allergic reaction, abrogating anaphylaxis [102].

### 3.6.2   GLYCA: A POTENT BIOMARKER OF CARDIOVASCULAR DISEASE

GlycA, a favorable biomarker of systemic inflammation, has been associated with enhanced glycan intricacy and circulating acute-phase protein levels in patients fighting autoimmune diseases such as RA, psoriasis, and lupus. Along with GlycA, glycoproteins such as hsCRP (C-reactive protein) and fibrinogen also serve as biomarkers of acute or chronic inflammation or infection of CVD (cardiovascular disease) [103]. GlycA acts as an NMR biomarker because of the protons from the N-acetyl glucosamine residues on the glycan portions of acute-phase proteins such as α1-acid glycoprotein, haptoglobin, α1-antitrypsin, and α1-antichymotrypsin [104]. Sialic acid, a terminal carbohydrate residue of glycoconjugates, which is directly proportional to tumor necrosis factor (TNFα) and IL-6 (Interleukin) have been linked to CVD [105]. Moreover, leukocytes, and neutrophils, which secrete α1-acid glycoprotein and haptoglobin, elevated in response to increased glycA may be attributed as a reliable source for increased glycA [105]. A high level of GlycA is proportional to the severity of disease in patients with RA, systemic lupus erythematous, and psoriasis. GlycA is a more reliable and early predictor of cardiovascular and other autoimmune diseases since it combines the glycosylated levels and the serum concentration of most frequent acute-phase proteins.

### 3.6.3  GLYCOSYLATION AND AUTOIMMUNE DISEASES

RA is a common autoimmune disorder associated with inflammation of the joints, a major symptom in the middle and upper aged woman. Individuals with RA produce a group of autoantibodies (auto-Abs), particularly the IgM called rheumatoid factors, which binds to the Fc region of IgG. These auto-Abs are complementary to the Fc region of IgG. This binding generates IgM-IgG complexes that are deposited in the joints. These immune complexes can activate the complement cascade and further promote type III hypersensitive reaction, which leads to chronic inflammation of the joints. Parekh and his co-workers have shown that abnormal N-glycosylation of IgG is responsible for RA [118]. Further studies have revealed that people with RA have reduced galactosylation levels of serum IgG [106]. Alterations have been reported in Fc portion of IgG because of reduced galactosyltransferase in RA patients [107]. Assessment of CRP and ESR (erythrocyte sedimentation rate) has been used as a standard method to diagnose the diseases besides both CRP and ESR diagnosis have certain limitations [108]. Apart from RA, concentration of ESR may also be altered in other diseases such as chronic kidney disease, pregnancy, anemia, etc., [109]. Moreover, concentration of CRP remains variable which makes it less trustworthy for RA diagnosis [110]. Recently, GlycA has been used as a potential biomarker for assessing and diagnosing autoimmune disease like RA and SLE (Systemic Lupus Erythematous) activity [111, 112]. Positive correlation of GlycA concentrations has been associated with RA and SLE disease activity [113].

### 3.6.4  GLYCOSYLATION AND CANCER

Cumulative evidence has given an indication of the association of aberrant glycosylation with tumor growth and progression. A major hallmark of malignant cancer cells is their incapability of cell adhesion. Cell membrane being composed of a myriad of glycoproteins is intimately involved in cell-cell and cell-extracellular matrix (ECM) adhesion. Any alteration in glyco-sylation of the protein components of the cell membrane such as distorted glycan structures, elevated functional activity of branched N-glycans, terminal sialylated glycan motifs and changed fucozylationrenders them incapable of cell adhesion [114]. Cancer cells display increased sialylation of the glycans which intensified negative charge due to increased sialylation resulting in the disruption of cell-cell adhesion due to electrostatic repulsion [115]. In context to this, cancer-associated sTn-antigen reduces cell adhesion in prostate cancer whereas sialyltransferase ST6GAL1 inhibits cell adhesion in breast cancer cells [116, 117]. One of the major problems encountered during cancer treatment is that most of the cancers are detectable only when they are at their advanced stage which makes treatment next to impossible. In this regard, glycobiomarkers have shown a promising future providing early disease diagnosis which can help in reducing cancer-related deaths. Two of the most well-known tumor suppressor genes, p53, and Rb (retinoblastoma), carry glycosylation sites at Ser149, whose activity is under the stringent control of O-GlcNAc [118]. Slight modification in its glycosylation, may lead to the amplification of pro-oncogenic activity. Moreover, there are compelling evidences that abnormal extracellular glycosylation as well as glycosylation of cytosolic proteins obstructs signaling cascades in cancer cell-cycle regulation, cellular energetics and their adaptation and stability to surrounding environment [119].

Evidence suggests that interactions between tumor specific glycans and lectins on immune cells can interfere with the anti-tumor response of the immune system [120]. Modification of cell surface glycans can stimulate siglecs, a transmembrane receptor protein present on immune cells, which promotes cancer progression by absconding immune surveillance in cancer [121]. Glycosylated IgG can work as a potential biomarker for cancer prog-nosis because of its function in tumor immune check [122]. Targeting altered glycosylation using anticancer vaccines for altered glycans can be a good therapeutic approach [123].

There are different types of cancer reported, which are complicated malignant diseases with different causative agents, etiology, or molecular profiles [124]. Oligosaccharide changes responsible for cancer are related to tumor prevention or cell proliferation [125]. Glycosyltransferases (GTs) are responsible for the changes in the oligosaccharide structures exposed to the lumen of the ER. The regulated expression of these enzymes at the transcriptional level makes them the biomarkers for a particular type of cancer [126]. For example, several fucosyltransferases (FUT) attach fucose at cell surface glycan residues in a α2-3 and/or 4 linkages of the N- and O-linked glycan structures, which cause expression of cancer-associated blood group Lewis antigens Lex/Ley and Lea/Leb. FUT3 is the major fucosyltransferase responsible for synthesizing SLea [127].

### 3.6.5   IGG AS GLYCOBIOMARKER FOR GASTRIC CANCER

IgG is the most abundant immunoglobulin present in serum. Being a glycoprotein, it serves as an attractive approach to study protein-specific glycosylation patterns in various diseases [129]. Research study pinpoints that aberrant glycosylation of the Fc region in IgG has been found to be linked with a diverse range of biological and physiological processes especially in autoimmune disease as well as cancer of the ovary and gastric type [129]. Thus IgG can be used as protein-specific glycan biomarkers for diagnosing cancer [130]. Ozcan et al. in their study have revealed the association of non-galactosylated bi-antennary glycans along with altered glycosylation pattern of Fc region of IgG with gastric cancer [130].

It has been reported that IgG which lacks glycosylation in its Fc region binds with its receptor with a much lower affinity mainly because of its altered confirmation and thus displays reduced functional activity [131]. It has been observed that removal of galactose from IgG glycans reduces the activity of antibodies, promoting chronic inflammation and thus altered immune system in gastric cancer. However, it is still unknown that whether cancer promotes inflammation or inflammation leads to cancer [132]. Future research is highly warranted for determining the predictive role of altered glycosylation patterns of IgG in cancer.

### 3.6.6   GLYCOSYLATION AND SYSTEMIC LUPUS ERYTHEMATOSUS (SLE)

SLE is a systemic autoimmune disease with myriad of symptoms ranging from kidney failure, arthritis, and erythema, loss of tolerance and development of an autoreactive immune response, including autoimmune cells producing pathogenic autoantibodies, mainly of the IgG1 and IgG3 subclasses. There have been innumerable reports supporting the fact that abnormal glycosylation is linked to chronic inflammation in SLE [133]. Research studies have reported that deficiency of a-mannosidase II (aM-II) promotes SLE progression since deficiency of aM-II disrupts normal branching in N-glycans which functions to halt autoimmune disease pathogenesis [134]. Glycosylation analysis may lead to the development of improved diagnostic methods and thus help to clarify the carbohydrate-related pathogenic mechanism of inflammation in SLE. Mass spectrometry (MS) and liquid chromatography/mass spectrometry (LC/MS) have further supported aberrant glycosylations in SLE by comparing mass spectra or chromatograms between case and control samples [135, 136]. Furthermore, increased oligosaccharides without galactose (Gal) on IgG rendering them pro-inflammatory, is prominently found in SLE and RA [137, 138]. The Fc portion of IgG harboring N-glycans mediates IgG effector functions. However, these on lacking terminal galactose make IgG pro-inflammatory, activating complement pathways. Conversely, the addition of galactose as well the sialylation of IgG has been known to decrease the inflammatory response of IgG [139, 140].

### 3.6.7  ENGINEERED GLYCANS AND GLYCAN MIMICS AS THERAPEUTIC AGENTS

The majority of glycan-based drugs so far have been discovered as carbohydrate-binding proteins (lectin) or glycosidase [141]. The antiviral compound Zanamivir (Relenza) is the most powerful medicine arising from engineered sugar moieties [142]. The binding of the viral hemagglutinin (HA) to sialic acid-containing glycans in the life cycle of the influenza virus on host cell surfaces serve as substrates for viral neuraminidase (NA), thereby helping in virus release and maturation [143] (Figure 3.4).

Both Zanamivir and Oseltamivir function effectively by binding with nano-molar affinity to viral NA and stopping viral budding entry into cells [144]. Resistance acquired to oseltamivir has provided the driving force for designing novel drugs that target the influenza virus. Recently, researchers have designed a covalent difluorinated NA inhibitor [145]. Miglitol (Glyzet) and Acarbose (Precose, Glucobay) are potent drugs for type II diabetes mellitus which control blood sugar by inhibiting glucosidases and amylases in the gut. For treating the lysosomal storage disease type 1, a gaucher disease

(GD), miglustat, an imino sugar (Zavesca, N-butyl-deoxynojirimycin), is a drug developed by Actelion Pharmaceuticals, National Taiwan University. Miglustat was first synthesized by Butters and Dwek, who found that N-alkylated analogs of the natural product deoxynojirimycin, which was the inhibitor of glucosyl-transferase involved in glucosylceramide biosynthesis [146].

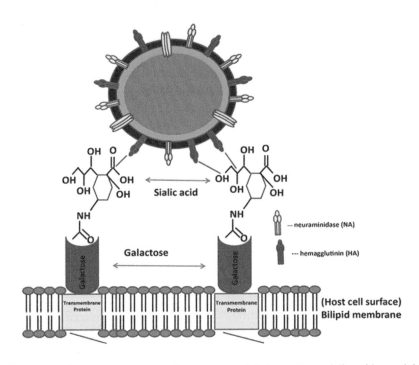

**FIGURE 3.4**   Binding mechanism of viral hemagglutinin (HA) to sialic acid-containing glycans on host cell surfaces, results in the release of virus in host cell and thus helping it to mature and further complete its life cycle in the host.

GD is a genetic disorder caused by the mutations in the GBA1 gene (beta-glucocerebrosidase) that results in the reduction of the activity of the β-glucocerebrosidase (GC) enzyme. Gaucher patients develop high levels of sphingolipid called glucosylceramide in different cells and other body parts. Treatment with miglustat (OGT 918, N-butyl-deoxynojirimycin), a medicine manufactured by Oxford GlycoSciences (UK-based Biopharmaceutical company), is used primarily to treat type I Gaucher disease (GD1). It is marketed under the trade name Zavesca and it was found not satisfactory as it enhances the formation of sphingolipid. Later, placental derived enzyme replacement therapy with commercially available Velaglucerase alfa, Taliglucerase alfa, Imiglucerase, and Eliglustat were found promising in GD1 treatment.

Many factors that are involved in bacterial adhesion on host cells are lectins and thus much work has been done towards creating glycan-based molecules to inhibit this interaction [147, 148]. Adhesion of *Pseudomonas aeruginosa* inside the cells is enhanced by the formation of lectins, such as PA-IL and PA-IIL (also known as LecA and LecB) formed in the cytoplasm of planktonic cells. Glycosaminoglycans (GAGs) are a class of highly charged linear polysaccharides found on the cell surface that play an important role in developmental physiology [149]. The invention of structure of the most significant clinical GAG, heparin, in 1982 [150] resulted in the formation of the active heparin pentasaccharide [151] from which first synthetically defined small molecule heparin, Fondaparinux, (a synthetic heparin analog) was made. The advantage of this small heparin is its excellent bioavailability and a longer circulating half-life of 18 hrs making it ideal for its administration under the skin for the disease treatment [152]. The advantage of fondaparinux is its increased potential and lesser risk of heparin-induced thrombocytopenia [153]. People have reported the chemical and enzymatic formation of a pentasaccharide heparin [154]. The significance of forming therapeutically defined heparin was considered in 2008 when over-sulfated chondroitin sulfate (CS) toxicity led to metabolic disorder [155]. A relatively similar technique has been used for the synthesis of a 3-O-sulfated heparin octasaccharide, which combines with the herpes simplex virus type 1 (HSV-1) thus blocking it in the cell [156]. The invention of selectins in 1980s was a breakthrough in the field of glycobiology that inspired several aspects in the field of medical biology.

The selectins aid in the migration of leukocytes to sites of inflammation by the attachment of the sialyl Lewis x (sLex) and sialyl Lewis A (sLe$^a$) carbohydrate moieties [157, 158, 219, 220]. Apart from migration of leukocytes to the inflammation sites [159], sLex is also important in metastasis of tumor and acts as ligand for human sperm binding to the zona pellucida of ova at the time of fertilization [160]. Many clinical trials have been stopped in making of small molecule glycan structures such as Cylexin (CY-1503) to inhibit the selectins, [161]. The fucosylated mimics of the Lewis structures give promising leads in clinical trials for treating asthma [162] and sickle cell [163]. A nuclear magnetic resonance (NMR) based fragment screen to design ligands with a sLex scaffold attached to a second site ligand has been reported [164]. Thus this technique managed E-selectin inhibitors with lower affinities, although they're *in vivo* trials are still under process. The glycosylation of proteins inside the cell with N-acetylglucosamine termed O-GlcNAcylation has a significant role in the regulation of cancer, diabetes, obesity, and Alzheimer's disease [165]. Recently, a highly selective and low nM inhibitor of O-linked N acetyl glucosamine (OGA), named as thiamet-G has been reported [166]. Treatment of rodents with thiamet-G enhanced O-GlcNAcylation levels in the brain, which reduced tau accumulation and loss of neuronal cells in a mouse model of Alzheimer's disease (AD) [167]. Novel methods using carbohydrate scaffolds are employed in order to screen the glycan chemical space for potent inhibitors of lectins involved in disease states [168, 169]. Moreover, many natural products are glycosylated, which affects their efficiency and specificity, and these are promising products in the field of therapeutics [170].

## 3.7    CARBOHYDRATE BASED VACCINES

Great efforts are being made in vaccines derived from carbohydrates thereby bringing carbohydrate chemistry into clinically significant platforms [171, 172]. Some recently utilized vaccines are composed of glycans including vaccinations against *Neisseria meningitides* (Menactra), *Streptococcus pneumonia* (Prevnar), *Haemophilus influenzae* type B (Hib; Hiberix, Comvax), and *Salmonella typhii* [173]. Advancements in protein conjugation led to expanded techniques available for the generation of carbohydrate vaccines specifically attached to protein carriers [174]. The most successful synthetically defined vaccine is the formation of the Cuban Hib vaccine, which is the first clinically approved fully synthetic carbohydrate vaccine. The vaccine is mainly based on the structure of the capsular polysaccharide antigen from Hib [175]. The Seeberger research group at the Department of Chemistry, MIT USA developed an antimalarial vaccine by synthesizing glycosylphosphatidylinositol (GPI) automatically from the causative agent of malaria, *Plasmodium falciparum* [176, 177]. They have also successfully developed a vaccine against the *Bacillus anthracis* spore tetra-saccharide, used to form antibodies against anthrax for imaging and detection purposes [178]. The glycan attached to tetanus toxoid resulted in a better serum anti-lipopolysaccharide (LPS) antibody response in mice in comparison to shorter synthetic O-SP (O-specific polysaccharide) sequences [179]. As a result, the immunized mice produced glycoconjugate induced anti-LPS antibodies which gave them protection against SF2a (*Shigella flexneri* 2a) infection. The preclinical data also suggest that these results are even applicable to human trials [180]. A polymer derived from the *Candida albicans* cell wall having the β-mannantrisaccharide epitope attached to chicken serum albumin has greater IgG formation in mice than the trisaccharide-tetanus toxoid conjugate vaccine alone [181]. Bundle and co-workers utilized such antibodies for identifying a minimal disaccharide epitope vaccine that protected rabbits from the fungal infection of *C. albicans* [182].

## 3.8   CONCLUSION AND FUTURE PERSPECTIVE

It has become evident that like genome and proteome, the role of glycome is equally important in the majority of the metabolic processes. In addition to these, glycans have been known to regulate various immunological pathways in multiple ways, such as GnT5 glycosyltransferase and ST6Gal1 regulating TCR and BCR signaling, respectively. Both N- and O-linked glycans have been known to affect the Ag processing and its presentation to antibodies. Aberrant glycosylation is likely to alter the normal innate and adaptive immune responses eventually causing diverse immune disorders like auto-immune diseases or congenital disorders of immune system. Specific lack of galactose and sialic acid from the N-glycans attached to IgG results in intense inflammatory response leading to RA. Moreover, abnormal glycosylation of IgG in its $F_c$ region has been linked to several diseases namely SLE, activation of complement cascade, cancer, etc. IgG, being the most abundant in serum as well as its association with several diseases, is now being used as a glycobiomarker. Thus, deciphering the information contained in protein-glycan interaction is extremely important which may be quite helpful in future. Interestingly, engineered glycans as well as carbohydrate-based vaccines are being used to take care of the problems arising due to altered glycosylation. More importantly, conjugation of carbohydrate vaccine with a protein carrier is proving beneficial in vaccine production.

Inhibitors of glycan-binding receptors are being considered as a potential target for treating cancer. To achieve this, different lectin blockers for inhibiting the interaction of selectins, siglecs, and galectins with their glycan ligands are being developed. However, these are still under clinical trials due to certain issues related to their bioavailability and susceptibility of lectin blockers [183]. In order to avoid this, polysaccharides as well as peptide inhibitors are being considered as alternate lectin inhibitors. Moreover, these ligands stress the importance that dietary carbohydrate compounds may serve as therapeutic agents. Liver metastasis and ovarian cancer are being treated through anti-E-selectin and anti-L-selectin mAbs respectively [184]. P-selectin Glycoprotein Ligand-1 (PSGL-1) in conjugation with IgG is known to inhibit leukocyte rolling and adhesion. Recent advancements in glycobiology have led to the development of anti-sense nucleotides where in small-interfering RNAs as well as mAbs for inhibiting lectins like those that Gal-3 and Gal-1 are being increasingly used for inhibiting tumor progression [185]. Despite these developments, deciphering the glycome code to specifically design selective inhibitors of lectin-glycan interactions with maximum bioavailability is highly warranted. Future research should be focused in this direction for development of highly selective inhibitor drug with maximum bioavailability and less toxicity.

## KEYWORDS

- **B-cell receptor**
- **galectin**
- **glycans**
- **glycosylation**
- **lectin**
- **vaccine**

## ACKNOWLEDGMENTS

Support by enhanced seed grant EF/2019-20/QE04-02 (to NKP) from Manipal University Jaipur, Rajasthan, India is gratefully acknowledged.

## REFERENCES

1. Lloyd, D. H., Viac, J., Werling, D., Rème, C. A., & Gatto, H., (2007). Role of sugars in surface microbe-host interactions and immune reaction modulation. *Vet. Dermatol., 18*, 197–204.
2. Patel, P., & Kearney, J. F., (2016). Immunological outcomes of antibody binding to glycans shared between microorganisms and mammals. *J. Immunol., 197*, 4201–4209.
3. Wolfert, M. A., & Boons, G., (2013). Adaptive immune activation: Glycosylation does matter. *Nat. Chem. Biol., 9*, 776–784.

4. Lyons, J. J., Milner, J. D., & Rosenzweig, S. D., (2015). Glycans instructing immunity: The emerging role of altered glycosylation in clinical immunology. *Front Pediatr., 3*, 54.

5. Leppanen, A., White, S. W., Helin, J., McEver, R. P., & Cummings, R. D., (2000). Binding of glycosulfopeptides to P-selectin requires stereospecific contributions of individual tyrosine sulfate and sugar residues. *Journal of Biological Chemistry, 275,* 39569–39578.

6. Marth, J. D., (2008). A unified vision of the building blocks of life. *Nat. Cell Biol., 10,* 1015, 1016.

7. Smyth, S., & Heron, A., (2006). Diabetes and obesity: The twin epidemics. *Nat. Med., 12,* 75–80.

8. Landsteiner, K., (1900). *Zur Kenntnis der antifermentativen, lytischen and agglutinierenden Wirkungen des Blutserums und der Lymphe Zentralbl Bakteriol, 27,* 357–362.

9. Kabat, E. A., & Leskowitz, S., (1955). Immunochemical studies on blood groups. XVII structural units involved in blood group A and B specificity. *J. Am. Chem. Soc., 77,* 5159–5164.

10. Morgan, W. T. J., & Watkins, W. M., (1953). The inhibition of the hemagglutinins in plant seeds by human blood group substances and simple sugars. *J. Exp. Pathol., 34,* 94–103.

11. Watkins, W. M., & Morgan, W. T., (1955). Inhibition by simple sugars of enzymes which decompose the blood-group substances. *Nature, 175,* 676–677.

12. Watkins, W. M., (2001). The ABO blood group system: Historical background. *Transfus. Med., 11,* 243–265.

13. McLean, J., (1916). The thromboplastic action of cephalin. *Am. J. Physiol., 41,* 250–257.

14. Lever, R., Mulloy, B., & Page, C. P., (2012). *Heparin: A Century of Progress.* Springer, Berlin.

15. Lindahl, U., Bäckström, G., Höök, M., Thunberg, L., Fransson, L. A., & Linker, A., (1979). Structure of the antithrombin binding site in heparin. *Proc. Natl. Acad. Sci. U.S.A., 76,* 3198–3202.

16. Dochez, A. R., & Avery, O. T., (1917). The elaboration of specific soluble substance by pneumococcus during growth. *J. Exp. Med., 26,* 477–493.

17. Heidelberger, M., & Avery, O. T., (1923). The soluble specific substance of pneumococcus. *J. Exp. Med., 38,* 73–79.

18. Francis, T. W. S., (1930). Cutaneous reactions in pneumonia. The development of antibodies following the intradermal injection of type-specific polysaccharide. *J. Exp. Med., 52,* 573–585.

19. Heidelberger, M., Dilapi, M. M., Siegal, M., & Walter, A. W., (1950). Persistence of antibodies in human subjects injected with pneumococcal polysaccharides. *The Journal of Immunology, 65,* 535–541.

20. Schatz, A., Bugle, E., & Waksman, S. A., (1944). Streptomycin, a substance exhibiting antibiotic activity against gram-positive and gram-negative bacteria. *Proc. Soc. Exp. Biol. Med., 55,* 66–69.

21. Wang, L. X., & Lomino, J. V., (2012). Emerging technologies for making glycan-defined glycoproteins. *ACS Chem. Biol., 7,* 110–122.

22. Reivich, M., Kuhl, D., Wolf, A., Greenberg, J., Phelps, M., Ido, T., Casella, V., et al., (1979). The [18F]fluorodeoxyglucose method for the measurement of local cerebral glucose utilization in man. *Circ. Res., 44,* 127–137.

23. Ametamey, S. M., Honer, M., & Schubiger, P. A., (2008). Molecular imaging with PET. *Chem. Rev., 108*, 1501–1516.

24. Laughlin, S. T., & Bertozzi, C. R., (2009). Imaging the glycome. *Proc. Natl. Acad. Sci. U.S.A, 106*, 12–17.

25. Varki, A., (2017). Biological roles of glycans. *Glycobiology, 27*, 3–49.

26. Apweiler, R., Hermjakob, H., & Sharon, N., (1999). On the frequency of protein glycosylation, as deduced from analysis of the SWISS-PROT database. *Biochim. Biophys. Acta, Gen. Subj., 1473*, 4–8.

27. Varki, A., (1993). Biological roles of oligosaccharides: All of the theories are correct. *Glycobiology, 3*, 97–130.

28. Helenius, A., & Aebi, M., (2004). Roles of N-linked glycans in the endoplasmic reticulum. *Annu. Rev. Biochem., 73*, 1019–1049.

29. Greene, E. R., Himmel, M. E., Beckham, G. T., & Tan, Z., (2015). Glycosylation of cellulases: Engineering better enzymes for biofuels. *Adv. Carbohydr. Chem. Biochem., 72*, 63–112.

30. Pol-Fachin, L., Verli, H., & Lins, R. D., (2014). Extension and validation of the GROMOS 53A6glyc parameter set for glycoproteins. *J. Comput. Chem., 35*, 2087–2095.

31. Kaushik, S., Mohanty, D., & Surolia, A., (2011). Role of glycosylation in structure and stability of erythrina corallodendron lectin (EcorL): A molecular dynamics study. *Protein Sci. 20*, 465–481.

32. Dessen, A., Gupta, D., Sabesan, S., Brewer, C. F., & Sacchettini, J. C., (1995). X-ray crystal structure of the soybean agglutinin cross-linked with a biantennary analog of the blood group I carbohydrate antigen. *Biochemistry, 34*, 4933–4942.

33. Halder, S., Surolia, A., & Mukhopadhyay, C., (2015). Impact of glycosylation on stability, structure, and unfolding of soybean agglutinin (SBA): An insight from thermal perturbation molecular dynamics simulations. *Glycoconj. J., 32*, 371–384.

34. Barb, A. W., Meng, L., Gao, Z., Johnson, R. W., Moremen, K. W., & Prestegard, J. H., (2012). NMR characterization of immunoglobulin G Fc glycan motion on enzymatic sialylation. *Biochemistry, 51*, 4618–4626.

35. Dall'Olio, F., (1996). Protein glycosylation in cancer biology: An overview. *Clin. Mol. Pathol, 49*, M126–M135.

36. Yadav, S. C., Prasanna, K. N. K., & Jagannadham, M. V., (2010). Deglycosylated milin unfolds via inactive monomeric intermediates. *Eur. Biophys. J., 39*, 1581–1588.

37. Gao, J., Bosco, D. A., Powers, E. T., & Kelly, J. W., (2009). Localized thermodynamic coupling between hydrogen bonding and microenvironment polarity substantially stabilizes proteins. *Nat. Struct. Mol. Biol., 16*, 684–690.

38. Cummings, R. D., (2009). The repertoire of glycan determinants in the human glycome. *Mol. Biosyst., 5*, 1087–1104.

39. Nairn, A. V., & Moremen, K. W., (2009). Glycotranscriptomics. In: Cummings, R., & Pierce, J. M., (eds.), *Handbook of Glycomics* (pp. 95–136).

40. Spiro, R. G., (2002). Protein glycosylation: Nature, distribution, enzymatic formation, and disease implications of glycopeptide bonds. *Glycobiology, 12*, 43R–56R.

41. Rao, R. S. P., & Wollenweber, B., (2010). Do N-glycoproteins have preference for specific sequons? *Bioinformation, 5*, 208–212.

42. Murray, A. N., Chen, W., Antonopoulos, A., Hanson, S. R., Wiseman, R. L., Dell, A., et al., (2015). Enhanced aromatic sequons increase oligosaccharyltransferase glycosylation efficiency and glycan homogeneity. *Chem. Biol., 22*, 1052–1062.

43. Jenny, L. J., Mark, B. J., Sean, O. R., & Brian, A. C., (2013). The regulatory power of glycans and their binding partners in immunity. *Trends in Immunology, 34*, 290–298.

44. Chen, W., Enck, S., Price, J. L., Powers, D. L., Powers, E. T., Wong, C., et al., (2013). Structural and energetic basis of carbohydrate-aromatic packing interactions in proteins. *J. Am. Chem. Soc., 135*, 9877–9884.

45. Van, D. S. P., Rudd, P. M., Dwek, R. A., & Opdenakker, G., (1998). Concepts and principles of O-linked glycosylation. *Crit. Rev. Biochem. Mol. Biol., 33*, 151–208.

46. Peter-Katalinić, J., (2005). Methods in enzymology: O-glycosylation of proteins. *Methods Enzymol., 405*, 139–171.

47. Tran, D. T., & Ten, H. K. G., (2013). Mucin-type O-glycosylation during development. *J. Biol. Chem., 288*, 6921–6929.

48. Orlean, P., (1990). Dolichol phosphate mannose synthase is required *in vivo* for glycosyl phosphatidylinositol membrane anchoring, O-mannosylation, and N-glycosylation of protein in *Saccharomyces cerevisiae*. *Mol. Cell. Biol., 10*, 5796–5805.

49. Xu, C., & Ng, D. T. W., (2015). O-mannosylation: The other glycan player of ER quality control. *Semin. Cell Dev. Biol., 41*, 129–134.

50. Bennett, E. P., Mandel, U., Clausen, H., Gerken, T. A., Fritz, T. A., Tabak, L. A., et al., (2012). Control of mucin-type O-glycosylation: A classification of the polypeptide GalNAc-transferase gene family. *Glycobiology, 22*, 736–756.

51. Hooper, L. V., Midtvedt, T., & Gordon, J. I., (2002). How host-microbial interactions shape the nutrient environment of the mammalian intestine. *Annu. Rev. Nutr., 22*, 283–307.

52. Midtvedt, T., (1974). Microbial bile acid transformation. *Am. J. Clin. Nutr., 27*, 1341–1347.

53. O'Keefe, S. J., (2008). Nutrition and colonic health: The critical role of the microbiota. *Curr. Opin. Gastroenterol., 24*, 51–58.

54. Ley, R. E., Peterson, D. A., & Gordon, J. I., (2006). Ecological and evolutionary forces shaping microbial diversity in the human intestine. *Cell, 124*, 837–848.

55. Eckburg, P. B., Bik, E. M., Bernstein, C. N., Purdom, E., Dethlefsen, L., Sargent, M., Gill, S. R., et al., (2005). Diversity of the human intestinal microbial flora. *Science, 308*, 1635–1638.

56. Cerdeno-Tarraga, A. M., Patrick, S., Crossman, L. C., Blakely, G., Abratt, V., Lennard, N., Poxton, I., Duerden, B., Harris, B., Quail, M. A., et al., (2005). Extensive DNA inversions in the B. fragilis genome control variable gene expression. *Science, 307*, 1463–1465.

57. Xu, J., Mahowald, M. A., Ley, R. E., Lozupone, C. A., Hamady, M., Martens, E. C., Henrissat, B., Coutinho, P. M., Minx, P., Latreille, P., et al., (2007). Evolution of symbiotic bacteria in the distal human intestine. *PLoS Biol., 5*, e156.

58. Coyne, M. J., & Comstock, L. E., (2008). Niche-specific features of the intestinal Bacteroidales. *J. Bacteriol., 190*, 736–742.

59. Bry, L., Falk, P. G., Midtvedt, T., & Gordon, J. I., (1996). A model of host-microbial interactions in an open mammalian ecosystem. *Science, 273*, 1380–1383.

60. Coyne, M. J., Reinap, B., Lee, M. M., & Comstock, L. E., (2005). Human symbionts use a host-like pathway for surface fucosylation. *Science, 307*, 1778–1781.

61. Girish, K. S., & Kemparaju, K., (2007). The magic glue hyaluronan and its eraser hyaluronidase: A biological overview. *Life Sci., 80*, 1921–1943.

62. Robinson, M. J., Sancho, D., Slack, E. C., LeibundGut-Landmann, S., & Reis e Sousa, C., (2006). Myeloid C-type lectins in innate immunity. *Nat. Immunol., 7*, 1258–1265.

63. Van, K. Y., & Rabinovich, G. A., (2008). Protein-glycan interactions in the control of innate and adaptive immune responses. *Nature Immunol., 9*, 593–601.

64. Baum, L. G., et al., (2003). Amelioration of graft versus host disease by galectin-1. *Clin. Immunol., 109*, 295–307.

65. Figdor, C. G., Van, K. Y., & Adema, G. J., (2002). C-type lectin receptors on dendritic cells and Langerhans cells. *Nat. Rev. Immunol., 2*, 77–84.

66. Blixt, O., Collins, B. E., Van, D. N. I. M., Crocker, P. R., & Paulson, J. C., (2003). Sialoside specificity of the Siglec family assessed using novel multivalent probes: Identification of potent inhibitors of myelin-associated glycoprotein. *J. Biol. Chem., 278*, 31007–31019.

67. Magesh, S., Ando, H., Tsubata, T., Ishida, H., & Kiso, M., (2011). High-affinity ligands of siglec receptors and their therapeutic potentials. *Curr. Med. Chem., 18*, 3537–3550.

68. Crocker, P. R., (2005). Siglecs in innate immunity. *Curr. Opin. Pharmacol., 5*, 431–437.

69. Barrionuevo, P., et al., (2007). A novel function for galectin-1 at the crossroad of innate and adaptive immunity: Galectin-1 regulates monocyte/macrophage physiology through a nonapoptotic ERK-dependent pathway. *J. Immunol., 178*, 436–445.

70. Collins, B. E., & Paulson, J. C., (2004). Cell surface biology mediated by low affinity multivalent protein-glycan interactions. *Current Opinion in Chemical Biology, 8*, 617–625.

71. Ko, H. L., Beuth, J., Solter, J., et al., (1987). *In vitro* and *in vivo* inhibition of lectin mediated adhesion of pseudomonas aeruginosa by receptor blocking carbohydrates. *Infection, 15*, 237–240.

72. Rebiere-Huet, J., Di Martino, P., & Hulen, C., (2004). Inhibition of pseudomonas aeruginosa adhesion to fibronectin by PA-IL and monosaccharides: Involvement of a lectin-like process. *Canadian Journal of Microbiology, 50*, 303–312.

73. King, S. S., Young, D. A., Nequin, L. G., et al., (2000). Use of specific sugars to inhibit bacterial adherence to equine endometrium *in vitro*. *American Journal of Veterinary Research, 61*, 446–449.

74. Rapoport, E. M., Kurmyshkina, O. V., & Bovin, N. V., (2008). Mammalian galectins: Structure, carbohydrate specificity, and functions. *Biochemistry (Mosc), 73*, 393–405.

75. Demetriou, M., Granovsky, M., Quaggin, S., & Dennis, J. W., (2001). Negative regulation of T-cell activation and autoimmunity by Mgat5 N-glycosylation. *Nature, 409*, 733–739.

76. Hennet, T., Chui, D., Paulson, J. C., & Marth, J. D., (1998). Immune regulation by the ST6Gal sialyltransferase. *Proc. Natl. Acad. Sci. U.S.A, 95*, 4504–4509.

77. Zajonc, D. M., & Kronenberg, M., (2007). CD1 mediated T cell recognition of glycolipids. *Curr. Opin. Struct. Biol., 17*, 521–529.

78. Zhao, X. J., & Cheung, N. K., (1995). GD2 oligosaccharide: Target for cytotoxic T lymphocytes. *J. Exp. Med., 182*, 67–74.

79. Wells, L., Vosseller, K., & Hart, G. W., (2001). Glycosylation of nucleocytoplasmic proteins: Signal transduction and O-GlcNAc. *Science, 291*, 2376–2378.

80. Moody, A. M., et al., (2001). Developmentally regulated glycosylation of the CD8αβcoreceptor stalk modulates ligand binding. *Cell, 107*, 501–512.

81. Hanisch, F. G., & Ninkovic, T., (2006). Immunology of O-glycosylated proteins: Approaches to the design of aMUC1 glycopeptide-based tumor vaccine. *Curr. Protein Pept. Sci., 7,* 307–315.

82. Ninkovic, T., & Hanisch, F. G., (2007). O-glycosylated human MUC1 repeats are processed *in vitro* by immunoproteasomes. *J. Immunol., 179,* 2380–2388.

83. Purcell, A. W., Van, D. I. R., & Gleeson, P. A., (2008). Impact of glycans on T-cell tolerance to glycosylated self-antigens. *Immunol. Cell Biol., 86,* 574–579.

84. Bäcklund, J., et al., (2002). Predominant selection of T cells specific for the glycosylated collagen type IIepitope (263–270) in humanized transgenic mice and in rheumatoid arthritis. *Proc. Natl. Acad. Sci. U.S.A, 99,* 9960–9965.

85. Andersson, I. E., et al., (2011). Design of glycopeptides used to investigate class II MHC binding and T-cell responses associated with autoimmune arthritis. *PLoS One, 6,* e17881.

86. Tarp, M. A., & Clausen, H., (2008). Mucin-type O-glycosylation and its potential use in drug and vaccine development. *Biochim. Biophys. Acta., 1780,* 546–563.

87. Steentoft, C., et al., (2011). Mining the O-glycoproteome using zinc-finger nuclease glycoengineered simple cell lines. *Nat. Methods, 8,* 977–982.

88. Wright, A., & Morrison, S. L., (1997). Effect of glycosylation on antibody function: Implications for genetic engineering. *Trends Biotech., 15,* 26–32.

89. Parekh, R. B., et al., (1985). Association of rheumatoid arthritis and primary osteoarthritis with changes in the glycosylation pattern of total serum IgG. *Nature, 316,* 452–457.

90. Hoffmeister, K. M., et al., (2003). Glycosylation restores survival of chilled blood platelets. *Science, 301,* 1531–1534.

91. Gayle, E. R., Beryl, E. M., Robert, B. S., Paul, B. M., Raymond, A. D., & Pauline, M. R., (2002). Glycosylation and the complement system. *Chem. Rev., 102,* 305–319.

92. Dahl, M. R., Thiel, S., Matsushita, M., Fujita, T., Willis, A. C., Christensen, T., Vorup-Jensen, T., & Jensenius, J. C., (2001). *Immunity, 15,* 127.

93. Hirani, S., Lambris, J. D., & Muller-Eberhard, H. J., (1986). *Biochem. J., 233,* 613.

94. Hase, S., Kikuchi, N., Ikenaka, T., & Inoue, K., (1985). *J. Biochem. (Tokyo), 98,* 863.

95. Hartmann, S., & Hofsteenge, J., (2000). Properdin, the positive regulator of complement, is highly C-mannosylated. *J. Biol. Chem., 275,* 28569–28574.

96. Marth, J. D., & Grewal, P. K., (2008). Mammalian glycosylation in immunity. *Nat. Rev. Immunol., 8,* 874–887.

97. Wagner, K. W., Punnoose, E. A., Januario, T., Lawrence, D. A., Pitti, R. M., Lancaster, K., Lee, D., et al., (2007). Death-receptor O-glycosylation controls tumor-cell sensitivity to the proapoptotic ligand Apo2L/TRAIL. *Nat. Med., 13,* 1070–1077.

98. Keppler, O. T., Peter, M. E., Hinderlich, S., Moldenhauer, G., Stehling, P., Schmitz, I., Schwartz-Albiez, R., et al., (1999). Differential sialylation of cell surface glycoconjugates in a human B lymphoma cell line regulates susceptibility for CD95 (APO-1/Fas)-mediated apoptosis and for infection by a lymphotropic virus. *Glycobiology, 9,* 557–569.

99. Mullen, T. D., & Obeid, L. M., (2012). Ceramide and apoptosis: Exploring the enigmatic connections between sphingolipid metabolism and programmed cell death. *Anticancer Agents in Medicinal Chemistry, 12,* 340–363.

100. Galli, S. J., & Tsai, M., (2012). IgE and mast cells in allergic disease. *Nat. Med., 18,* 693–704.

101. Plomp, R., Hensbergen, P. J., Rombouts, Y., Zauner, G., Dragan, I., Koeleman, C. A. M., Deelder, A. M., & Wuhrer, M., (2014). Site-specific N-glycosylation analysis of human immunoglobulin E. *J. Proteome Res., 13*, 536–546.

102. Dombrowicz, D., Brini, A. T., Flamand, V., Hicks, E., Snouwaert, J. N., Kinet, J. P., & Koller, B. H., (1996). Anaphylaxis mediated through a humanized high affinity IgE receptor. *J. Immunol., 157*, 1645–1651.

103. Connelly, M. A., Gruppen, E. G., Otvos, J. D., & Dullaart, R. P., (2016). Inflammatory glycoproteins in cardiometabolic disorders, autoimmune diseases, and cancer. *Clin. Chim. Acta, 459*, 177–186.

104. Otvos, J. D., Shalaurova, I., Wolak-Dinsmore, J., Connelly, M. A., Mackey, R. H., Stein, J. H., & Tracy, R. P., (2015). GlycA: A composite nuclear magnetic resonance biomarker of systemic inflammation. *Clin. Chem., 61*, 714–723.

105. Ritchie, S. C., Wurtz, P., Nath, A. P., Abraham, G., Havulinna, A. S., Fearnley, L. G., Sarin, A. P., Kangas, A. J., Soininen, P., Aalto, K., et al., (2015). The biomarker GlycA is associated with chronic inflammation and predicts long-term risk of severe infection. *Cell Syst., 1*, 293–301.

106. Soltys, A. J., Hay, F. C., Bond, A., Axford, J. S., Jones, M. G., Randen, I., et al., (1994). The binding of synovial tissue derived human monoclonal immunoglobulin M rheumatoid factor to immunoglobulin G preparations of differing galactose content. *Scand. J. Immunol., 40*, 135–143.

107. Mizuochi, T., Hamako, J., Nose, M., & Titani, K., (1990). Structural changes in the oligosaccharide chains of IgG in autoimmune MRL/Mp-lpr/lpr mice. *J. Immunol., 145*, 1794–1798.

108. Ercan, A., Barnes, M. G., Hazen, M., Tory, H., Henderson, L., Dedeoglu, F., et al., (2012). Multiple juvenile idiopathic arthritis subtypes demonstrate proinflammatory IgG glycosylation. *Arthritis Rheum, 64*, 3025–3033.

109. Mizuochi, T., Pastore, Y., Shikata, K., Kuroki, A., Kikuchi, S., Fulpius, T., et al., (2001). Role of galactosylation in the renal pathogenicity of murine immunoglobulin G3 monoclonal cryoglobulins. *Blood, 97*, 3537–3543.

110. Sibéril, S., De, R. C., Bihoreau, N., Fernandez, N., Meterreau, J. L., Regenman, A., et al., (2006). Selection of a human anti-RhD monoclonal antibody for therapeutic use: Impact of IgG glycosylation on activating and inhibitory Fc gamma R functions. *Clin. Immunol., 118*, 170–179, 751.

111. Otani, M., Kuroki, A., Kikuchi, S., Kihara, M., Nakata, J., Ito, K., et al., (2012). Sialylation determines the nephritogenicity of IgG3 cryoglobulins. *J. Am. Soc. Nephrol., 23*, 1869–1878.

112. Bakchoul, T., Walek, K., Krautwurst, A., Rummel, M., Bein, G., Santoso, S., et al., (2013). Glycosylation of autoantibodies: Insights into the mechanisms of immune thrombocytopenia. *Thromb. Haemost., 110*, 1259–1266.

113. Holland, M., Takada, K., Okumoto, T., Takahashi, N., Kato, K., Adu, D., et al., (2002). Hypogalactosylation of serum IgG in patients with ANCA-associated systemic vasculitis. *Clin. Exp. Immunol., 129*, 183–190.

114. Pinho, S. S., & Reis, C. A., (2015). *Nat. Rev. Cancer, 15*, 540–555.

115. Seidenfaden, R., Krauter, A., Schertzinger, F., Gerardy-Schahn, R., & Hildebrandt, H., (2003). Polysialic acid directs tumor cell growth by controlling heterophilic neural cell adhesion molecule interactions. *Mol. Cell Biol., 23*, 5908–5918.

116. Munkley, J., & Elliott, D. J., (2016). Sugars and cell adhesion: The role of ST6GalNAc1 in prostate cancer progression. *Cancer Cell and Microenvironment., 3,* e1174.

117. Lin, S., Kemmner, W., Grigull, S., & Schlag, P. M., (2002). Cell surface alpha 2, 6sialylation affects adhesion of breast carcinoma cells. *Exp. Cell Res., 276,* 101–110.

118. Fardini, Y., Dehennaut, V., Lefebvre, T., & Issad, T., (2013). O-GlcNAcylation: A new cancer hallmark? *Frontiers in Endocrinology, 4,* 99.

119. Slawson, C., & Hart, G. W., (2011). *Nat. Rev. Cancer, 11,* 678–684.

120. Rabinovich, G. A., & Toscano, M. A., (2009). Turning 'sweet' on immunity: Galectin-glycan interactions in immune tolerance and inflammation. *Nature Reviews Immunology, 9,* 338–352.

121. Dimitroff, C. J., (2015). Galectin-binding O-glycosylations as regulators of malignancy. *Cancer Res., 75,* 3195–3202.

122. Ruhaak, L. R., & Lebrilla, C. B., (2015). Glycans in the immune system and the altered glycan theory of autoimmunity: A critical review. *Journal of Autoimmunity, 57,* 1–13.

123. Slovin, S. F., Ragupathi, G., Adluri, S., Ungers, G., Terry, K., Kim, S., Spassova, M., et al., (1999). Carbohydrate vaccines in cancer: Immunogenicity of a fully synthetic Globo H hexasaccharide conjugates in man. *Proc. Natl. Acad. Sci. U.S.A, 96,* 5710–5715.

124. Vogelstein, B., et al., (2013). Cancer genome landscapes. *Science, 339,* 1546–1558.

125. Couldrey, C., & Green, J. E., (2000). Metastases: The glycan connection. *Breast Cancer Res., 2,* 321–323.

126. Meany, D. L., & Chan, D. W., (2011). Aberrant glycosylation associated with enzymes as cancer biomarkers. *Clin. Proteomics, 8,* 7.

127. Kannagi, R., (2004). Molecular mechanism for cancer-associated induction of sialyl Lewis X and sialyl Lewis an expression-the Warburg effect revisited. *Glycoconj. J., 20,* 353–364.

128. Huhn, C., Selman, M. H., Ruhaak, L. R., Deelder, A. M., & Wuhrer, M., (2009). IgG glycosylation analysis. *Proteomics, 9,* 882–913.

129. Kodar, K., Stadlmann, J., Klaamas, K., Sergeyev, B., & Kurtenkov, O., (2012). Immunoglobulin G Fc N-glycan profiling in patients with gastric cancer by LC-ESI-MS: Relation to tumor progression and survival. *Glycoconj. J., 29,* 57–66.

130. Ozcan, S., Barkauskas, D. A., Ruhaak, L. R., Torres, J., Cooke, C. L., An, H. J., et al., (2014). Serum glycan signatures of gastric cancer. *Cancer Prev. Res. (Phila), 7,* 226–235.

131. Walker, M. R., Lund, J., Thompson, K. M., & Jefferis, R., (1989). Aglycosylation of human IgG1 and IgG3 monoclonal antibodies can eliminate recognition by human cells expressing Fc gamma RI and/or Fc gamma RII receptors. *Biochem. J., 259,* 347–353.

132. Balkwill, F., Charles, K. A., & Mantovani, A., (2005). Smoldering and polarized inflammation in the initiation and promotion of malignant disease. *Cancer Cell, 7,* 211–227.

133. Elliott, M. A., Elliott, H. G., Gallagher, K., McGuire, J., Field, M., & Smith, K. D., (1997). Investigation into the concanavalin A reactivity, fucosylation and oligosaccharide microheterogeneity of alpha 1-acid glycoprotein expressed in the sera of patients with rheumatoid arthritis. *J. Chromatogr. B. Biomed. Sci. Appl., 688,* 229–237.

134. Chui, D., Sellakumar, G., Green, R., et al., (2001). Genetic remodeling of protein glycosylation *in vivo* induces autoimmune disease. *Proc. Natl. Acad. Sci. U.S.A, 98,* 1142–1147.

135. Faid, M., & Zinedine, A., (2007). Isolation and characterization of strains of bifidobacteria with probiotic properties *in vitro*. *World Journal of Dairy and Food Sciences, 2,* 28–34.

136. Miyamoto, S., (2006). Clinical applications of glycomic approaches for the detection of cancer and other diseases. *Curr. Opin. Mol. Ther., 8,* 507–513.

137. Tomana, M., Schrohenloher, R. E., Reveille, J. D., Arnett, F. C., & Koopman, W. J., (1992). Abnormal galactosylation of serum IgG in patients with systemic lupus erythematosus and members of families with high frequency of autoimmune diseases. *Rheumatol. Int., 12,* 191–194.

138. Arnold, J. N., Wormald, M. R., Sim, R. B., Rudd, P. M., & Dwek, R. A., (2007). The impact of glycosylation on the biological function and structure of human immunoglobulins. *Annu. Rev. Immunol., 25,* 21–50.

139. Zak, I., Lewandowska, E., & Gnyp, W., (2000). Selectin glycoprotein ligands. *Acta Biochim. Pol., 47,* 393–412.

140. Axford, J. S., (1999). Glycosylation and rheumatic disease. *Biochim. Biophys. Acta, 1455,* 219–229.

141. Asano, N., (2003). Glycosidase inhibitors: Update and perspectives on practical use. *Glycobiology, 13,* 93R–104R.

142. Von, I. M., Wu, W. Y., Kok, G. B., Pegg, M. S., Dyason, J. C., Jin, B., Van, P. T., Smythe, M. L., White, H. F., Oliver, S. W., et al., (1993). Rational design of potent sialidase-based inhibitors of influenza virus replication. *Nature, 363,* 418–423.

143. Shriver, Z., Raman, R., Viswanathan, K., & Sasisekharan, R., (2009). Context-specific target definition in influenza a virus hemagglutinin-glycan receptor interactions. *Chem. Biol., 16,* 803–814.

144. Matrosovich, M. N., Matrosovich, T. Y., Gray, T., Roberts, N. A., & Klenk, H. D., (2004). Neuraminidase is important for the initiation of influenza virus infection in human airway epithelium. *J. Virol., 78,* 12665–12667.

145. Kim, J. H., Resende, R., Wennekes, T., Chen, H. M., Bance, N., Buchini, S., Watts, A. G., Pilling, P., Streltsov, V. A., Petric, M., et al., (2013). Mechanism-based covalent neuraminidase inhibitors with broad-spectrum influenza antiviral activity. *Science, 340,* 71–75.

146. Abian, O., Alfonso, P., Velazquez-Campoy, A., Giraldo, P., Pocovi, M., & Sancho, J., (2011). Therapeutic strategies for gaucher disease: Miglustat (NB-DNJ) as a pharmacological chaperone for glucocerebrosidase and the different thermostability of velaglucerase alfa and imiglucerase. *Mol. Pharm., 8,* 2390–2397.

147. Ernst, B., & Magnani, J. L., (2009). From carbohydrate leads to glycomimetic drugs. *Nat. Rev. Drug Discov., 8,* 661–677.

148. Imberty, A., Chabre, Y. M., & Roy, R., (2008). Glycomimetics and glycodendrimers as high affinity microbial antiadhesins. *Chemistry, 14,* 7490–7499.

149. Sasisekharan, R., Raman, R., & Prabhakar, V., (2006). Glycomics approach to structure-function relationships of glycosaminoglycans. *Annu. Rev. Biomed. Eng., 8,* 181–231.

150. Thunberg, L., Bäckström, G., & Lindahl, U., (1982). Further characterization of the antithrombin-binding sequence in heparin. *Carbohydr. Res., 100,* 393–410.

151. Linhardt, R. J., & Claude, S., (2003). Hudson Award address in carbohydrate chemistry. Heparin: Structure and activity. *J. Med. Chem., 46,* 2551–2564.

152. Petitou, M., & Van, B. C. A. A., (2004). A synthetic antithrombin III binding pentasaccharide is now a drug! What comes next? *Angew. Chem. Int. Ed. Engl., 43,* 3118–3133.

153. Maccarana, M., & Lindahl, U., (1993). Mode of interaction between platelet factor 4 and heparin. *Glycobiology, 3*, 271–277.

154. Xu, Y., Masuko, S., Takieddin, M., Xu, H., Liu, R., Jing, J., Mousa, S. A., et al., (2011). Chemoenzymatic synthesis of homogeneous ultralow molecular weight heparins. *Science, 334*, 498–501.

155. Liu, H., Zhang, Z., & Linhardt, R. J., (2009). Lessons learned from the contamination of heparin. *Nat. Prod. Rep., 26*, 313–321.

156. Copeland, R., Balasubramaniam, A., Tiwari, V., Zhang, F., Bridges, A., Linhardt, R. J., Shukla, D., & Liu, J., (2008). Using a 3-O-sulfated heparin octasaccharide to inhibit the entry of herpes simplex virus type 1. *Biochemistry, 47*, 5774–5783.

157. Bevilacqua, M., Butcher, E., Furie, B., Furie, B., Gallatin, M., Gimbrone, M., Harlan, J., Kishimoto, K., Lasky, L., McEver, R., et al., (1991). Selectins: A family of adhesion receptors. *Cell, 67*, 233.

158. Lasky, L. A., (1995). Selectin-carbohydrate interactions and the initiation of the inflammatory response. *Annu. Rev. Biochem., 64*, 113–139.

159. Imhof, B. A., & Aurrand-Lions, M., (2004). Adhesion mechanisms regulating the migration of monocytes. *Nat. Rev. Immunol., 4*, 432–444.

160. Pang, P. C., Chiu, P. C. N., Lee, C. L., Chang, L. Y., Panico, M., Morris, H. R., Haslam, S. M., et al., (2011). Human sperm binding is mediated by the sialyl-Lewis(x) oligosaccharide on the zona pellucida. *Science, 333*, 1761–1764.

161. Kerr, K. M., Auger, W. R., Marsh, J. J., Comito, R. M., Fedullo, R. L., Smits, G. J., Kapelanski, D. P., et al., (2000). The use of cylexin (CY-1503) in prevention of reperfusion lung injury in patients undergoing pulmonary thromboendarterectomy. *Am. J. Respir. Crit. Care Med., 162*, 14–20.

162. Kogan, T. P., Dupré, B., Bui, H., McAbee, K. L., Kassir, J. M., Scott, I. L., Hu, X., et al., (1998). Novel synthetic inhibitors of selectin-mediated cell adhesion: Synthesis of 1, 6-bis[3-(3-carboxymethylphenyl)-4-(2-alpha-D-mannopyranosyloxy) phenyl]hexane (TBC1269). *J. Med. Chem., 41*, 1099–1111.

163. Chang, J., Patton, J. T., Sarkar, A., Ernst, B., Magnani, J. L., & Frenette, P. S., (2010). GMI-1070, a novel pan-selectin antagonist, reverses acute vascular occlusions in sickle cell mice. *Blood, 116*, 1779–1786.

164. Egger, J., Weckerle, C., Cutting, B., Schwardt, O., Rabbani, S., Lemme, K., & Ernst, B., (2013). Nanomolar E-selectin antagonists with prolonged half-lives by a fragment-based approach. *J. Am. Chem. Soc., 135*, 9820–9828.

165. Bond, M. R., & Hanover, J. A., (2013). O-GlcNAc cycling: A link between metabolism and chronic disease. *Annu. Rev. Nutr., 33*, 205–229.

166. Yuzwa, S. A., Macauley, M. S., Heinonen, J. E., Shan, X., Dennis, R. J., He, Y., Whitworth, G. E., et al., (2008). A potent mechanism-inspired O-GlcNAcase inhibitor that blocks phosphorylation of tau *in vivo*. *Nat. Chem. Biol., 4*, 483–490.

167. Yuzwa, S. A., Shan, X., Macauley, M. S., Clark, T., Skorobogatko, Y., Vosseller, K., & Vocadlo, D. J., (2012). Increasing OGlcNAc slows neurodegeneration and stabilizes tau against aggregation. *Nat. Chem. Biol., 8*, 393–399.

168. Diot, J. D., Garcia, M. I., Twigg, G., Ortiz, M. C., Haupt, K., Butters, T. D., Kovensky, J., & Gouin, S. G., (2011). Amphiphilic 1-deoxynojirimycin derivatives through click strategies for chemical chaperoning in N370S Gaucher cells. *J. Org. Chem., 76*, 7757–7768.

169. Weïwer, M., Chen, C. C., Kemp, M. M., & Linhardt, R. J., (2009). Synthesis and biological evaluation of non-hydrolyzable 1, 2, 3-triazole-linked sialic acid derivatives as neuraminidase inhibitors. *Eur. J. Org. Chem.*, 2611–2620.

170. Ostash, B., Yan, X., Fedorenko, V., & Bechthold, A., (2010). Chemoenzymatic and bioenzymatic synthesis of carbohydrate containing natural products. *Top. Curr. Chem.*, *297*, 105–148.

171. Boltje, T. J., Buskas, T., & Boons, G. J., (2009). Opportunities and challenges in synthetic oligosaccharide and glycoconjugate research. *Nat. Chem., 1,* 611–622.

172. Seeberger, P. H., & Werz, D. B., (2007). Synthesis and medical applications of oligosaccharides. *Nature, 446,* 1046–1051.

173. Morelli, L., Poletti, L., & Lay, L., (2011). Carbohydrates and immunology: Synthetic oligosaccharide antigens for vaccine formulation. *Eur. J. Org. Chem.*, 5723–5777.

174. Grayson, E. J., Bernardes, G. J. L., Chalker, J. M., Boutureira, O., Koeppe, J. R., & Davis, B. G., (2011). A coordinated synthesis and conjugation strategy for the preparation of homogeneous glycoconjugate vaccine candidates. *Angew. Chem. Int. Ed. Engl., 50,* 4127–4132.

175. Verez-Bencomo, V., Fernández-Santana, V., Hardy, E., Toledo, M. E., Rodríguez, M. C., Heynngnezz, L., Rodriguez, A., Baly, A., Herrera, L., Izquierdo, M., et al., (2004). A synthetic conjugate polysaccharide vaccine against hemophilus influenzae type b. *Science, 305,* 522–525.

176. Hewitt, M. C., Snyder, D. A., & Seeberger, P. H., (2002). Rapid synthesis of a glycosylphosphatidylinositol-based malaria vaccine using automated solid-phase oligosaccharide synthesis. *J. Am. Chem. Soc., 124,* 13434–13436.

177. Schofield, L., Hewitt, M. C., Evans, K., Siomos, M. A., & Seeberger, P. H., (2002). Synthetic GPI as a candidate antitoxic vaccine in a model of malaria. *Nature, 418,* 785–789.

178. Tamborrini, M., Werz, D. B., Frey, J., Pluschke, G., & Seeberger, P. H., (2006). Anti-carbohydrate antibodies for the detection of anthrax spores. *Angew. Chem. Int. Ed. Engl., 45,* 6581–6582.

179. Belot, F., Wright, K., Costachel, C., Phalipon, A., & Mulard, L. A., (2004). Blockwise approach to fragments of the O-specific polysaccharide of shigella flexneri serotype 2a: Convergent synthesis of a decasaccharide representative of a dimer of the branched repeating unit. *J. Org. Chem. 69,* 1060–1074.

180. Phalipon, A., Tanguy, M., Grandjean, C., Guerreiro, C., Bélot, F., Cohen, D., Sansonetti, P. J., & Mulard, L. A., (2009). A synthetic carbohydrate-protein conjugate vaccine candidate against *Shigella flexneri* 2a infection. *J. Immunol., 182,* 2241–2247.

181. Lipinski, T., Kitov, P. I., Szpacenko, A., Paszkiewicz, E., & Bundle, D. R., (2011). Synthesis and immunogenicity of a glycopolymer conjugate. *Bioconjug. Chem., 22,* 274–281.

182. Bundle, D. R., Nycholat, C., Costello, C., Rennie, R., & Lipinski, T., (2012). Design of a Candida albicans disaccharide conjugate vaccine by reverse engineering a protective monoclonal antibody. *ACS Chem. Biol., 7,* 1754–1763.

183. Chen, Z. P., Jing, Y. M. S., Song, B. M. S., Han, Y. M. S., & Chu, Y. P., (2009). Chemically modified heparin inhibits *in vitro* L-selectin-mediated human ovarian carcinoma cell adhesion. *Int. J. Gynecol. Cancer, 19,* 540–546.

184. Brodt, P., Fallavollita, L., Bresalier, R. S., Meterissian, S., Norton, C. R., & Wolitzky, B. A., (1997). Liver endothelial E-selectin mediates carcinoma cell adhesion and promotes liver metastasis. *Int. J. Cancer, 71,* 612–619.

185. Mirandola, L., Yu, Y., Cannon, M. J., Jenkins, M. R., Rahman, R. L., Nguyen, D. D., et al., (2014). Galectin-3 inhibition suppresses drug resistance, motility, invasion, and angiogenic potential in ovarian cancer. *Gynecol. Oncol., 135,* 573–579.

# CHAPTER 4

# Role of Glycans in Neurodegeneration

ABHAI KUMAR[1] and SMITA SINGH[2]

[1]*Interdisciplinary School of Life Sciences, Institute of Science, Banaras Hindu University, Varanasi, Uttar Pradesh, India, E-mail: singhabhai2000@gmail.com*

[2]*Center of Advanced Studies in Botany, Institute of Science, Banaras Hindu University, Varanasi, Uttar Pradesh, India*

## ABSTRACT

Among the major classes of macromolecules, glycans are the most diversified macromolecules, playing broad-spectrum roles in the development, growth, functioning, or survival of the organism. The role of glycans has been significantly recognized in neural cell development and the adult nervous system in the last decade. The diversity in the structure of these glycans and their involvement in diverse functions such as cell migration, neurite outgrowth and fasciculation, synapses formation, stabilization, and modulation of synaptic efficacy increase their importance in the nervous system. The importance of glycan in neurological functioning is well evident; inborn error in glycosylation results in psychomotor difficulties, mental retardation, and other neurodegenerative disorders. The current chapter documents present an understanding of the functional roles of these glycans in neurodegeneration to explore their roles in pathological manifestation of disease and intervention in disease treatment.

## 4.1 INTRODUCTION

Neurodegenerative diseases are most predominantly found in the elderly population all over the world. The underlying causes of neurodegeneration and effective therapeutic intervention are lacking [2]. Neurodegenerative diseases are identified as protein misfolding diseases, proteinopathies, and protein conformational disorders. The loss of neurons and synaptic alterations are typical symptoms of neurodegeneration. The defects in posttranslational modification of proteins result in misfolding, which in turn cause protein aggregation and accumulation, leading to neuronal damage [2]. In the brain, some native proteins (prion, tau, β-amyloid, α-synuclein, and Huntington) undergo conformational changes via genetic and environmental factors wherein secondary structures of protein convert from α-helix/random coil to β-sheet. Consequently, neurotoxic misfolded protein aggregates are deposited in central nervous systems (CNSs) and the brain leading to neurodegenerative diseases [2]. Among others, the four most important neurodegenerative diseases are prion diseases, Alzheimer's disease (AD), Parkinson's disease (PD), and Huntington's disease (HD). One of the most vital components which are altered in neurodegeneration diseases includes the chain of monosaccharides, which differ in length from a few to several hundred, named as Glycans. The role of glycans in neural cell interaction is well identified in the developing and developed a nervous system in the last decade [2]. The glycans having complex structures are involved in diverse functions such as cell migration, neurite outgrowth and fasciculation, synapse formation and stabilization, and modulation of synaptic efficacy [2]. Understanding of disease pathology in past decades has shifted from omics generated complicated data to cell-cell interaction based on complex polysaccharide present on the extracellular matrix (ECM) of cell [2]. Among these, the most extensively studied complex polysaccharides are the glycosaminoglycans (GAGs). An emerging paradigm in the extracellular modulation of biological function is the specific interactions between GAGs and numerous proteins at the cell-ECM interface. Despite the numerous known biological roles of GAGs, it has been challenging to decode their structure-function relationships. The biosynthesis of GAGs is non-template driven and involves several enzymes and their tissue-specific isoforms [2]. The reports have shown that O-GlcNAcylation is linked with neurogenesis and neuronal morphology. O-GlcNAc transferase expression, enriched at neuronal synapses [2, 4], has been found to dynamically modify various neuronal proteins related to synaptic function, learning, memory, and neurodegeneration [2, 4]. A recent neuroproteomics study has identified about 249 O-GlcNacylated proteins in the mouse cerebral cortex [2]. Several pre- and postsynaptic proteins, including synapsin, piccolo, bassoon, and SHANK2 proteins, have been found to be extensively O-GlcNAcylated [2]. The current chapter highlights on the classification of glycans and their role in the modulation of structural and regulatory proteins governing different biological functions in neuronal cells.

## 4.2   CLASSIFICATION OF GLYCANS

There are different classes of glycans, mainly N-linked glycans (NLGs), O-linked glycans, glycolipids, O-GlcNAc, and GAGs. The attachment of glycan on asparagine residue of protein is known as N-linked glycosylation, whereas linkage of glycans on serine or threonine residue through N-acetyl-galactosamine on protein is known as O-linked glycosylation. The process of removal of glycans from protein is known as deglycosylation and is used for the study of peptide and/or glycan portion of a glycoprotein. The functional role of glycans strongly correlates with their structures. The structure of glycan in specific glycoproteins requires detailed information for interpretation of glycan chains. The chemical and enzymatic methods are well established for oligosaccharide removal from proteins. The $\beta$-elimination with mild alkali or mild hydrazinolysis results in degradation of protein, while enzymatic methods are more sensitive and result in complete sugar removal with no protein degradation [1].

### 4.2.1   STRUCTURAL AND MODULATORY ROLE OF GLYCANS IN NEURODEGENERATIVE FUNCTION

The interaction of the protein with each other is strongly modulated by the process of glycosylation. The process of glycosylation regulates the binding abilities of growth factor receptors (EGFs) in the Golgi apparatus (GA), preventing the interaction of EGF with growth factor synthesized in the same cell. The cellular functions of proteins in cells are regulated at a structural and functional level by glycans and glycan-binding proteins (GBPs). There are two major groups of GBPs: (i) proteins which recognize glycans from the same organism are called intrinsic GBPs, and (ii) proteins which recognize glycans from different organisms are called extrinsic GBPs. The cell-cell interactions, recognition of extracellular molecules, and glycans on the same cell are mainly mediated by intrinsic GBPs. The interactions with pathogenic microbial adhesins, agglutinins, or toxins are mediated by extrinsic GBPs. The role of extrinsic GBPs in symbiotic relationships is also well-established [1]. The extent and type of glycosylation on growth factors and hormones play an important modulatory role in their activity. The homophilic binding between neuronal cells is facilitated by polysialic acid present on neural cell adhesion molecule (NCAM). The length of the polysialic acid chain increases in the case of embryonic state or in other neuronal plasticity preventing homophilic binding. The glycocalyx on the surface of all eukaryotic cells and the polysaccharide coats of various prokaryotes provides physical resistance to external stress [2]. The proteoglycans play an important role in the maintenance of tissue structure, porosity, and integrity. Proteoglycan has a specific site for attachment with glycans, which play an important role in overall organization of the matrix. The glycans form a dense layer of the glycocalyx on underlying glycoproteins providing a protecting shield to underlying polypeptide from recognition by protease or antibodies [3].

The tuning effects of glycans on receptor-ligand functions are partial, but their overall effect on the biological outcome can be very drastic. There are instances where polysialic acid of embryonic NCAM interferes with the interactions of other unrelated receptor-ligand pair. The tyrosine phosphorylation activity of endogenous EGF and insulin receptor is modulated by endogenous cell-surface ganglioside [4]. Glycosylation is an important mechanism which brings functional diversity to basic receptor-ligand interactions, as compared to basic gene product derived from the genome. There are certain exceptions to many receptors whose ligand binding did not necessitate glycosylation, and many peptide ligands do not need glycosylation for binding with their receptors [4].

Glycans also act as a protective storage depot for biologically important molecules such as heparin-binding growth factors, which are attached to glycosaminoglycan (GAG) chains of the extracellular matrix, and stimulate adjacent cells. These functions of glycans prevent the diffusion of factors away from the site, protect proteolytic degradations, increase their sustainability inside the cell, and regulate their release under specific conditions. Similarly, the functions of secretory protein present in secretory granules are modulated by glycosaminoglycan chains [4]. The role of GAGs in the formation of amyloid fibril has been explained by various hypotheses. However, the exact mechanism of GAGs involvement in amyloidogenesis is still largely unexplored.

## 4.3   ROLE OF GLYCOSAMINOGLYCANS (GAGS) IN AMYLOID DEPOSITION

The presence of a significant amount of polysaccharides in eukaryotic tissues compromises glycosaminoglycan, mainly GAGs in the human body.

The molecular weight of this heteropolysaccharide around 10–100 kDa and is composed of long-chain disaccharides repeating units [5]. Hyaluronic acid is a non-sulfated form of GAGs, and sulfated form includes heparin sulfate (HS), dermatan sulfate (DS), chondroitin sulfate (CS), keratin sulfate and keratin. The viscosity of the solution is increased due to the presence of negative charge and extended conformation of GAGs. The GAGs are present on the extracellular matrix of multicellular organisms as free macromolecules or attached to proteoglycan through covalent linkages. The involvement of heparin, heparan sulfate (HS), and chondroitin-4-sulfate in amyloidosis are well-reported [6]. Several reports provide strong evidence on the association of GAGs with amyloid fibril formation and stabilization in human tissues [1, 6, 7]. The recent studies on neurodegenerative diseases like Alzheimer's, Parkinson's, and Dementia with Lewy bodies (DLB) have a significant amount of proteinaceous deposits containing glycosaminoglycan family (GAGs) in tissues [8, 9]. The most abundant GAGs found among all glycosaminoglycan family were heparin sulfate in all neurodegenerative disorders [8, 9]. The level of deposition and distribution of amyloid fibril in humans is mainly regulated by extracellular matrix components, mainly glycoproteins and GAGs, which are variably distributed and may regulate the amount and tissue distribution of amyloid deposition in the cell [9, 10]. Several reports emphasized that GAGs catalyzed reactions like amyloid fibril aggregation, nucleation, amyloid formation, and stabilization [11, 12]. *In vitro* studies suggested active interaction between GAGs highly sulfated form heparin sulfate with Aβ peptide in the formation of amyloid fibril formation [12]. The influence on fibril formation and peptide conformation was investigated by understanding the interaction between GAGs with amyloid-beta (Aβ) isoforms [12–14]. Studies indicated that the initial folded state of amyloid fibril can be destabilized by GAGs and induces the transition to aggregation form. Further, it can also induce the assembly of unfolded monomer into amyloid oligomers.

## 4.4 INHIBITION OF HEPARIN SULFATE BIOSYNTHESIS

The role of heparin sulfate in the regulation of amyloid deposition was observed for the first time when functional inhibition of its biosynthesis resulted in attenuation of amyloid deposition in an animal model [15–17]. The coiling of Aβ peptide is accelerated into amyloidogenic β-sheet form in the presence of heparin. Furthermore, this acceleration of amyloidogenic β-sheet transition leads to the formation of amyloid fibrils resulting from enhanced nucleation of Aβ1–42. The role of GAGs in amyloid formation was further confirmed when incubation of GAG with Aβ1–42 fibrils increases lateral aggregation and increases the accumulation of fibril [18–20]. The role of sulfate moiety on GAGs has a significant role in Aβ fibril formation enhancement; the removal of sulfate moiety from GAGs prevents the promoting effect of Aβ fibril formation [21, 22]. Studies done on low-molecular-weight heparins (LMWHs) have antagonistic effects as compared to their higher molecular weight forms; the β-plated structures of the protein are blocked by LMWHs and therefore will provide a therapeutic approach in preventing the interaction between Aβ peptides and proteoglycans [23]. Inhibitory activity of heparin on APP-cleaving enzyme 1 (BACE-1) suggested that BACE-1 is very crucial for amyloidogenic processing of APP and formation of Aβ peptide [23]. The toxic level of insoluble and soluble assemblies of Aβ, mainly Aβ42 dimers and trimers, result in cellular toxicity, and affect the cognitive function in neurons [24]. The presence of HS on the cell surface and their role in Aβ toxicity and cell internalization was explained by Sandwall and coworkers [25]. Cells mutant for HS expression were resistant to toxicity caused by Aβ and did not internalize Aβ peptide. Besides this, cells over-expressing heparinase were resistant to Aβ40 toxicity, and the addition of heparin also provided prevention from cellular toxicity caused by Aβ [25]. The role of heparin sulfate in the induction of fibril formation in tau protein, α-synuclein, gelsolin, β2-microglobulin acyl-phosphatase, and islet amyloid polypeptide (IAPP) variant has been reported [26, 27]. The random coil state of IAPP is induced to a helical state in the presence of heparin. In fact, the expression of GAGs on the cell membrane and the presence of heparin in cell media might prevent from IAPP induced cell toxicity [28]. There is a sufficient body of scientific evidence to prove that GAGs accelerate the amyloid fibril formation process and prevent the cell from cytotoxic prefibrillar oligomeric state leading to cell toxicity. GAGs also convert prion protein from PrPC to PrPSC [29, 30]. Heparin is very effective among GAGs to play a significant role in fibril formation probably because of high sulfate content, electrostatic interactions which are crucial for heparin to bind with amyloid fibril, removal of sulfate with magnesium, calcium ions significantly suppressed the interaction between amyloid and heparin and thereby indicating their electrostatic nature [32]. Heparin induces amyloid formation in the cell by the interaction of amyloid binding sites with heparin, but the mechanism of amyloid aggregation formation through heparin in the cell is still unknown [34]. The role of heparin sulfate in the destabilization of folded forms of amyloid and induction in the formation of the aggregate form of amyloids, along with unfolded form into amyloid oligomers, was reported by Motamedi-Shad and coworkers [34]. The heparin sulfate follows two independent pathways in the conversion of acylphosphatase into amyloidogenic form; fast phase pathway in which interaction of heparin sulfate with few protein molecules convert them into β-sheet oligomers, and slow phase in which normal aggregation of protein occurs which cannot interact with heparin sulfate [35]. The high density of sulfate on the surface of GAGs and their polymeric nature play a very significant role in amyloidogenesis. Sulfated GAGs accelerated transthyretin (TTR) amyloidogenesis and did not influence the initial steps of cascade involved in TTR amyloidogenesis. The increase in aggregation of TTR oligomers is through quaternary structure formation induced by electrostatic interaction of the polymeric surface of GAGs with TTR oligomers. Thus binding of heparin to amyloidogenic protein increases the aggregate formation and fibrillation process [36].

### 4.4.1  STRUCTURAL MOTIF RECOGNITION OF HEPARIN BY PROTEINS

Heparin and heparin sulfate bear a common binding site in protein, as evident from structural and NMR spectroscopy [37]. The non-sulfated carbohydrates are more likely to align with the side chain of asparagine, aspartate, glutamate, arginine, histidine, and tryptophan residues [38]. The aliphatic moiety of side chains, mainly alanine, glycine, isoleucine, and glycans, mediate cellular functions wherein leucine is inserted inside the protein and is not accessible for sugar-binding. The increase in solvent accessibility of sugar residue is through binding of the hydrophobic face of sugar with the indole ring of tryptophan residue [38]. The presence of negatively charged carboxylate and sulfate group on the surface of the heparin/heparin sulfate chain provides strong ionic pairing with a positively charged amino acid of proteins. The basic amino acids arginine, lysine, and histidine bind with GAGs through ionic interaction [39]. The presence of the guanidino group in arginine increases its binding with GAGs as compared with lysine because the guanidino group forms hydrogen bond along with electrostatic interaction with the sulfate group. The interaction of GAG with protein is determined by the ratio of arginine to lysine [40].

The interaction between GAGs and protein is also influenced by hydrogen bonds, hydrophobic interactions, and van der Waals forces due to carbohydrate backbone. The amino acid such as asparagine and glutamine present on the heparin-binding domain form hydrogen bonds with protein. The binding of heparin-binding protein with heparin sulfate/heparin is enhanced by the presence of serine and glycine [41]. Several workers have identified that proteins possess a high positive charge domain for binding with heparin/heparin sulfate, apolipoprotein B-100 and low-density lipo-protein (LDL) have five to seven positive charge regions that bind with heparin-binding domain [41–43]. Cardin and Weintraub found that α-helical or β-sheet domains of heparin-binding proteins contain repeated repeats of arginine and lysine with a very rare occurrence of histidine. The non-basic amino acid such as asparagine, serine, alanine, glycine, isoleucine, leucine, and tyrosine are also low in occurrence as compared with basic amino acids [44]. However, many studies reported that the binding of heparin with proteins does not require consensus sequences of basic amino acids [59, 60]. The heparin-binding ability is also influenced by the distance between two amino acids of 20 Å [45]. The CPC clip motif defines the binding site of heparin with chemokines and human amyloid β protein [46]. The space between two amino acids will be 20 Å when 13 residues of basic amino acids are present in α-helical conformation and 7 residues in β-strand conforma-tion. The clips like the structure of the CPC motif have arginine and lysine as cationic residues and asparagine, glutamine, threonine, tyrosine, and serine as polar residues; the distance between α-carbons and side-chain remain fixed, which form a clip-like structure where heparin binds. The proteins in Protein Data Bank show structural similarity with this structure motif [46].

## 4.5   PROTEIN FOLDING ANOMALIES IN NEURODEGENERATIVE DISORDERS

The addition of glycan to the protein chain regulates it's proper folding and stability in the endoplasmic reticulum (ER) and its localization to a specific site of action, whereas proteins that are not folded properly due to wrong glycosylation are directed to the proteasome degradation pathway. The biological functions of certain glycoproteins are not altered by glycosylation [47]. The glycosylation of amyloid protein leads to their active aggregation and deposition in brain tissues that leads to neurodegeneration [47]. The assembly of two to three 3 nm protofilament twisted around each other gives rise to unbranched amyloid fibril, which is 7 to 10 nm in diameter [47]. The β-sheet structures are predominantly found in amyloid fibril, which plays an important role in its oligomerization [48]. The formation of amyloidogenic peptides occurs through the mechanism of nucleation and polymerization. The formation of Aβ peptide and IAPP occurs through polymerization, incorporating nucleation and extension of the amyloid fibril. The oligomeric complexes are formed and converted into fibril, and these oligomeric complexes induce the conversion of random coil conformation of the amyloidogenic polypeptide into β-sheet and further lead to amyloid formation. The formation of amyloid fibril follows other mechanisms in which monomer directs the conversion of native state to prefibrillar state, which induces other monomers to follow the same process of fibril formation [49–52]. Several studies have been conducted to understand the cytotoxicity of these prefibrillar assemblies at different transition stages into mature fibrils, but still, the mechanism by which these fibrils exert their immediate toxicity is unknown [53]. The disease-specific amyloids have unique polypeptides, and their aggregation into prefibrillar aggregates, which are highly ordered tissue deposits, are still unknown [53]. Many forms of functional amyloid are also produced in nature, and often these systems require careful control of their assembly to avoid the potentially toxic effects. The best-characterized functional amyloid system is the *Escherichia coli* bacterial curli system [54]. Results obtained from various studies have strongly emphasized that abnormal folding of polypeptides are promoted and influenced by glycans, which convert them into amyloidogenic intermediates which are rich in β-sheets and provide a structural template for their self-assembly [54]. The role of GAGs and heparin-sulfate in the amyloidogenesis process is well established as they increase aggregation, insolubility of fibrils, and prevent them from proteolysis [55].

## 4.5.1 THERAPEUTIC ROLE OF GLYCOCONJUGATES IN NEURODEGENERATIVE DISORDERS

The role of glycoconjugates and polysaccharides in different biological activities is well-established [56–58]. The cytokines and neurotrophic factors influence neurodegenerative disease, while the immunomodulatory and neuroprotective effect of these factors by polysaccharides and their synthetic derivatives have been reported [59]. The studies from animal models and cell culture studies suggest that aggregation of proteins either in the form of senile plaques; neurofibrillary tangles, or Lewy bodies are associated with activation of the pro-apoptotic pathway, oxidative stress, and neuroinflammation

[60, 61]. The roles of antioxidants and anti-inflammatory drugs have posed significant progress in treatment and prevention in the progression of these neurodegenerative disorders. Previous research reports suggested the role of Vitamin E in improving cognitive performance and prevention from neuro-degeneration [62]. The prevention from neurodegeneration by antioxidants has been shown in different experimental models [63–65]. The late-stage prevention from neurodegeneration can be attained by intake of a diet rich in carotenoids and antioxidants such as Vitamin C and E [66–68]. The cognitive performance was improved in age-related cognitive disorders, and the risk of AD reduces among individuals found using non-steroid anti-inflammatory drugs (NSAID) [69]. However, contradictory results were also reported in the usage of NSAID in prevention from progression to AD [70, 71]. The inhibition of neuroinflammation is still the main target, to slow down the progression of neurodegeneration, because NSAID targets complement-mediated inflammation and hence can also target various infection based neuro-inflammation which involves complement-mediated activation [72, 73]. The neurodegenerative diseases are age-associated and involve a number of risk factors, such as smoking, diabetes mellitus, and hypertension. Similarly, PD is associated with pesticide exposure and hyper-homocysteinemia [74]. The neurodegeneration leading to cognitive impairment is influenced by various overlapping factors that determine disease pathology and progression. The antioxidants and NSAID mainly target these glycoproteins at the biochemical and molecular level. Further, the role of proteoglycans in the identification of the neurodegeneration process as a biomarker is also complementing the field of biomarker-based early diagnosis and treatment of these neurodegenerative disorders [74].

## 4.6   CONCLUSION AND FUTURE CHALLENGES

The challenges in understanding the interaction of GAGs with proteins from structural and functional aspects are high. The specific interaction of GAGs, mainly heparin with other polypeptides, is still unclear, and previous studies found that GAG induced non-toxic fibril protein formation while in matured cells, GAG modified fibril protein causes cell damage and inflammation [53]. The role of exogenous heparin in attenuating the oligomer formation in cells prevents their interaction with the cell membrane [52]. Therefore, the use of heparin in therapeutic targets of amyloidogenesis and amyloid fibril formation can provide a future therapeutic target for neurodegenerative diseases.

The role of heparin sulfate in internalization and propagation of α-synuclein, tau, and their aggregation into extracellular space and entry into neighboring cell to fibril formation can be exploited for developing mechanism-based inhibition of neuro-fibrillary tangles involved in the process of neurodegeneration. The paradoxical role of GAGs in amyloidosis is through the interaction of soluble proteins into amyloid aggregates resulting in disease pathogenesis, and on the other side, it induces proteins to non-toxic fibril form, which can be utilized as a therapeutic target in patients suffering from amyloid involving neurodegenerative disorder.

## KEYWORDS

- **Alzheimer's disease**
- **endoplasmic reticulum**
- **extracellular matrix**
- **glycan-binding proteins**
- **glycosaminoglycans**
- **growth factor receptor**

## REFERENCES

1. Abe, A., & Shayman, J. A., (1998). Purification and characterization of 1-o-aceylceramide synthase a novel phospholipase A2 with transacylase activity. *J. Bio. Chem., 273*, 8467–8474.
2. Varki, A., & Lowe, J. B., (2009). Biological roles of glycans. *Essentials of Glycobiology* (2nd edn.). Cold Spring Harbor (NY): Cold Spring Harbor Laboratory Press, Chapter 6.
3. Berger, E. G., Buddecke, E., Kamerling, J. P., Kobata, A., Paulson, J. C., & Vliegenthart, J. F. G., (1982). Structure, biosynthesis, and functions of glycoprotein glycans. *Experientia, 38*, 1129–1162.
4. Rademacher, T. W., Parekh, R. B., & Dwek, R. A., (1988). Glycobiology. *Annu. Rev. Biochem., 57*, 785–838.
5. Ancsin, J. B., (2003). Amyloidogenesis: Historical and modern observations point to heparan sulfate proteoglycans as a major culprit. *Amyloid, 10*, 67–79.
6. Diaz-Nido, J., Wandossel, F., & Avila, J., (2002). Glycosaminoglycans and beta-amyloid, prion, and tau peptides in neurodegenerative diseases. *Peptides, 23*, 1323–1332.
7. Gruys, E., Ultee, A., & Upragarin, N., (2006). Glycosaminoglycans are part of amyloid fibrils: Ultrastructural evidence in avian AA amyloid stained with cuprolinic blue and labeled with immunogold. *Amyloid, 13*, 13–19.

8. Young, I. D., Ailles, L., Narindrasorasak, S., Tan, R., & Kisilevsky, R., (1992). Localization of the basement membrane heparan sulfate proteoglycan in islet amyloid deposits in type II diabetes mellitus. *Arch. Pathol. Lab. Med., 116*, 951–954.

9. Snow, A. D., Wight, T. N., Nochlin, D., Koike, Y., Kimata, K., De Armond, S. J., & Prusiner, S. B., (1990). Immunolocalization of heparan sulfate proteoglycans to the prion protein amyloid plaques of gerstmann-straussler syndrome, Creutzfeldt-Jakob disease and scrapie. *Lab. Investig., 63*, 601–611.

10. Papy-Garcia, D., Christophe, M., Huynh, M. B., Fernando, S., Ludmilla, S., Sepulveda-Diaz, J. E., & Raisman-Vozari, R., (2011). Glycosaminoglycans, protein aggregation, and neurodegeneration. *Curr. Protein Pept. Sci., 12*, 258–268.

11. Zhu, M., Souillac, P. O., Ionesco-Zanetti, C., Carter, S. A., & Fink, A. L., (2002). Surface-catalyzed amyloid fibril formation. *J. Biol. Chem., 277*, 50914–50922.

12. Salmivirta, M., Lidholt, K., & Lindahl, U., (1996). Heparan sulfate: A piece of information. *FASEB J., 10*, 1270–1279.

13. Castillo, G. M., Lukito, W., Wight, T. N., & Snow, A. D., (1999). The sulfate moieties of glycosaminoglycans are critical for the enhancement of beta-amyloid protein fibril formation. *J. Neurochem., 72*, 1681–1687.

14. Castillo, G. M., Ngo, C., Cummings, J., Wight, T. N., & Snow, A. D., (1997). Perlecan binds to the beta-amyloid proteins (A beta) of Alzheimer's disease, accelerates A beta fibril formation, and maintains A beta fibril stability. *J. Neurochem., 69*, 2452–2465.

15. Kisilevsky, R., Szarek, W. A., Ancsin, J. B., Elimova, E., Marone, S., Bhat, S., & Berkin, A., (2004). Inhibition of amyloid A amyloidogenesis *in vivo* and in tissue culture by 4-deoxy analogues of peracetylated 2-acetamido-2-deoxy-alpha- and beta-d-glucose: Implications for the treatment of various amyloidosis. *Am. J. Pathol., 164*, 2127–2137.

16. Elimova, E., Kisilevsky, R., Szarek, W. A., & Ancsin, J. B., (2004). Amyloidogenesis recapitulated in cell culture: A peptide inhibitor provides direct evidence for the role of heparan sulfate and suggests a new treatment strategy. *FASEB J., 18*, 1749–1751.

17. Li, J. P., Galvis, M. L., Gong, F., Zhang, X., Zcharia, E., Metzger, S., Vlodavsky, I., Kisilevsky, R., & Lindahl, U., (2005). *In vivo* fragmentation of heparan sulfate by heparanase overexpression renders mice resistant to amyloid protein A amyloidosis. *Proc. Natl. Acad. Sci. U.S.A, 102*, 6473–6477.

18. Fraser, P. E., Nguyen, J. T., Chin, D. T., & Kirschner, D. A., (1992). Effects of sulfate ions on Alzheimer beta/A4 peptide assemblies: Implications for amyloid fibril-proteoglycan interactions. *J. Neurochem., 59*, 1531–1540.

19. Fraser, P. E., Darabie, A. A., & McLaurin, J. A., (2001). Amyloid-beta interactions with chondroitin sulfate-derived monosaccharides and disaccharides. Implications for drug development. *J. Biol. Chem., 276*, 6412–6419.

20. McLaurin, J., & Fraser, P. E., (2000). Effect of amino-acid substitutions on Alzheimer's amyloid-beta peptide-glycosaminoglycan interactions. *Eur. J. Biochem., 267*, 6353–6361.

21. Valle-Delgado, J. J., Alfonso-Prieto, M., De Groot, N. S., Ventura, S., Samitier, J., Rovira, C., & Fernàndez-Busquets, X., (2010). Modulation of Abeta42 fibrillogenesis by glycosaminoglycan structure. *FASEB J., 24*, 4250–4261.

22. Ariga, T., Miyatake, T., & Yu, R. K., (2010). Role of proteoglycans and glycosaminoglycans in the pathogenesis of Alzheimer's disease and related disorders: Amyloidogenesis and therapeutic strategies: A review. *J. Neurosci. Res., 88*, 2303–2315.

23. Scholefield, Z., Yates, E. A., Wayne, G., Amour, A., McDowell, W., & Turnbull, J. E., (2003). Heparan sulfate regulates amyloid precursor protein processing by BACE1, the Alzheimer's beta-secretase. *J. Cell Biol., 163*, 97–107.

24. Bergamaschini, L., Rossi, E., Vergani, C., De Simoni, M. G., (2009). Alzheimer's disease: Another target for heparin therapy. *Sci. World J., 9*, 891–908.

25. Sandwall, E., O'Callaghan, P., Zhang, X., Lindahl, U., Lannfelt, L., & Li, J. P., (2010). Heparan sulfate mediates amyloid-beta internalization and cytotoxicity. *Glycobiology, 20*, 533–541.

26. Walsh, D. M., Klyubin, I., Shankar, G. M., Townsend, M., Fadeeva, J. V., Betts, V., Podlisny, M. B., Cleary, J. P., Ashe, K. H., Rowan, M. J., et al., (2005). The role of cell-derived oligomers of abeta in Alzheimer's disease and avenues for therapeutic intervention. *Biochem. Soc. Trans., 33*, 1087–1090.

27. Gouras, G. K., Tsai, J., Naslund, J., Vincent, B., Edgar, M., Checler, F., Greenfield, J. P., Haroutunian, V., Buxbaum, J. D., Xu, H., et al., (2000). Intraneuronal A beta 42 accumulation in human brain. *Am. J. Pathol., 156*, 15–20.

28. De Carufel, C. A., Nguyen, P. T., Sahnouni, S., & Bourgault, S., (2013). New insights into the roles of sulfated glycosaminoglycans in islet amyloid polypeptide amyloidogenesis and cytotoxicity. *Biopolymers, 100*, 645–655.

29. Wirths, O., Multhaup, G., & Bayer, T. A., (2004). A modified beta-amyloid hypothesis: Intraneuronal accumulation of the beta-amyloid peptide-the first step of a fatal cascade. *J. Neurochem., 91*, 513–520.

30. Goedert, M., Jakes, R., Spillantini, M. G., Hasegawa, M., Smith, M. J., & Crowther, R. A., (1996 and 2015). Assembly of microtubule-associated protein tau into Alzheimer-like filaments induced by sulfated glycosaminoglycans. *Nature, 383*, 550–553; *Molecules, 20*, 2525.

31. Paudel, H. K., & Li, W., (1999). Heparin-induced conformational change in microtubule-associated protein Tau as detected by chemical cross-linking and phosphopeptide mapping. *J. Biol. Chem., 274*, 8029–8038.

32. Cohlberg, J. A., Li, J., Uverskky, V. N., & Fink, A. L., (2002). Heparin and other glycosaminoglycans stimulate the formation of amyloid fibrils from alpha synuclein *in vitro. Biochemistry, 41*, 1502–1511.

33. Suk, J. Y., Zhang, F., Balch, W. E., Linhardt, R. J., & Kelly, J. F., (2006). Heparin accelerates gelsolin amyloidogenesis. *Biochemistry, 45*, 2234–2242.

34. Motamedi-Shad, N., Monsellier, E., & Chiti, F., (2009). Amyloid formation by the model protein muscle acylphosphatase is accelerated by heparin and heparan sulfate through a scaffolding-based mechanism. *J. Biochem., 146*, 805–814.

35. Bourgault, S., Solomon, J. P., Reixach, N., & Kelly, J. W., (2011). Sulfated glycosaminoglycans accelerate transthyretin amyloidogenesis by quaternary structural conversion. *Biochemistry, 50*, 1001–1015.

36. Relini, A., De Stefano, S., Torrassa, S., Cavalleri, O., Rolandi, R., Gliozzi, A., Giorgetti, S., Raimondi, S., Marchese, L., Verga, L., et al., (2008). Heparin strongly enhances the formation of beta2-microglobulin amyloid fibrils in the presence of type I collagen. *J. Biol. Chem., 283*, 4912–4920.

37. Calamai, M., Kumita, J. R., Mifsud, J., Parrini, C., Ramazzotti, M., Ramponi, G., Taddei, N., et al., (2006). Nature and significance of the interactions between amyloid fibrils and biological polyelectrolytes. *Biochemistry, 45*, 12806–12815.

38. Meng, F., Abedini, A., Song, B., & Raleigh, D. P., (2007). Amyloid formation by pro-islet amyloid polypeptide processing intermediates: Examination of the role of protein heparan sulfate interactions and implications for islet amyloid formation in type 2 diabetes. *Biochemistry, 46,* 12091–12099.

39. McLaughlin, R. W., De Stigter, J. K., Sikkink, L. A., Baden, E. M., & Ramirez-Alvarado, M., (2006). The effects of sodium sulfate, glycosaminoglycans, and Congo red on the structure, stability, and amyloid formation of an immunoglobulin light-chain protein. *Protein. Sci., 15,* 1710–1722.

40. Madine, J., & Middleton, D. A., (2010). Comparison of aggregation enhancement and inhibition as strategies for reducing the cytotoxicity of the aortic amyloid polypeptide medin. *Eur. Biophys. J., 39,* 1281–1288.

41. Cardin, A. D., Randall, C. J., Hirose, N., & Jackson, R. L., (1987). Physical-chemical interaction of heparin and human plasma low-density lipoproteins. *Biochemistry, 26,* 5513–5518.

42. Wong, C., Xiong, L. W., Horiuchi, M., Raymond, L., Wehrly, K., Chesebro, B., & Caughey, B., (2001). Sulfated glycans and elevated temperature stimulate PrP(Sc)-dependent cell-free formation of protease-resistant prion protein. *EMBO J., 20,* 377–386.

43. Weisgraber, K. H., Rail, S. C. Jr., Mahley, R. W., Milne, R. W., Marcel, Y. L., & Sparrow, J. T., (1986). Human apolipoprotein E. determination of the heparin binding sites of apolipoprotein E3. *J. Biol. Chem., 261,* 2068–2076.

44. Cardin, A. D., & Weintraub, H. J., (1989). Molecular modeling of protein-glycosaminoglycan interactions. *Arteriosclerosis, 9,* 21–32.

45. Vieira, T. C., Cordeiro, Y., Caughey, B., & Silva, J. L., (2014). Heparin binding confers prion stability and impairs its aggregation. *FASEB J., 28,* 2667–2676.

46. Narindrasorasak, S., Lowery, D., Gonzalez-DeWhitt, P., Poorman, R. A., Greenberg, B., & Kisilevsky, R., (1991). High affinity interactions between the Alzheimer's beta-amyloid precursor proteins and the basement membrane form of heparan sulfate proteoglycan. *J. Biol. Chem., 266,* 12878–12883.

47. Brunden, K. R., Richter-Cook, N. J., Chaturvedi, N., & Frederickson, R. C., (1993). pH-dependent binding of synthetic beta-amyloid peptides to glycosaminoglycans. *J. Neurochem., 61,* 2147–2154.

48. Caughey, B., Brown, K., Raymond, G. J., Katzenstein, G. E., & Thresher, W., (1994). Binding of the protease-sensitive form of PrP (prion protein) to sulfated glycosaminoglycan and Congo red. *J. Virol., 68,* 2135–2141.

49. Warner, R. G., Hundt, C., Weiss, S., & Turnbull, J. E., (2002). Identification of the heparan sulfate binding sites in the cellular prion protein. *J. Biol. Chem., 277,* 18421–18430.

50. Park, K., & Verchere, C. B., (2001). Identification of a heparin-binding domain in the N-terminal cleavage site of pro-islet amyloid polypeptide. Implications for islet amyloid formation. *J. Biol. Chem., 276,* 16611–16616.

51. Ohashi, K., Kisilevsky, R., & Yanagishita, M., (2002). Affinity binding of glycosaminoglycans with beta(2)-microglobulin. *Nephron., 90,* 158–168.

52. Solomon, J. P., Bourgault, S., Powers, E. T., & Kelly, J. W., (2011). Heparin binds 8 kDa gelsolin cross-β-sheet oligomers and accelerates amyloidogenesis by hastening fibril extension. *Biochemistry, 50,* 2486–2498.

53. Sasisekharan, R., Raman, R., & Prabhakar, V., (2006). Glycomis approach to structure-function relationships of glycosaminoglycans. *Annu. Rev. Biomed. Eng., 8,* 181–231.

54. Malik, A., & Ahmad, S., (2007). Sequence and structural features of carbohydrate binding in proteins and assessment of predictability using a neural network. *BMC Struct. Biol., 7*, 1.

55. Shionyu-Mitsuyama, C., Shirai, T., Ishida, H., & Yamane, T., (2003). An empirical approach for structure-based prediction of carbohydrate-binding sites on proteins. *Protein. Eng. Des. Sel., 16*, 467–478.

56. Taroni, C., Jones, S., & Thornton, J. M., (2000). Analysis and prediction of carbohydrate binding sites. *Protein. Eng. Des. Sel., 13*, 89–98.

57. Fromm, J. R., Hileman, R. E., Caldwell, E. E. O., Weiler, J. M., & Linhardt, R. J., (1997). Pattern and spacing of basic amino acids in heparin binding sites. *Arch. Biochem. Biophys., 343*, 92–100.

58. Hileman, R. E., Fromm, J. R., Weiler, J. M., & Linhardt, R. J., (1998). Glycosaminoglycan-protein interactions: Definition of consensus sites in glycosaminoglycan binding proteins. *BioEssays., 20*, 156–167.

59. Agostinho, P., Cunha, R. A., & Oliveira, C., (2010). Neuroinflammation, oxidative stress and the pathogenesis of Alzheimer's disease. *Curr. Pharm. Des., 16*, 2766–2778.

60. Gao, H. M., Kotzbauer, P. T., Uryu, K., Leight, S., Trojanowski, J. Q., & Lee, V. M., (2008). Neuroinflammation and oxidation/nitration of alpha-synuclein linked to dopaminergic neurodegeneration. *J. Neurosci., 28*, 7687–7698.

61. Engelhart, M. J., Geerlings, M. I., Ruitenberg, A., et al., (2002). Dietary intake of antioxidants and risk of Alzheimer's disease. *JAMA, 287*, 3223–3229.

62. Sun, A. Y., Wang, Q., Simonyi, A., & Sun, G. Y., (2008). Botanical phenolics and brain health. *Neuromolecular Med., 10*, 259–274.

63. Szekely, C. A., & Zandi, P. P., (2010). Non-steroidal anti-inflammatory drugs and Alzheimer's disease: The epidemiological evidence. *CNS Neurol. Disord. Drug Targets, 9*, 132–139.

64. Wang, Q., Xu, J., Rottinghaus, G. E., et al., (2002). Resveratrol protects against global cerebral ischemic injury in gerbils. *Brain Res., 958*, 439–447.

65. Baird, A., Schubert, D., Ling, N., & Guillemin, R., (1988). Three-dimensional structure of human basic fibroblast growth factor. *Proc. Natl. Acad. Sci. U.S.A, 85*, 2324–2328.

66. Margalit, H., Fischer, N., & Ben-Sasson, S. A., (1993). Comparative analysis of structurally defined heparin binding sequences reveals a distinct spatial distribution of basic residues. *J. Biol. Chem., 268*, 19228–19231.

67. Torrent, M., Nogués, M. V., Andreu, D., & Boix, E., (2012). The "CPC Clip Motif": A conserved structural signature for heparin-binding proteins. *PLoS One, 7*, e42692.

68. Vilasi, S., Sarcina, R., Maritato, R., De Simone, A., Irace, G., & Sirangelo, I., (2011). Heparin induces harmless fibril formation in amyloidogenic W7FW14F apomyoglobin and amyloid aggregation in wild-type protein *in vitro*. *PLoS One, 6*, e22076.

69. Campioni, S., Mannini, B., Pensalfini, A., Zampagni, M., Parrini, C., Evangelisti, E., Relini, A., Stefani, M., Dobson, C. M., Cecchi, C., et al., (2010). A causative link between the structure of aberrant protein oligomers and their toxicity. *Nat. Chem. Biol., 6*, 140–147.

70. Yu, M. S., Lai, C. S., Ho, Y. S., et al., (2007). Characterization of the effects of anti-aging medicine fructuslycii on beta-amyloid peptide neurotoxicity. *Int. J. Mol. Med., 20*, 261–268.

71. Ho, Y. S., Yu, M. S., Yang, X. F., So, K. F., Yuen, W. H., & Chang, R. C. C., (2010). Neuroprotective effects of polysaccharides from wolfberry, the fruits of Lyciumbarbarum,

against homocysteine-induced toxicity in rat cortical neurons. *J. Alzheimers Dis., 19,* 813–827.

72. Guillozet-Bongaarts, A. L., Garcia-Sierra, F., Reynolds, M. R., et al., (2005). Tau truncation during neurofibrillary tangle evolution in Alzheimer' disease. *Neurobiol. Aging., 26,* 1015–1022.

73. Hunot, S., Vila, M., Teismann, P., et al., (2004). JNK-mediated induction of cyclooxygenase 2 is required for neurodegeneration in a mouse model of Parkinson's disease. *Proc. Natl. Acad. Sci., 101,* 665–670.

74. Tang, B. L., & Chua, C. E., (2008). SIRT1 and neuronal diseases. *Mol. Aspects Med., 29,* 187–200.

# CHAPTER 5

# Glycome in Metastasis: Glycan Remodeling and Tumor Progression

AYYAGARI ARCHANA,[1] DURGASHREE DUTTA,[2] SAFIKUR RAHMAN,[3] and RINKI MINAKSHI[1]

[1]Department of Microbiology, Swami Shraddhanand College, University of Delhi, New Delhi – 110 036, India

[2]Department of Biochemistry, Jan Nayak Chaudhary Devilal Dental College, Sirsa, Haryana, India

[3]Department of Medical Biotechnology, Yeungnam University, Gyeongsan – 712-749, South Korea

## ABSTRACT

The immense role of glycosylation in protein structure stabilization, as well as proper biological functioning, is well documented. Glycosylation happens to be the most common and crucial post-translational modification that is carried out just after protein synthesis by specific enzymes in the endoplasmic reticulum (ER) and Golgi apparatus (GA). Lectins are host cell surface proteins that identify and bind to their target glycans. Moreover, mammalian cells turning cancerous are reported to be associated with changes in glycosylation profile on their surface, indicating how powerfully glycosylation drives the development and angiogenesis of tumors, culminating in metastasis. Glycosylation regulates antibody functioning in a big way. The presence and quantity of fucose and galactose on the Fc region of immunoglobulins modulate their binding affinity to their own receptors located on immune cells. A few adhesion molecules of tumor cells, such as mucins, selectins, cadherins, integrins, laminins, etc., have been discussed here in terms of their altered glycome. A unique feature associated with metastasis is the development of neo-vasculature from preexisting blood vessels. Aberrations in glycosylation patterns have been identified among the causes of angiogenesis of metastatic tumors. Equipped with adequate knowledge of the altered glycome of various types of cancers, their reliable diagnosis, prevention, and therapy is fast becoming a possibility. Glycome remodeling enzymes could be a potential route of preventing as well as treating tumors at large. The present was designed to review the role of glycans in metastasis with the intention of highlighting the possible role of glycans in the therapeutic intervention of cancer. Future insights in this context have also been included in the chapter.

## 5.1   INTRODUCTION

The importance of carbohydrates in the biological world doesn't need an introduction and has been observed as a well-established field. Right from the formation of backbone in DNA and RNA, carbohydrates share closeness to proteins and lipids in the cell. The whole array of important signaling cascade depends on the cell surface receptors, which are protein conjugated with carbohydrates (glycoconjugates). The polymers of oligosaccharides with heterogenous substituting residues form a repertoire of glycans, and the entire set of all glycans present in an organism or cell is termed as its glycome. The machinery of glycosylation is a part of post-translational modification, which takes place in the endoplasmic reticulum (ER) and Golgi apparatus (GA), using nucleotide sugar residue donor and two important enzymes, glycosyltransferases and glycosidases. Identical proteins can be linked to different glycan residues, thereby giving rise to structurally variant glycoforms. Thus, glycome adds to the diversity of proteome. Glycans are inevitable for the proper functioning of the multicellular organization because studies have shown that the absence of glycans during embryo development may even prove lethal [1]. Additionally, various prokaryotic and eukaryotic pathogens show the expression of glycoconjugates on their cell surface, which are immunogenic in the mammalian host body.

The glycan-binding receptors on the cell surface are called lectins, which identify their specific ligands like glycans-conjugated to protein or lipid. Mammalian cells possess lectins, which are proteins that specifically bind to certain carbohydrate molecules. Purified lectins have been used *in vitro* to study the process of glycosylation. In the living system, lectins are responsible for conducting cell-to-cell interactions, and hence their binding to their specific glycans controls cell trafficking [2]. The immune cells are decorated by a number of such lectins that form the basis of the immune response. The constitution of all the human membrane-bound, as well as secreted proteins, has reported the presence of glycans [3]. More than half of all the cellular protein repertoire is represented by glycoproteins [4].

Glycans show their functional presence in effectively all the major processes of life like fertilization, development of embryo, proliferation of cells, immune system cells, and aging. The glycans help in the proper functioning of protein as they help in maintaining proper structural integrity, solubility, and function of a protein. These aspects are of crucial importance in the pathophysiology of various diseases. The involvement of more than hundreds of genes in the biosynthesis of glycans renders them prone to alterations during various pathogenesis. Proteins get glycosylated through a complex but systematic sequence of enzymatic steps that have been well characterized. The enzymes carrying out glycosylation of proteins are termed Glycosyltransferases, and their functioning rests on the availability of the exact monosaccharides needed for the same [5, 6]. The carbohydrate residues that get attached to the proteins stabilize them as well as may play a pivotal role in helping them efficiently perform their assigned biological functions. As a matter of fact, the solubility, half-life, affinity, and specificity of all proteins depend a great deal upon their glycosylation patterns. It has been shown that proteins with similar amino acid sequences may give rise to more than one glycoforms, possessing mutually distinct key properties, such as their localization, folding, stability, and ability to bind their ligands [2]. This leads to alterations in protein folding and routing, cell-to-cell interaction, differentiation, and immune response [7–9]. Accordingly, the molecular changes in protein glycosylation patterns may be useful in differentiating and identifying the diseased status by comparing the same with the patterns of the same proteins found in healthy individuals. This may serve as a simple, yet powerful tool used for disease diagnosis [10, 11]. Furthermore, since cell-to-cell contact and communication plays a crucial role in cancer, alterations in glycosylation patterns of cell surface proteins on cancerous cells can potentially affect the interactions of malignant cells with their environment [9, 12].

The lag behind the knowledge about glycans in diseases has been covered well with the advancements in the skills for studying the complexity and regulation of glycan functioning. The transformations during the establishment of malignancy unequivocally accompany deviations in the glycosylation patterns on proteins that represent a conspicuous glycoprofile of a particular cancer type. The role of glycosylation has been satisfactorily established in the proliferation, invasion, and angiogenesis of a tumor [13]. This chapter focuses on the role played by glycans in the metastatic journey of cancer.

## 5.2   THE CHEMICAL IDENTITY OF GLYCANS

Glycans are formed by the covalent bonding of oligosaccharides and poly-saccharides, either existing as free complexes or conjugated with proteins or lipids (glycoconjugates). The repertoire of these glycans constitutes an enormous number of molecules, thereby forming various families of glycoconjugates. They are the products of a post-translational modification. The primary types of glycosylation events occurring in the cell include the following (Figure 5.1) [14]:

1. **N-Linked Glycosylation:** The addition of oligosaccharides on the asparagine residues (on the consensus Asn-X-Ser/Thr motif) on the glycoproteins and this event occurs in the ER, e.g., cadherins, integrins, ICAM1, cathepsins.
2. **O-Linked Glycosylation:** The addition of N-acetyl galactosamine to serine and/or threonine residues, followed by further attachment of residues like galactose, N-acetyl-D-glucosamine, and/or sialic acid and this event occurs in the GA, e.g., integrins, CD44, selectins, mucin.

Apart from these, the other important glycans, which have been exten-sively studied in the case of tumor metastasis, are:

i.   Glycosaminoglycans (GAG): glycans with repeating units of an amino sugar;
ii.  Glycoshingolipid: glycans linked to ceramide.

The diversity witnessed among glycans ranges from the variation in the composition of monosaccharides like galactose (Gal), N-acetylgalactos-amins (GalNAc), in the linkage pattern like between 1st and 3rd carbons or 1st and 4th carbons, in the nature of branching, in the presence of substitution groups like sulfation and in the nature of conjugates like protein or lipid. The core structure of N-glycans consistently harbors the pentasaccharide, Man3GlcNAc2. The classification of N-glycans includes the following types [9]:

a. **Mannose-Type:** Core attached to only mannosyl (Man) residues.
b. **Complex-Type:** Branches (two: biantennary to four: tetraantennary) of N-acetylglucosamine (GlcNAc), fucose (Fuc), galactose (Gal), sialic acid (NeuNAc), and sulfates.
c. **Hybrid Type:** Two types of mannose branches, only Man on Man α 6 arm and more branching on Man α 1–3 arm.
d. **Poly-N-Acetyllactosamine Type:** Repeating residues of Galβ1-4GlcNAcβ1-3.

The tumor microenvironment usurps several events like the activation of surface receptors, upregulation of cellular adhesion molecules, and manipulates the cellular motility whereby the behavior of the neoplastic cells is controlled. The role of these glycosylation pathways has been studied in tumor progression where the aberrations are generally encountered due to either the truncation of normal oligosaccharides like the Tn antigen or the formation of rare sequences like the Lewis[x/a] structures. The significance of such aberrations is tremendous in tumor progression that arises due to dysregulation of glycosyltransferases.

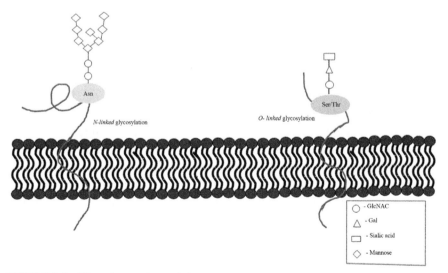

**FIGURE 5.1**   The primary glycosylation types. In *N-linked* glycosylation, oligosaccharides are added on the asparagine residues (on the consensus Asn-X-Ser/Thr motif) of the glycoprotein. In *O-linked* glycosylation, N-acetyl galactosamine is added to serine and/or threonine residues of glycoprotein.

## 5.3   GLYCOSYLATION REGULATES ANTIBODY FUNCTION

A plethora of glycoproteins constitutes the major players of our innate as well as adaptive immune response to all antigenic challenges. Of these, immunoglobulins of five distinct classes and their subclasses are the major components, which take care of the adaptive immunity. It is a well-established fact that they are glycoproteins, and that without their specified carbohydrate content, they wouldn't be able to perform their respective assigned functions in imparting the much-needed immunity. The most important effector molecules of our immune system are the five classes of immunoglobulins, which are essentially glycosylated [15]. As much as 15% of their weight is estimated to be their glycan content, without which they would be likely to suffer a partial or total loss in their functionality. Of this, immunoglobulin G (IgG) deserves the maximum importance due to its highest abundance and greatest involvement in imparting immunity, and consequently has been extensively studied. The two amino termini of Y-shaped immunoglobulins are the antigen-binding or variable regions, imparting specificity, a key factor that determines their specific binding to their immunogen epitopes. On the other hand, their carboxy-terminal, termed Fc portion, possesses a constant composition, and is recognized by its receptors on the surface of phagocytes, hence playing the lead role in the clearance of immune complexes. Each heavy chain of this immunoglobulin is bestowed with a single bi-antennary N-glycan that is covalently attached as an integral component of the Fc portion of the antibody. It is evident that its effector functions get substantially altered if the sugars binding to the antibody are changed during its glycosylation step [16].

Although the total sugar profile of the entire set of immunoglobulins found in any individual, particularly in homeostatic conditions, is found to be rather constant at any given point in time [17], it appears to be subject to gradual alteration with the advancing age of the individual [18, 19], and may even start changing at a fast pace if the individual is in a disturbed homeostasis state [20]. At the level of population studies, it may be safely inferred that immunoglobulin glycome undergoes a high-level variation all the time. This not only provides an evolutionary advantage but also mirrors alterations in environmental stimuli [18, 21].

The majority of human IgG in circulation contains a fucose sugar molecule attached to the first N-acetylglucosamine of the sugar portion of this glycoprotein [21]. This manifests in a decline in the efficiency of the Fc portion of the antibody to bind to its receptors located on leukocytes such as macrophages and natural killer (NK) cells [22, 23]. The occurrence of this phenomenon is interpreted as a safety regulation that controls excessive ADCC (antibody-dependent cell-mediated cytotoxicity), which, if uncontrolled, could play havoc in the system [24]. Also, with this clue in hand, the monoclonal and chimeric antibodies, synthesized for therapeutic applications of cancer, are deliberately designed without the fucose residue, so that they can bring about the required ADCC in an amplified manner [23, 25, 26]. Further evidence of the part played by the glycome in the immune response is that it is the sugar-sugar interaction of the Fc portion of immunoglobulin and its receptor that facilitates their mutual affinity and binding [27, 28]. Similar to fucose, the sugar galactose is also a component of the immunoglobulin glycome. Its content alters with advancing age and changing the hormonal profile of an individual, but may also shift rather swiftly during the episodes of inflammation [20]. A decline in galactosylation of IgG has been found associated with the pathogenesis of a number of autoimmune and/or inflammatory diseases. Immuno-complexes with higher galactose content opposes the participation of complement split product C5a in inflammation [29, 30]. Glycans of IgG with terminal sialic acid have been shown to convert the immunoglobulin role in the inflammation from positive to inhibitory in some studies on mice. This may be due to the fact that high sialic acid content on the Fc portion of the antibody makes it recognizable to the CD 209 receptor, which in turn inhibits inflammation. Such immunomodulatory mechanisms of heavily sialylated IgG that downregulates inflammation may be credited for keeping autoimmune and chronically inflammatory diseases at bay among immunocompetent individuals [30, 31].

## 5.4 THE N-ACETYLGLUCOSAMINYLTRANSFERASES

The oligosaccharides on the cell surface glycoproteins are remodeled by the action of enzyme, glycosyltransferases. The glycosyltransferases are often accompanied with glycosidases, which reside in the nucleus, ER, GA, and cytosol. Changes in the function of these glycosyltransferases are responsible for the altered sugar chains on the surface glycoproteins in cancer [32, 33]. The post-translational modification of protein involves the enzymatic addition or subtraction of N-glycans on the core amino acid motif of the parent peptide chain. The glycosyltransferases majorly involved in the biosynthetic pathway of N-glycan are N-acetyl-glucosaminyltransferase III (GnT-III), N-acetyl-glucosaminyltransferase V (GnT-V) and $\alpha$-1,6-fucozyltransferase (Fut8) (Figure 5.2).

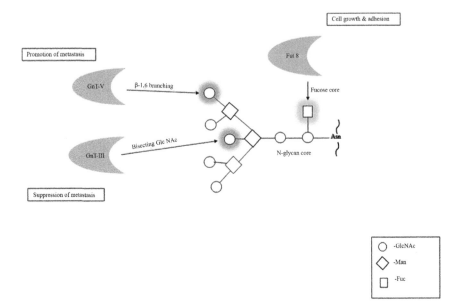

**FIGURE 5.2**    The major N-acetylglucosaminyltransferases. GnT-III catalyzes the addition of N-acetylglucosamine (GlcNAc) through β-1,4 linkage on to β-linked mannose present on the trimannosyl core of sugar chain leading to the bisection of GlcNAc linkage. This shows role in the suppression of metastasis. GnT-V catalyzes of addition of GlcNAc on α-1,6 mannose through β-1,6-linkage by GnT-V, giving rise to branching structure, β-1,6-GlcNAc. This promotes metastasis. Fut8 transfers fucose (Fuc) residue on the innermost GlcNAc of the core glycan via α-1,6-linkage, which gives rise to 'core fucose' structure. It supports cell growth and adhesion.

## 5.4.1    GNT-III

The gene, MGAT3 (β-1,4-mannosyl glycoprotein β-1,4-N-acetylglucosaminyltransferase), finds its cytogenetic location on 22q13.1 chromosome, codes for GnT-III [34, 35]. The catalysis involves the addition of N-acetyl-glucosamine (GlcNAc) through β-1,4 linkage on to β-linked mannose present on the trimannosyl core of sugar chain leading to the bisection of GlcNAc linkage. This bisection halts additional processing and elaboration of N-glycans that favors impediment of metastasis [36, 37]. The tumor suppression effect of GlcNAc bisection has been studied in many model cases of metastasis, like lung and breast neoplasia [37, 38]. However, some studies have proven to antagonize roles of GnT-III where it has been observed to augment the progression of cancer [39].

### 5.4.2   GNT-V

The gene MGAT5 ($\alpha$-1,6-mannosyl glycoprotein $\beta$-1,6-N-acetylglucosaminyltransferase), is located on 2q21.3-q21.3 chromosome, code for GnT-V [40]. The catalysis of addition of GlcNAc on $\alpha$-1,6 mannose through $\beta$-1,6-linkage by GnT-V, gives rise to branching structure, $\beta$-1,6-GlcNAc [41]. We have augmenting evidences in support of the uncontrolled upregulation of MGAT5 gene and its aberrant enzymatic activity that prop metastasizing neoplasm [42]. Report favors the prognostic association of GnT-V in malignancies [43].

### 5.4.3   FUT8

The cytogenetic location of FUT8 gene is on 14q23.3 [44]. The transfer of fucose (Fuc) residue on the innermost GlcNAc of the core glycan via $\alpha$-1,6-linkage is catalyzed by Fut8, which gives rise to 'core fucose' structure [45]. Fut8 has also been reported to be aberrantly upregulated in hepato-carcinogenesis [46]. In disease models of lung cancer, core fucose has been shown to be required for cell proliferation that is dependent on growth factors [47]. In several lines of cancer like hepatoma, lung, breast, and prostate, the unregulated upregulation of FUT8 has been reported [42]. The use of core fucose in antibody cancer therapy involving the mechanism of antibody-dependent cellular cytotoxicity (ADCC) has been well documented [42].

## 5.5   THE SIALYLTRANSFERASE FAMILY

The increased level of sialic acid on the N-glycans present on tumor surface is a part of the aberrant glycosylation feature observed in neoplasm [48, 49]. The class of glycosyltransferases which perform the role in the addition of sialic acid onto other sugar moieties are termed sialyltranferases (STs). Many forms of cancer have shown the overexpression of STs [50]. The hypersialylation of adhesion molecules has been well studied in case of multiple myeloma and breast cancer, where the sialylated ligands are important for homing and metastasis of tumor cells [51]. Notably, sialyl Lewis X (SLe$^x$) and its isomer sialyl Lewis A (SLe$^a$) are significant in metastasis.

## 5.6 MULTIPLE FACTORS REGULATE BRANCHING OF N-GLYCANS IN CANCER

### 5.6.1 EPIGENETICS

The phenotypic expression of aberrant glycosylation pattern seen in cancer has been linked to epigenetic changes. The methylation of CpG Island (upstream promoter region) has been known to silence the transcription of genes by binding to chromatin silencing proteins like histone deacetylase (HDACs). Some genes involved in the branching of N-glycan have been observed to be either silenced by methylation or triggered by hypomethylation. In the case of ovarian cancer, the MGAT3 promoter has been reported to be hypomethylated and similar studies have been reported for other cancer types [52, 53]. It has also been demonstrated in ovarian cancer epithelial cells that methylation causes a reduction in the formation of core fucose and an increase in branching as well as sialylation due to alterations in MGAT5 [54].

### 5.6.2 GENES AND TRANSCRIPTION FACTORS

The involvement of some crucial transcription factors provoking the upregulation of genes that promote tumorigenesis has been studied. The best documented is the oncogenic stimulation of MGAT5 through the Src-Raf-Ets2 pathway [55]. Relatedly, Her-2/neu oncogene has been linked with the upregulation of MGAT5 through Ras-Raf-Ets pathway [56]. The transcription factor, HNF1 has been found to negatively regulate the FUT8 gene [57].

### 5.6.3 REGULATORY MIRNA

The short noncoding RNA (miRNA) cause negative regulation of gene expression. The miRNA-198 has been shown to target FUT8 in colorectal cancer. The regulatory miRNA causes the suppression of FUT8 resulting in the inhibition of cancer proliferation and invasion [58]. Similarly, miRNA-122 and miRNA-34a have been found to suppress FUT8 in hepatocarcinoma [59].

### 5.6.4    REPERTOIRE OF NUCLEOTIDE SUGARS

The glycosylation status is very much dependent upon the concentration of donor nucleotide sugar substrates like UDP-GlcNAc, which depend on the metabolism of cell. It is well established that the metabolism of sugar and energy is transformed in cancer [60]. Quantification experiments have shown that the concentrations of nucleotide sugars vary between cancer of breast and pancreas [61].

### 5.6.5    THE SUBCELLULAR LOCATIONS OF GLYCOSYLTRANSFERASES

The GA is known to localize the N-glycan branching enzymes. It has been shown in breast malignancy that there is relocation of GalNAc-Ts into the ER which enhances progression and invasiveness of the tumor [62]. In human hepatoma cells, the regulation of GnT-III localization and activity has been observed to be affected by caveolin-1 [63, 64].

## 5.7    ROLE OF GLYCANS IN METASTASIS

The pathophysiology of cancer phenotype not only includes modifications in the expression of gene, but also the transformed cell undergoes major alterations itself as well as affects its microenvironment like the extracellular matrix. The affect percolates to post-translational modifications of protein as well as lipids that the cancer usurps to establish its domain. The progression and development of cancer in its secondary form is termed metastasis, which requires the completion of a multitude of cascading events. The phenotype of malignancy requires an active communication between the tumor and its microenvironment. This association has been strongly recognized to be modulated by the process of glycosylation. Normally the carbohydrates present on the cell surface act as adhesion molecules that serve two important functions-recognition of cell and modulation of receptors for social function of the cell in a tissue. The glycans perform crucial functions in the progress of cancer through an array of processes like inflammation, cell-to-cell adhesion, cell to matrix association and signaling [65]. Malignancy is breaking the rules of this cellular society where aberration in glycosylation status of the cell makes it insidious.

The steps of metastasis can be summarized as follows [66, 67]:

1. **Invasion of Neighboring Tissue:** The primary lesion leads to degradation and penetration of the basal membrane to spread to the surrounding tissues.
2. **Entering Circulatory System:** When the tumor gets increased vascularization, they gain access into the circulatory system.
3. **Endothelial Attachment:** The spreading tumor gets attached to the distant endothelium.
4. **Encroachment of Endothelium:** The tumor gets penetrance into the basal membrane of the endothelium where they colonize resulting into a metastatic focus.

The uncontrolled growth of a malignant tumor is accompanied by the loss of contact inhibition, independence on the growth factors, higher cellular mobility, and exploitation of hydrolases activity for solubilization of basement membrane. Additionally, the ratio of matrix metalloproteinases (MMP) to tissue inhibitors of metalloproteinases (TIMP) plays an important parameter for invasion ability of tumor. The alterations in the level of glycosylation of adhesion molecules present on the cell surfaces have been widely documented in the pathology of malignancy. The adhesion molecules like selectins, integrins, fibronectins, and mucins show alteration in their glycosylation status in various malignant pathologies. These glycosylation variations from the normal tissues have been well exploited in the prognosis of malignancies [67–69]. The metastasizing tumor interacts with the endothelium through integrins and lectins via the mechanism of tethering and rolling [70, 71]. The selectin-ligand interaction is greatly favored by moieties like SLe$^x$ and SLe$^a$ [71].

### 5.7.1   THE 'SUGARINESS' OF TUMOR CELLS

The post-translational modifications of protein comprise N-glycosylation as one of the most important aspect of soluble as well as membrane glycoproteins. The addition of oligosaccharides on proteins gives them unique properties due to their types of linkages and highly branched nature. This gives them the extraordinary property of storing more diverse set of information as compared to other biomolecules [72]. The coding template doesn't control the addition of these sugar chains on the glycoproteins rather the variations in their numbers completely depend upon the change in physiological properties of the cell. Glycocode undergoes major change during metastasis as compared to the normal cells. One of the significant phenotypic illustration of tumor involves an increase in the effective mass of the composite molecules formed by N-glycans on the cell surface, which is due to the rise in branching of the trimannosyl core (after the additions of poly-N-acetyl-lactosaminoglycan and sialic acid). It has been reported that there is an increase in $\beta$1,6-linked branching on gp130, the cell surface glycoprotein, which augments the metastatic potential [73]. This branching has also been implicated in the facilitation of the invasive potential of tumors through the basement membrane [74–77]. The cases of human neoplasia like breast and colon as well as melanoma have been studied to show consistent rise in malignancy with increasing $\beta$1,6-linked branching [76]. Colon carcinoma studies disclosed that higher number of poly-N-acetyl-lactosaminyl side chains and sialic acid addition supported metastasis [78].

The aberrations in glycosylation induced by tumor follow two mechanisms postulated by Hakomori and Kannagi-the incomplete synthesis process observes impairment in synthesis of normal glycans (e.g., truncated structures sialyl Tn, STn on breast cancer) and the neo-synthesis process shows induction of cancer-associated carbohydrate determinants (e.g., SL$^x$, and SL$^a$) [65].

The aberrations in glycosylation most significantly noticed in cancer includes following [74, 79–83]:

a.  Formation of $\beta$1-6 GlcNAc on N-linked structures;
b.  O-glycosylation;
c.  addition of a variety of fucozyl- and sialyl-linkages;
d.  Accumulation of ganglio- and globo-structures.

The accumulation of altered glycosphingolipid has been reasoned behind change in the nature of interaction between cell and its basal membrane that ultimately provokes aberrant mobility of the transformed cells [84, 85]. The metastatic potential gets fueled up with overall rise in sialic acid additions to the cell surface [73]. The SLe$^x$ moiety load on the colorectal carcinoma was found to be directly related to the metastasis of the later to the liver [86]. The constant additions of sugar moieties on the cell surface needs to reach a threshold density before the glycosylation could initiate its metastatic effect [75]. In cases of high-grade tumors of neuroblastoma and gliomas, increased sialylation causes accumulation of polysialic acid in neural cell adhesion molecule 1 (NCAM1) [65]. Reports show overexpression of gangliosides in breast cancer, melanoma, and neuroblastoma that aid in tumor migration [65]. The extravasation of transformed cells into new metastatic foci is accompanied by lymphocytic infiltration, which is visible as inflammation

[87]. Inflammation is initiated by a group of cell surface lectins known as selectins [88–90]. All of these work by selecting a glycoprotein or glycolipid ligands on the cell surface and binding to them in a calcium dependent manner [91]. They play a crucial role in attracting leukocytes to the inflammation site, leading to rolling, adhesion, and extravasation into tissues as well as signal transduction. Since selectins play such a pivotal role in inflammation and immune surveillance, they are certainly a significant component of a successful immune response against antigenic challenges. They have been reported to be playing significantly involved in metastasis [92] and maternal-fetal interactions [93].

## 5.8    THE FOCI OF GLYCOSYLATION IN MALIGNANCY

The cells in tissues are decorated with a variety of adhesion molecules that play a very crucial role in the interaction of these cells with their neighboring cells. The glycosylation status of these proteins is important in the behavior of tumor whereby they form the foci of glycosylation to promote malignancy (Figure 5.3). Some of the important adhesion molecules are discussed herewith:

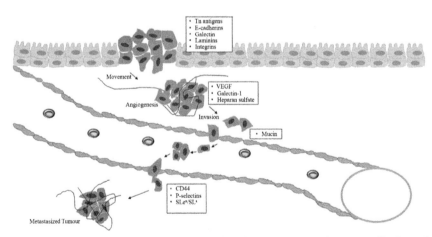

**FIGURE 5.3**  Molecules aiding in metastasis. The movement of tumor cells from the primary niche is initiated by Tn antigens, E-cadherins, galectin, laminins, and integrins. The molecules favoring angiogenesis are VEGF (vascular endothelial growth factor), galectin-1 and heparan sulfate. The invasive capacity of transformed cells is augmented by aberrantly expressed mucins. After crossing the vasculature, migratory tumor cells establish their new niche at some secondary site that is favored by CD44, P-selectins, and SLe[x]/SLe[a].

### 5.8.1   MUCINS

Mucins are huge glycoproteins with higher content of O-linked oligosaccharides. They have been reported to be the carriers of aberrant glycosylation in metastatic tumors [94, 95]. The anti-adhesion property of mucin help in the dissemination of tumor cell from its primary focus of origin to distant niches [94, 96]. In one study on rat mammary tumor, it was shown that NK cell-mediated lysis of tumor was halted due to elevated levels of sialomucin [97]. Tumor cells affect their mucin expression by increasing SLe$^x$ content, in order to establish a favorable tumor microenvironment, which has been shown in the cell lines of colonic carcinoma [98]. The aberrant glycosylation of mucin in pancreatic cancer has been reported to enhance metastatic potential as well as induce genes for multidrug resistance [99, 100].

### 5.8.2   SELECTINS

The adhesion molecules present on the leukocytes, platelets, and the endothelial cells are the selectins, which mediate the linkage of vascular cells on the endothelium. Selectins are calcium-dependent N-glycoprotein lectins, which form a family of three members:

1.  **E-Selectins:** They are expressed on endothelial cells upon induction with TNFα and IL-1α.
2.  **P-Selectins:** They are found in the storage granules of platelets (α-granules) and endothelium (Weibel-Palade bodies), which upon activation get translocated to the cell surface.
3.  **L-Selectins:** They are constitutively expressed on leukocytes that mediate faster leukocytic rolling on the endothelium.

The ligands of all the selectins are the glycoproteins present on cell surfaces. One of the most described selectin ligands is PSGL-1 that interacts with all the three members. The implication of selectins in extravasation of tumor cells through tethering and rolling on the activated ligands of endothelial cells is crucial for metastasis. All this is enhanced under the effect of TNFα [101]. The direct involvement of E-selectin ligand in the metastatic phenotype of colon carcinoma has been shown to be linked to SLe$^a$ levels [102]. Similar results of stronger E-selectin interactions have been witnessed in metastatic breast cancer cell line [103]. We have accumulating results that advocate the induction of E-selectin expression directly or indirectly by tumor cells. Either tumor cells stimulate the endothelial cells directly by secreting IL-α or indirectly by evoking mononuclear leukocytes that secrete IL-1β leading to induction of E-selectin expression [104]. The activated platelets express P-selectins that aid in the extravasation of tumor cells and this becomes a key step in the promotion of succeeding colonization to distant tissues [105]. In leukemic cell lines, the homing of leukocytes in tissues is greatly influenced by L-selectins [106]. It has been displayed well in transgenic mice that the immune surveillance was cleverly curtailed by tumor cells through the repression of L-selectin expression favoring tumor progression [107].

Any discrepancies in the selection component of the host system may render the individual prone to a number of diseases such as atherosclerosis, deep venous thrombosis, and tumor metastasis [108]. The importance and role played by the sugar component of selectins can't be overemphasized. It is a well-established fact that the fucose content of selectins has a significant role in the proper adhesion of leukocytes to the sides of blood capillaries prior to their extravasation into the tissues at the site of inflammation [109]. The patients of primary immunodeficiency called Leukocyte adhesion deficiency type II (LAD II) provide an elegant evidence of the outcome of defective fucose processing, resulting in dysfunctional selectin ligands on the host cell surface [110]. Such spontaneous models were witnessed to be frequently contracting bacterial infections [111, 112]. It is highly likely that any variations in the glycosylation pattern of proteins involved in immunity of human beings may also affect the working of selectins and thus influencing various disease mechanisms [18].

### 5.8.3   PSGL-1

The P-selectin glycoprotein ligand-1 (PSGL-1) is a homodimer transmembrane protein expressed on lymphoid and myeloid cells as well as platelets. PSGL-1 is a ligand for all three selectins, P-, E-, and L-showing variable binding affinity [113]. The binding of selectins with PSGL-1 is mediated by specific enzymes that are expressed constitutively by myeloid and T-cell progenitors [114, 115]. The selectin binding property of PSGL-1 is activated in only proliferating and differentiating T-cells [116, 117]. The ligand harbors SLe$^x$ that mediates its interaction with the selectins. The regulatory role-played by PSGL-1 in rolling and tethering of various immune cells like plasma B-cells, macrophages, and T-cells is inevitable for the migration of these cells to the inflammation foci [118]. Thus PSGL-1 actively promotes lymphocytic infiltration. In addition, PSGL-1 acts as an inhibitory checkpoint which expedites exhaustion of T-cells in tumor [119].

### 5.8.4   CADHERINS

The cell-to-cell adhesion receptors are cadherins, which are type I transmembrane glycoproteins. The ectodomain facilitates the binding to neighboring call cadherin while the endodomain in the cytoplasm interacts through special proteins (catenins) with actin cytoskeleton, signaling, and trafficking complexes [120]. Cadherins are also known to play a role in metastasis [121–123]. The cadherin family includes N-cadherin present in neural and muscle tissue, E-cadherin in epithelial cells, P-cadherin in the placenta, and L-cadherin present in the liver [124, 125]. The N-glycan aberration commonly observed on E-cadherin is the rise in β-1,6-branching during malignancy [126]. This leads to the internalization of E-cadherin into the cytoplasm resulting in the interruption of cell-to-cell contact for easing invasion and metastasis [127].

### 5.8.5   GALECTINS

The S-type lectins are the Galectins that take part in the cell to cell as well as a cell to matrix association through the recognition of N-acetyllactosamine residues. These residues are common components of the glycoproteins in the extracellular matrix (ECM) like fibronectin and laminin. Galectins are non-glycosylated protein which has been known to be expressed in metastatic cells [128–130].

### 5.8.6   INTEGRINS

The integrins form the largest group of glycoprotein adhesion molecules that have been widely studied for their presence on the tumor cell surface as receptors, which facilitate interaction between the cell and ECM proteins like laminin and fibronectin [131]. The metastatic potential of aberrantly expressed integrin levels has been proved in prostate, colon, and breast cancer [131]. The epithelial cell integrins show N-glycosylation changes like a rise in -1,6-branching that encourage invasion [73]. One study in melanoma cells depicted that the removal of α-2, 8-linked sialic acid from integrin led to inhibition of cell interaction with fibronectin [132]. The experiment in gastric cancer cells has shown that a rise in N-glycan branching facilitates laminin-integrin clustering to help in cell motility as well as adhesion [133].

### 5.8.7   LAMININS

The laminins are a part of the basement membrane proteins, which show association with integrins, a feature commonly exploited by invading tumor cells [134]. It has been verified that the modulation of laminin and integrin interaction through N-glycans promotes cell migration [131].

### 5.8.8   HYALURONAN

The ER and GA facilitate the assembly of glycans, but hyaluronan is a large free glycosaminoglycan that polymerizes at the plasma membrane destined to be secreted into the ECM. The presence of hyaluronan in the ECM of the tumor that binds to CD44 (glycosaminoglycan) receptor is well documented [13]. Cancerous cells display spliced variants of CD44 molecules that have an aberrant association with hyaluronan, playing promotional role in metastasis [135, 136].

### 5.8.9   HEPARAN SULFATE (HS)

Heparan sulfate (HS) are GAGs that are part of proteoglycans with discrete negatively charged sulfate groups assisting in cell to matrix interactions. They also act as co-receptors and aid in the stowage and protection of ligands. In pancreatic cancer cells, a GPI-anchored HS proteoglycan known as glypican-1, mediates ligand-receptor complexes [137]. However, a mutational study within GLYPICAN-3 gene in Simpson-Golabi-Behmel syndrome, patients was reported with malignancies [138].

## 5.9   ROLE OF GLYCANS IN ANGIOGENESIS

The development of neo-vasculature from the preexisting blood vessels is a crucial phenotype of angiogenesis during metastasis. The trigger for angiogenesis originates from factors like vascular endothelial growth factor (VEGF), which enhances the endothelial metabolic rate for supporting genesis of new vasculature [64]. The regulatory network controlling the remodeling of glycosylation of cell surface receptors like VEGF receptor 2 and Notch in the endothelial cells, thus plays an essential role in the angiogenesis required for tumor metastasis [139]. The substantial role played by Galectin-1 in the angiogenesis of tumor has been well reported [140, 141]. The β1,6-linked branching supports binding Galectin-1 on endothelial cells [142]. HS proteoglycans have been well noticed to be associated with growth factors supporting angiogenesis [143]. There are many instances of aberration in glycosylation pattern imposed during angiogenesis of metastatic tumor. The presence of lactate stimulates hyper-O-GlcNAcylation that promotes angiogenesis during metastasis (Warburg Effect) [144]. The notch receptors show modifications through glycosyltransferases that cause extensions on O-fucose thereby effecting angiogenesis [144].

## 5.10   THERAPEUTIC INTERVENTIONS ON TUMOR GLYCOME

The ever-evolving field of glycobiology has given us tremendous data that we are using to understand the mechano-chemical properties of tumor in metastasis. The glycome studies help in deciphering the consequences of molecular and cellular trafficking in the cancer metastasis.

We are enriched with the knowledge about neo-molecules that are potentially used as biomarkers of various cancers. Some of the most noticeable biomarkers used in the diagnosis of patients with specific cancer are prostate-specific cancer (PSA), carcinoma antigen 125 (CA125 or MUC16, in ovarian cancer), Sle$^a$ (in colon cancer), and aberrantly glycosylated MUC1 (in breast cancer) [65]. In hepatocellular carcinoma, higher fucozylation of alpha-fetoprotein has been approved by FDA to be a significant early tumor marker [145]. The introduction of new techniques like high-throughput mass spectrometry (MS) has contributed highly to the analysis of aberrations in glycan, useful in the diagnosis of cancer [146, 147]. Patients with gastric cancer and intestinal metaplasia have shown serum STn [148]. The role of circulating exosomes, which are characteristic of certain glyco-conjugates, in early cancer detection is marvelous. As in the case of pancreatic cancer, the detection of exosomes possessing the proteoglycan, glypican-1, helps in early diagnosis [149]. Use of IgG autoantibodies in the detection of aberrantly glycosylated MUC1-STn in colorectal cancer was shown to be 95% specific [150].

The targeting of glycome remodeling enzymes as well as the molecules aiding in metastasis can help us in stopping the insidious journey of tumor. The use of definite GAGs like heparin have been used to attenuate metastasis in animal models [151]. Anti-Galectin-1 monoclonal antibodies were documented to prevent angiogenesis in refractory tumors showing anti-VEGF feature [142]. In Phase II clinical trial of ovarian cancer, targeting of MUC16 (CA125) with antibody is under consideration whereas in the Phase III trial of breast carcinoma, MUC1 is being targeted through vaccine therapy [152, 153]. The competitive inhibitor of GnT-V, Swainsonine, has been proved to reduce tumor invasion in Phase I trials [154]. The abrogation of angiogenesis by polymers of sulfonic acid is well-documented [155].

## 5.11   CONCLUSION AND FUTURE PERSPECTIVES

The normal functioning of a cell is dependent upon the regulated modulation of protein glycosylation, which in turn serve a wide array of functional role in the physiology of a cell. The regulatory control of glycosylation is beyond the master template, DNA. Therefore, a comprehensive learning about the molecular aspects of glycan right from the beginning of study of glycosyltransferases to its controlling parameters is the need of this time. Though lots of studies on the same are available, much more is required for a more convincing and reliable grasp on the subject. A better understanding of the processes involved in various levels of glycan remodeling and their exploitation for progression by tumor can really help us in deciphering the mysteries behind refractory tumors. The glycome forms the basis of cellular communication thereby establishing a cellular society. The advent of metastasis exploits this aspect by altering the cell surface receptors and components of cellular microenvironment. A handful of cases of diagnosis as well as immunotherapies of various cancers based on the cancer glycome have been recently appearing. However, what is already demonstrated in this field is just the tip of the huge iceberg yet to be discovered. We need more molecular paradigms in glycobiology that can provide higher degrees of understandings in cancer glycomics.

## KEYWORDS

- **antibody-dependent cell-mediated cytotoxicity**
- **endoplasmic reticulum**
- **extracellular matrix**
- **galactose N-acetylgalactosamins**
- **glycosaminoglycans**
- **histone deacetylase**
- **matrix metalloproteinases**

## REFERENCES

1. Marek, K. W., Vijay, I. K., & Marth, J. D., (1999). A recessive deletion in the GlcNAc-1-phosphotransferase gene results in peri-implantation embryonic lethality. *Glycobiology, 9*, 1263–1271.

2. Dwek, R. A., (1996). Glycobiology: Toward understanding the function of sugars. *Chem. Rev., 96*, 683–720.

3. Moremen, K. W., Tiemeyer, M., & Nairn, A. V., (2012). Vertebrate protein glycosylation: Diversity, synthesis and function. *Nat. Rev. Mol. Cell Biol., 13*, 448–462.

4. Apweiler, R., Hermjakob, H., & Sharon, N., (1999). On the frequency of protein glycosylation, as deduced from analysis of the SWISS-PROT database. *Biochim. Biophys. Acta, 1473*, 4–8.

5. Spiro, R. G., (2002). Protein glycosylation: Nature, distribution, enzymatic formation, and disease implications of glycopeptide bonds. *Glycobiology, 12*, 43R–56R.

6. Kuzmanov, U., Kosanam, H., & Diamandis, E. P., (2013). The sweet and sour of serological glycoprotein tumor biomarker quantification. *BMC Med., 11*, 31.

7. Crocker, P. R., & Feizi, T., (1996). Carbohydrate recognition systems: Functional triads in cell-cell interactions. *Curr. Opin. Struct. Biol., 6*, 679–691.

8. Feizi, T., (2000). Carbohydrate-mediated recognition systems in innate immunity. *Immunol. Rev., 173*, 79–88.

9. Helenius, A., & Aebi, M., (2001). Intracellular functions of N-linked glycans. *Science, 291*, 2364–2369.

10. Hakomori, S., (1996). Tumor malignancy defined by aberrant glycosylation and sphingo(glyco)lipid metabolism. *Cancer Res., 56*, 5309–5318.

11. Kobata, A., (1998). A retrospective and prospective view of glycopathology. *Glycoconj. J., 15*, 323–331.

12. Rudd, P. M., Wormald, M. R., Stanfield, R. L., Huang, M., Mattsson, N., Speir, J. A., Digennaro, J. A., et al., (1999). Roles for glycosylation of cell surface receptors involved in cellular immune recognition. *J. Mol. Biol., 293*, 351–366.

13. Fuster, M. M., & Esko, J. D., (2005). The sweet and sour of cancer: Glycans as novel therapeutic targets. *Nat. Rev. Cancer, 5*, 526–542.

14. Varki, A., & Freeze, H. H., (1994). The major glycosylation pathways of mammalian membranes: A summary. In: Maddy, A. H., & Harris, J. R., (eds.), *Subcellular Biochemistry, Membrane Biogenesis* (Vol. 22, pp. 71–100).

15. Rudd, P. M., Elliott, T., Cresswell, P., Wilson, I. A., & Dwek, R. A., (2001). Glycosylation and the immune system. *Science, 291*, 2370–2376.

16. Schwab, I., & Nimmerjahn, F., (2013). Intravenous immunoglobulin therapy: How does IgG modulate the immune system? *Nat. Rev. Immunol., 13*, 176–189.

17. Gornik, O., Wagner, J., Pucic, M., Knezevic, A., Redzic, I., & Lauc, G., (2009). Stability of N-glycan profiles in human plasma. *Glycobiology, 19*, 1547–1553.

18. Knezevic, A., Polasek, O., Gornik, O., Rudan, I., Campbell, H., Hayward, C., Wright, A., et al., (2009). Variability, heritability, and environmental determinants of human plasma N-glycome. *J. Proteome Res., 8*, 694–701.

19. Kristic, J., Vuckovic, F., Menni, C., Klaric, L., Keser, T., Beceheli, I., Pucic-Bakovic, M., et al., (2014). Glycans are a novel biomarker of chronological and biological ages. *J. Gerontol. A Biol. Sci. Med. Sci., 69*, 779–789.

20. Novokmet, M., Lukic, E., Vuckovic, F., Ethuric, Z., Keser, T., Rajsl, K., Remondini, D., et al., (2014). Changes in IgG and total plasma protein glycomes in acute systemic inflammation. *Sci. Rep., 4*, 4347.

21. Pucic, M., Knezevic, A., Vidic, J., Adamczyk, B., Novokmet, M., Polasek, O., Gornik, O., et al., (2011). High throughput isolation and glycosylation analysis of IgG-variability and heritability of the IgG glycome in three isolated human populations. *Mol. Cell Proteomics, 10*, M111 010090.

22. Niwa, R., Hatanaka, S., Shoji-Hosaka, E., Sakurada, M., Kobayashi, Y., Uehara, A., Yokoi, H., et al., (2004). Enhancement of the antibody-dependent cellular cytotoxicity of low-fucose IgG1 Is independent of FcgammaRIIIa functional polymorphism. *Clin. Cancer Res., 10*, 6248–6255.

23. Iida, S., Misaka, H., Inoue, M., Shibata, M., Nakano, R., Yamane-Ohnuki, N., Wakitani, M., et al., (2006). Nonfucosylated therapeutic IgG1 antibody can evade the inhibitory effect of serum immunoglobulin G on antibody-dependent cellular cytotoxicity through its high binding to FcgammaRIIIa. *Clin. Cancer Res., 12*, 2879–2887.

24. Scanlan, C. N., Burton, D. R., & Dwek, R. A., (2008). Making autoantibodies safe. *Proc. Natl. Acad. Sci. U.S.A, 105*, 4081, 4082.

25. Shinkawa, T., Nakamura, K., Yamane, N., Shoji-Hosaka, E., Kanda, Y., Sakurada, M., Uchida, K., et al., (2003). The absence of fucose but not the presence of galactose or bisecting N-acetylglucosamine of human IgG1 complex-type oligosaccharides shows the critical role of enhancing antibody-dependent cellular cytotoxicity. *J. Biol. Chem., 278*, 3466–3473.

26. Preithner, S., Elm, S., Lippold, S., Locher, M., Wolf, A., Da Silva, A. J., Baeuerle, P. A., & Prang, N. S., (2006). High concentrations of therapeutic IgG1 antibodies are needed to compensate for inhibition of antibody-dependent cellular cytotoxicity by excess endogenous immunoglobulin G. *Mol. Immunol., 43*, 1183–1193.

27. Ferrara, C., Stuart, F., Sondermann, P., Brunker, P., & Umana, P., (2006). The carbohydrate at FcgammaRIIIa Asn-162. An element required for high affinity binding to non-fucosylated IgG glycoforms. *J. Biol. Chem., 281*, 5032–5036.

28. Ferrara, C., Grau, S., Jager, C., Sondermann, P., Brunker, P., Waldhauer, I., Hennig, M., et al., (2011). Unique carbohydrate-carbohydrate interactions are required for high

affinity binding between FcgammaRIII and antibodies lacking core fucose. *Proc. Natl. Acad. Sci. U.S.A, 108*, 12669–12674.

29. Malhotra, R., Wormald, M. R., Rudd, P. M., Fischer, P. B., Dwek, R. A., & Sim, R. B., (1995). Glycosylation changes of IgG associated with rheumatoid arthritis can activate complement via the mannose-binding protein. *Nat. Med., 1*, 237–243.

30. Mihai, S., & Nimmerjahn, F., (2013). The role of Fc receptors and complement in autoimmunity. *Autoimmun. Rev., 12*, 657–660.

31. Nimmerjahn, F., & Ravetch, J. V., (2008). Fcgamma receptors as regulators of immune responses. *Nat. Rev. Immunol., 8*, 34–47.

32. Taniguchi, N., Yoshimura, M., Miyoshi, E., Ihara, Y., Nishikawa, A., Kang, R., & Ikeda, Y., (1998). Gene expression and regulation of N-acetylglucosaminyltransferases III and V in cancer tissues. *Adv. Enzyme Regul., 38*, 223–232.

33. Gu, J., Sato, Y., Kariya, Y., Isaji, T., Taniguchi, N., & Fukuda, T., (2009). A mutual regulation between cell-cell adhesion and N-glycosylation: Implication of the bisecting GlcNAc for biological functions. *J. Proteome Res., 8*, 431–435.

34. Ihara, Y., Nishikawa, A., Tohma, T., Soejima, H., Niikawa, N., & Taniguchi, N., (1993). cDNA cloning, expression, and chromosomal localization of human N-acetylglucosaminyltransferase III (GnT-III). *J. Biochem., 113*, 692–698.

35. Pinho, S. S., Reis, C. A., Paredes, J., Magalhaes, A. M., Ferreira, A. C., Figueiredo, J., Xiaogang, W., et al., (2009). The role of N-acetylglucosaminyltransferase III and V in the post-transcriptional modifications of E-cadherin. *Hum. Mol. Genet., 18*, 2599–2608.

36. Schachter, H., (1986). Biosynthetic controls that determine the branching and microheterogeneity of protein-bound oligosaccharides. *Adv. Exp. Med. Bio., 205*, 53–85.

37. Yoshimura, M., Nishikawa, A., Ihara, Y., Taniguchi, S., & Taniguchi, N., (1995). Suppression of lung metastasis of B16 mouse melanoma by N-acetylglucosaminyltransferase III gene transfection. *Proc. Natl. Acad. Sci. U.S.A, 92*, 8754–8758.

38. Song, Y., Aglipay, J. A., Bernstein, J. D., Goswami, S., & Stanley, P., (2010). The bisecting GlcNAc on N-glycans inhibits growth factor signaling and retards mammary tumor progression. *Cancer Res., 70*, 3361–3371.

39. Yoshimura, M., Ihara, Y., Ohnishi, A., Ijuhin, N., Nishiura, T., Kanakura, Y., Matsuzawa, Y., & Taniguchi, N., (1996). Bisecting N-acetylglucosamine on K562 cells suppresses natural killer cytotoxicity and promotes spleen colonization. *Cancer Res., 56*, 412–418.

40. https://www.omim.org/Entry/601774 (accessed on 10 December 2020).

41. Dennis, J. W., Taniguchi, N., & Pierce, M., (2014). Mannosyl (Alpha-1, 6-)-glycoprotein beta-1, 6-N-acetyl-glucosaminyltransferase (MGAT5). In: Taniguchi, N., Honke, K., Fukuda, M., Narimatsu, H., Yamaguchi, Y., & Angata, T., (eds.), *Handbook of Glycosyltransferases and Related Genes* (pp. 233–246). (Tokyo: Springer Japan).

42. Kizuka, Y., & Taniguchi, N., (2016). Enzymes for N-glycan branching and their genetic and nongenetic regulation in cancer. *Biomolecules, 6*.

43. Hanashima, S., Manabe, S., Inamori, K., Taniguchi, N., & Ito, Y., (2004). Synthesis of a bisubstrate-type inhibitor of N-acetylglucosaminyltransferases. *Angew. Chem. Int. Ed. Engl., 43*, 5674–5677.

44. https://www.omim.org/entry/602589 (accessed on 10 December 2020).

45. Ihara, H., Tsukamoto, H., Gu, J., Miyoshi, E., Taniguchi, N., & Ikeda, Y., (2014). Fucosyltransferase 8. GDP-fucose N-glycan core α6-fucosyltransferase (FUT8). In: Taniguchi, N., Honke, K., Fukuda, M., Narimatsu, H., Yamaguchi, Y., & Angata, T.,

(eds.), *Handbook of Glycosyltransferases and Related Genes* (pp. 581–596). (Tokyo: Springer Japan).

46. Noda, K., Miyoshi, E., Uozumi, N., Gao, C. X., Suzuki, K., Hayashi, N., Hori, M., & Taniguchi, N., (1998). High expression of alpha-1-6 fucosyltransferase during rat hepatocarcinogenesis. *Int. J. Cancer, 75*, 444–450.

47. Liu, Y. C., Yen, H. Y., Chen, C. Y., Chen, C. H., Cheng, P. F., Juan, Y. H., Khoo, K. H., et al., (2011). Sialylation and fucosylation of epidermal growth factor receptor suppress its dimerization and activation in lung cancer cells. *Proc. Natl. Acad. Sci. U.S.A, 108*, 11332–11337.

48. Schultz, M. J., Swindall, A. F., & Bellis, S. L., (2012). Regulation of the metastatic cell phenotype by sialylated glycans. *Cancer Metastasis Rev., 31*, 501–518.

49. Bull, C., Stoel, M. A., Den, B. M. H., & Adema, G. J., (2014). Sialic acids sweeten a tumor's life. *Cancer Res., 74*, 3199–3204.

50. Harduin-Lepers, A., Krzewinski-Recchi, M. A., Colomb, F., Foulquier, F., Groux-Degroote, S., & Delannoy, P., (2012). Sialyltransferases functions in cancers. *Front Biosci. (Elite Ed.), 4*, 499–515.

51. Natoni, A., Macauley, M. S., & O'dwyer, M. E., (2016). Targeting selectins and their ligands in cancer. *Front Oncol., 6*, 93.

52. Anugraham, M., Jacob, F., Nixdorf, S., Everest-Dass, A. V., Heinzelmann-Schwarz, V., & Packer, N. H., (2014). Specific glycosylation of membrane proteins in epithelial ovarian cancer cell lines: Glycan structures reflect gene expression and DNA methylation status. *Mol. Cell Proteomics, 13*, 2213–2232.

53. Vojta, A., Samarzija, I., Bockor, L., & Zoldos, V., (2016). Glyco-genes change expression in cancer through aberrant methylation. *Biochim. Biophys. Acta, 1860*, 1776–1785.

54. Saldova, R., Dempsey, E., Perez-Garay, M., Marino, K., Watson, J. A., Blanco-Fernandez, A., Struwe, W. B., et al., (2011). 5-AZA-2'-deoxycytidine induced demethylation influences N-glycosylation of secreted glycoproteins in ovarian cancer. *Epigenetics, 6*, 1362–1372.

55. Buckhaults, P., Chen, L., Fregien, N., & Pierce, M., (1997). Transcriptional regulation of N-acetylglucosaminyltransferase V by the src oncogene. *J. Biol. Chem., 272*, 19575–19581.

56. Chen, L., Zhang, W., Fregien, N., & Pierce, M., (1998). The her-2/neu oncogene stimulates the transcription of N-acetylglucosaminyltransferase V and expression of its cell surface oligosaccharide products. *Oncogene, 17*, 2087–2093.

57. Lauc, G., Essafi, A., Huffman, J. E., Hayward, C., Knezevic, A., Kattla, J. J., Polasek, O., et al., (2010). Genomics meets glycomics-the first GWAS study of human N-Glycome identifies HNF1alpha as a master regulator of plasma protein fucosylation. *PLoS Genet., 6*, e1001256.

58. Wang, M., Wang, J., Kong, X., Chen, H., Wang, Y., Qin, M., Lin, Y., et al., (2014). MiR-198 represses tumor growth and metastasis in colorectal cancer by targeting fucosyl transferase 8. *Sci. Rep., 4*, 6145.

59. Bernardi, C., Soffientini, U., Piacente, F., & Tonetti, M. G., (2013). Effects of microRNAs on fucosyltransferase 8 (FUT8) expressions in hepatocarcinoma cells. *PLoS One, 8*, e76540.

60. Boroughs, L. K., & Deberardinis, R. J., (2015). Metabolic pathways promoting cancer cell survival and growth. *Nat. Cell Biol., 17*, 351–359.

61. Nakajima, K., Kitazume, S., Angata, T., Fujinawa, R., Ohtsubo, K., Miyoshi, E., & Taniguchi, N., (2010). Simultaneous determination of nucleotide sugars with ion-pair reversed-phase HPLC. *Glycobiology, 20,* 865–871.

62. Gill, D. J., Tham, K. M., Chia, J., Wang, S. C., Steentoft, C., Clausen, H., Bard-Chapeau, E. A., & Bard, F. A., (2013). Initiation of GalNAc-type O-glycosylation in the endoplasmic reticulum promotes cancer cell invasiveness. *Proc. Natl. Acad. Sci. U.S.A, 110,* E3152–3161.

63. Sasai, K., Ikeda, Y., Ihara, H., Honke, K., & Taniguchi, N., (2003). Caveolin-1 regulates the functional localization of N-acetylglucosaminyltransferase III within the Golgi apparatus. *J. Biol. Chem., 278,* 25295–25301.

64. Wang, Z., Dabrosin, C., Yin, X., Fuster, M. M., Arreola, A., Rathmell, W. K., Generali, D., et al., (2015). Broad targeting of angiogenesis for cancer prevention and therapy. *Semin. Cancer Biol., 35,* S224–S243.

65. Pinho, S. S., & Reis, C. A., (2015). Glycosylation in cancer: Mechanisms and clinical implications. *Nat. Rev. Cancer, 15,* 540–555.

66. Akiyama, S. K., Olden, K., & Yamada, K. M., (1995). Fibronectin and integrins in invasion and metastasis. *Cancer Metastasis Rev., 14,* 173–189.

67. Pantel, K., Schlimok, G., Angstwurm, M., Passlick, B., Izbicki, J. R., Johnson, J. P., & Riethmuller, G., (1995). Early metastasis of human solid tumors: Expression of cell adhesion molecules. *Ciba Found. Symp., 189,* 157–170.

68. Stetler-Stevenson, W. G., Aznavoorian, S., & Liotta, L. A., (1993). Tumor cell interactions with the extracellular matrix during invasion and metastasis. *Annu. Rev. Cell Biol., 9,* 541–573.

69. Ruoslahti, E., (1996). How cancer spreads. *Sci. Am., 275,* 72–77.

70. Tedder, T. F., Steeber, D. A., Chen, A., & Engel, P., (1995). The selectins: Vascular adhesion molecules. *FASEB J., 9,* 866–873.

71. Kansas, G. S., (1996). Selectins and their ligands: Current concepts and controversies. *Blood, 88,* 3259–3287.

72. Schachter, H., (1994). In: Fukuda, M., & Hindsgaul, O., (eds.), *Molecular Glycobiology* (pp. 88–162). Oxford University Press, New York.

73. Dennis, J. W., Laferte, S., Waghorne, C., Breitman, M. L., & Kerbel, R. S., (1987). Beta 1-6 branching of Asn-linked oligosaccharides is directly associated with metastasis. *Science, 236,* 582–585.

74. Dennis, J. W., & Laferte, S., (1988). Asn-linked oligosaccharides and the metastatic phenotype. In: Reading, C. L., Hakomori, S., & Marcus, D. M., (eds.), *Altered Glycosylation in Tumor Cells* (pp. 257–267). Alan R. Liss, New York.

75. Easton, E. W., Bolscher, J. G., & Van, D. E. D. H., (1991). Enzymatic amplification involving glycosyltransferases forms the basis for the increased size of asparagine-linked glycans at the surface of NIH 3T3 cells expressing the N-ras proto-oncogene. *J. Biol. Chem., 266,* 21674–21680.

76. Fernandes, B., Sagman, U., Auger, M., Demetrio, M., & Dennis, J. W., (1991). Beta 1-6 branched oligosaccharides as a marker of tumor progression in human breast and colon neoplasia. *Cancer Res., 51,* 718–723.

77. Yousefi, S., Higgins, E., Daoling, Z., Pollex-Kruger, A., Hindsgaul, O., & Dennis, J. W., (1991). Increased UDP-GlcNAc: Gal beta 1-3GalNAc-R (GlcNAc to GaLNAc) beta-1, 6-N-acetylglucosaminyltransferase activity in metastatic murine tumor cell lines. Control of polylactosamine synthesis. *J. Biol. Chem., 266,* 1772–1782.

78. Saitoh, O., Wang, W. C., Lotan, R., & Fukuda, M., (1992). Differential glycosylation and cell surface expression of lysosomal membrane glycoproteins in sublines of a human colon cancer exhibiting distinct metastatic potentials. *J. Biol. Chem., 267*, 5700–5711.

79. Matsuura, H., Takio, K., Titani, K., Greene, T., Levery, S. B., Salyan, M. E., & Hakomori, S., (1988). The oncofetal structure of human fibronectin defined by monoclonal antibody FDC-6. Unique structural requirement for the antigenic specificity provided by a glycosylhexapeptide. *J. Biol. Chem., 263*, 3314–3322.

80. Hakomori, S., (1989). Aberrant glycosylation in tumors and tumor-associated carbohydrate antigens. *Adv. Cancer Res., 52*, 257–331.

81. Dennis, J. W., (1992). *Oligosaccharides in Carcinogenesis and Metastasis* (pp. 1–3). In GlycoNews, II, Oxford GlycoSystems, Oxford.

82. Muramatsu, T., (1993). Carbohydrate signals in metastasis and prognosis of human carcinomas. *Glycobiology, 3*, 291–296.

83. Lowe, J. B., (1994). Specificity and expression of carbohydrate ligands. In: Wegner, C., (eds.), *Adhesion Molecules* (pp. 111–133). Academic Press, London.

84. Hakomori, S., Nudelman, E., Levery, S. B., & Kannagi, R., (1984). Novel fucolipids accumulating in human adenocarcinoma. I. Glycolipids with di- or trifucosylated type 2 chain. *J. Biol. Chem., 259*, 4672–4680.

85. Ladisch, S., Sweeley, C. C., Becker, H., & Gage, D., (1989). Aberrant fatty acyl alpha-hydroxylation in human neuroblastoma tumor gangliosides. *J. Biol. Chem., 264*, 12097–12105.

86. Irimura, T., Nakamori, S., Matsushita, Y., Taniuchi, Y., Todoroki, N., Tsuji, T., Izumi, Y., et al., (1993). Colorectal cancer metastasis determined by carbohydrate-mediated cell adhesion: Role of sialyl-LeX antigens. *Semin. Cancer Biol., 4*, 319–324.

87. Sackstein, R., (2011). The biology of CD44 and HCELL in hematopoiesis: The 'step 2-bypass pathway' and other emerging perspectives. *Curr. Opin. Hematol., 18*, 239–248.

88. Lasky, L. A., (1992). Selectins: Interpreters of cell-specific carbohydrate information during inflammation. *Science, 258*, 964–969.

89. Lasky, L. A., (1995). Selectin-carbohydrate interactions and the initiation of the inflammatory response. *Annu. Rev. Biochem., 64*, 113–139.

90. Austrup, F., Vestweber, D., Borges, E., Lohning, M., Brauer, R., Herz, U., Renz, H., et al., (1997). P- and E-selectin mediate recruitment of T-helper-1 but not T-helper-2 cells into inflamed tissues. *Nature, 385*, 81–83.

91. Mitoma, J., Bao, X., Petryanik, B., Schaerli, P., Gauguet, J. M., Yu, S. Y., Kawashima, H., et al., (2007). Critical functions of N-glycans in L-selectin-mediated lymphocyte homing and recruitment. *Nat. Immunol., 8*, 409–418.

92. St Hill, C. A., (2011). Interactions between endothelial selectins and cancer cells regulate metastasis. *Front Biosci. (Landmark Ed), 16*, 3233–3251.

93. Genbacev, O. D., Prakobphol, A., Foulk, R. A., Krtolica, A. R., Ilic, D., Singer, M. S., Yang, Z. Q., et al., (2003). Trophoblast L-selectin-mediated adhesion at the maternal-fetal interface. *Science, 299*, 405–408.

94. Hilkens, J., Ligtenberg, M. J., Vos, H. L., & Litvinov, S. V., (1992). Cell membrane-associated mucins and their adhesion-modulating property. *Trends Biochem. Sci., 17*, 359–363.

95. Kim, Y. S., Gum, J. Jr., & Brockhausen, I., (1996). Mucin glycoproteins in neoplasia. *Glycoconj. J., 13*, 693–707.

96. Ligtenberg, M. J., Buijs, F., Vos, H. L., & Hilkens, J., (1992). Suppression of cellular aggregation by high levels of episialin. *Cancer Res., 52*, 2318–2324.

97. Moriarty, J., Skelly, C. M., Bharathan, S., Moody, C. E., & Sherblom, A. P., (1990). Sialomucin and lytic susceptibility of rat mammary tumor ascites cells. *Cancer Res., 50*, 6800–6805.

98. Chachadi, V. B., Cheng, H., Klinkebiel, D., Christman, J. K., & Cheng, P. W., (2011). 5-Aza-2'-deoxycytidine increases sialyl Lewis X on MUC1 by stimulating beta-galactoside: alpha2, 3-sialyltransferase 6 gene. *Int. J. Biochem. Cell Biol., 43*, 586–593.

99. Andrianifahanana, M., Moniaux, N., Schmied, B. M., Ringel, J., Friess, H., Hollingsworth, M. A., Buchler, M. W., et al., (2001). Mucin (MUC) gene expression in human pancreatic adenocarcinoma and chronic pancreatitis: A potential role of MUC4 as a tumor marker of diagnostic significance. *Clin. Cancer Res., 7*, 4033–4040.

100. Nath, S., Daneshvar, K., Roy, L. D., Grover, P., Kidiyoor, A., Mosley, L., Sahraei, M., & Mukherjee, P., (2013). MUC1 induces drug resistance in pancreatic cancer cells via upregulation of multidrug resistance genes. *Oncogenesis, 2*, e51.

101. Heidemann, F., Schildt, A., Schmid, K., Bruns, O. T., Riecken, K., Jung, C., Ittrich, H., et al., (2014). Selectins mediate small cell lung cancer systemic metastasis. *PLoS One, 9*, e92327.

102. Ben-David, T., Sagi-Assif, O., Meshel, T., Lifshitz, V., Yron, I., & Witz, I. P., (2008). The involvement of the sLe-a selectin ligand in the extravasation of human colorectal carcinoma cells. *Immunol. Lett., 116*, 218–224.

103. Geng, Y., Yeh, K., Takatani, T., & King, M. R., (2012). Three to tango: MUC1 as a ligand for both E-selectin and ICAM-1 in the breast cancer metastatic cascade. *Front Oncol., 2*, 76.

104. Kannagi, R., Izawa, M., Koike, T., Miyazaki, K., & Kimura, N., (2004). Carbohydrate-mediated cell adhesion in cancer metastasis and angiogenesis. *Cancer Sci., 95*, 377–384.

105. Kim, Y. J., & Varki, A., (1997). Perspectives on the significance of altered glycosylation of glycoproteins in cancer. *Glycoconj. J., 14*, 569–576.

106. Stoolman, L. M., & Kaldjian, E., (1992). Adhesion molecules involved in the trafficking of normal and malignant leukocytes. *Invasion Metastasis, 12*, 101–111.

107. Onrust, S. V., Hartl, P. M., Rosen, S. D., & Hanahan, D., (1996). Modulation of L-selectin ligand expression during an immune response accompanying tumorigenesis in transgenic mice. *J. Clin. Invest., 97*, 54–64.

108. Bedard, P. W., & Kaila, N., (2010). Selectin inhibitors: A patent review. *Expert. Opin. Ther. Pat., 20*, 781–793.

109. Homeister, J. W., Thall, A. D., Petryniak, B., Maly, P., Rogers, C. E., Smith, P. L., Kelly, R. J., et al., (2001). The alpha(1, 3)fucosyltransferases FucT-IV and FucT-VII exert collaborative control over selectin-dependent leukocyte recruitment and lymphocyte homing. *Immunity, 15*, 115–126.

110. Smith, P. L., Myers, J. T., Rogers, C. E., Zhou, L., Petryniak, B., Becker, D. J., Homeister, J. W., & Lowe, J. B., (2002). Conditional control of selectin ligand expression and global fucosylation events in mice with a targeted mutation at the FX locus. *J. Cell Biol., 158*, 801–815.

111. Etzioni, A., Frydman, M., Pollack, S., Avidor, I., Phillips, M. L., Paulson, J. C., & Gershoni-Baruch, R., (1992). Brief report: Recurrent severe infections caused by a novel leukocyte adhesion deficiency. *N. Engl. J. Med., 327*, 1789–1792.

112. Dennis, J. W., Nabi, I. R., & Demetriou, M., (2009). Metabolism, cell surface organization, and disease. *Cell, 139*, 1229–1241.

113. Tinoco, R., Otero, D. C., Takahashi, A. A., & Bradley, L. M., (2017). PSGL-1: A new player in the immune checkpoint landscape. *Trends Immunol., 38*, 323–335.

114. Kieffer, J. D., Fuhlbrigge, R. C., Armerding, D., Robert, C., Ferenczi, K., Camphausen, R. T., & Kupper, T. S., (2001). Neutrophils, monocytes, and dendritic cells express the same specialized form of PSGL-1 as do skin-homing memory T-cells: Cutaneous lymphocyte antigen. *Biochem. Biophys. Res. Commun., 285*, 577–587.

115. Rossi, F. M., Corbel, S. Y., Merzaban, J. S., Carlow, D. A., Gossens, K., Duenas, J., So, L., et al., (2005). Recruitment of adult thymic progenitors is regulated by P-selectin and its ligand PSGL-1. *Nat. Immunol., 6*, 626–634.

116. Wagers, A. J., & Kansas, G. S., (2000). Potent induction of alpha(1,3)-fucosyltransferase VII in activated CD4+ T cells by TGF-beta 1 through a p38 mitogen-activated protein kinase-dependent pathway. *J. Immunol., 165*, 5011–5016.

117. Carlow, D. A., Williams, M. J., & Ziltener, H. J., (2005). Inducing P-selectin ligand formation in CD8 T cells: IL-2 and IL-12 are active *in vitro* but not required *in vivo*. *J. Immunol., 174*, 3959–3966.

118. Nunez-Andrade, N., Lamana, A., Sancho, D., Gisbert, J. P., Gonzalez-Amaro, R., Sanchez-Madrid, F., & Urzainqui, A., (2011). P-selectin glycoprotein ligand-1 modulates immune inflammatory responses in the enteric lamina propria. *J. Pathol., 224*, 212–221.

119. Tinoco, R., Carrette, F., Barraza, M. L., Otero, D. C., Magana, J., Bosenberg, M. W., Swain, S. L., & Bradley, L. M., (2016). PSGL-1 Is an immune checkpoint regulator that promotes T-cell exhaustion. *Immunity, 44*, 1190–1203.

120. Jeanes, A., Gottardi, C. J., & Yap, A. S., (2008). Cadherins and cancer: How does cadherin dysfunction promote tumor progression? *Oncogene, 27*, 6920–6929.

121. Frixen, U. H., Behrens, J., Sachs, M., Eberle, G., Voss, B., Warda, A., Lochner, D., & Birchmeier, W., (1991). E-cadherin-mediated cell-cell adhesion prevents invasiveness of human carcinoma cells. *J. Cell Biol., 113*, 173–185.

122. Oka, H., Shiozaki, H., Kobayashi, K., Inoue, M., Tahara, H., Kobayashi, T., Takatsuka, Y., Matsuyoshi, N., Hirano, S., Takeichi, M., et al., (1993). Expression of E-cadherin cell adhesion molecules in human breast cancer tissues and its relationship to metastasis. *Cancer Res., 53*, 1696–1701.

123. Gabbert, H. E., Mueller, W., Schneiders, A., Meier, S., Moll, R., Birchmeier, W., & Hommel, G., (1996). Prognostic value of E-cadherin expression in 413 gastric carcinomas. *Int. J. Cancer, 69*, 184–189.

124. Takeichi, M., (1991). Cadherin cell adhesion receptors as a morphogenetic regulator. *Science, 251*, 1451–1455.

125. Shiozaki, H., Oka, H., Inoue, M., Tamura, S., & Monden, M., (1996). E-cadherin mediated adhesion system in cancer cells. *Cancer, 77*, 1605–1613.

126. Jamal, B. T., Nita-Lazar, M., Gao, Z., Amin, B., Walker, J., & Kukuruzinska, M. A., (2009). N-glycosylation status of E-cadherin controls cytoskeletal dynamics through the organization of distinct beta-catenin- and gamma-catenin-containing AJs. *Cell Health Cytoskelet.*, 67–80.

127. Pinho, S. S., Figueiredo, J., Cabral, J., Carvalho, S., Dourado, J., Magalhaes, A., Gartner, F., et al., (2013). E-cadherin and adherens-junctions stability in gastric carcinoma: Functional implications of glycosyltransferases involving N-glycan branching

biosynthesis, N-acetylglucosaminyltransferases III and V. *Biochim. Biophys. Acta, 1830*, 2690–2700.

128. Inohara, H., & Raz, A., (1994). Effects of natural complex carbohydrate (citrus pectin) on murine melanoma cell properties related to galectin-3 functions. *Glycoconj. J., 11*, 527–532.

129. Nangiamakker, P., Thompson, E., Hogan, C., Ochieng, J., & Raz, A., (1995). Induction of tumorigenicity by galectin-3 in a nontumorigenic human breast-carcinoma cell-line. *Int. J. Oncol., 7*, 1079–1087.

130. Wang, L., Inohara, H., Pienta, K. J., & Raz, A., (1995). Galectin-3 is a nuclear matrix protein which binds RNA. *Biochem. Biophys. Res. Commun., 217*, 292–303.

131. Rambaruth, N. D., & Dwek, M. V., (2011). Cell surface glycan-lectin interactions in tumor metastasis. *Acta Histochem., 113*, 591–600.

132. Nadanaka, S., Sato, C., Kitajima, K., Katagiri, K., Irie, S., & Yamagata, T., (2001). Occurrence of oligosialic acids on integrin alpha 5 subunit and their involvement in cell adhesion to fibronectin. *J. Biol. Chem., 276*, 33657–33664.

133. Kariya, Y., Kato, R., Itoh, S., Fukuda, T., Shibukawa, Y., Sanzen, N., Sekiguchi, K., et al., (2008). N-Glycosylation of laminin-332 regulates its biological functions. A novel functions of the bisecting GlcNAc. *J. Biol. Chem., 283*, 33036–33045.

134. Colognato, H., & Yurchenco, P. D., (2000). Form and function: The laminin family of heterotrimers. *Dev. Dyn., 218*, 213–234.

135. Carter, W. G., & Wayner, E. A., (1988). Characterization of the class III collagen receptor, a phosphorylated, transmembrane glycoprotein expressed in nucleated human cells. *J Biol. Chem., 263*, 4193–4201.

136. Nehls, V., & Hayen, W., (2000). Are hyaluronan receptors involved in three-dimensional cell migration? *Histol. Histopathol., 15*, 629–636.

137. Kleeff, J., Ishiwata, T., Kumbasar, A., Friess, H., Buchler, M. W., Lander, A. D., & Korc, M., (1998). The cell-surface heparan sulfate proteoglycan glypican-1 regulates growth factor action in pancreatic carcinoma cells and is overexpressed in human pancreatic cancer. *J. Clin. Invest., 102*, 1662–1673.

138. DeBaun, M. R., Ess, J., & Saunders, S., (2001). Simpson Golabi Behmel syndrome: Progress toward understanding the molecular basis for overgrowth, malformation, and cancer predisposition. *Mol. Genet. Metab., 72*, 279–286.

139. Bousseau, S., Vergori, L., Soleti, R., Lenaers, G., Martinez, M. C., & Andriantsitohaina, R., (2018). Glycosylation as new pharmacological strategies for diseases associated with excessive angiogenesis. *Pharmacol. Ther.*

140. Thijssen, V. L., Postel, R., Brandwijk, R. J., Dings, R. P., Nesmelova, I., Satijn, S., Verhofstad, N., et al., (2006). Galectin-1 is essential in tumor angiogenesis and is a target for antiangiogenesis therapy. *Proc. Natl. Acad. Sci. U.S.A, 103*, 15975–15980.

141. Mathieu, V., De Lassalle, E. M., Toelen, J., Mohr, T., Bellahcene, A., Van, G. G., Verschuere, T., et al., (2012). Galectin-1 in melanoma biology and related neo-angiogenesis processes. *J. Invest. Dermatol., 132*, 2245–2254.

142. Croci, D. O., Cerliani, J. P., Dalotto-Moreno, T., Mendez-Huergo, S. P., Mascanfroni, I. D., Dergan-Dylon, S., Toscano, M. A., et al., (2014). Glycosylation-dependent lectin-receptor interactions preserve angiogenesis in anti-VEGF refractory tumors. *Cell, 156*, 744–758.

143. Fuster, M. M., & Wang, L., (2010). Endothelial heparan sulfate in angiogenesis. *Prog. Mol. Biol. Transl. Sci., 93*, 179–212.

144. Cheng, W. K., & Oon, C. E., (2018). How glycosylation aids tumor angiogenesis: An updated review. *Biomed. Pharmacother., 103*, 1246–1252.

145. Sato, Y., Nakata, K., Kato, Y., Shima, M., Ishii, N., Koji, T., Taketa, K., et al., (1993). Early recognition of hepatocellular carcinoma based on altered profiles of alpha-fetoprotein. *N. Engl. J. Med., 328*, 1802–1806.

146. Steentoft, C., Vakhrushev, S. Y., Vester-Christensen, M. B., Schjoldager, K. T., Kong, Y., Bennett, E. P., Mandel, U., et al., (2011). Mining the O-glycoproteome using zinc-finger nuclease-glycoengineered simple cell lines. *Nat. Methods, 8*, 977–982.

147. Campos, D., Freitas, D., Gomes, J., Magalhaes, A., Steentoft, C., Gomes, C., Vester-Christensen, M. B., et al., (2015). Probing the O-glycoproteome of gastric cancer cell lines for biomarker discovery. *Mol. Cell Proteomics, 14*, 1616–1629.

148. Gomes, C., Almeida, A., Ferreira, J. A., Silva, L., Santos-Sousa, H., Pinto-De-Sousa, J., Santos, L. L., et al., (2013). Glycoproteomic analysis of serum from patients with gastric precancerous lesions. *J. Proteome Res., 12*, 1454–1466.

149. Melo, S. A., Luecke, L. B., Kahlert, C., Fernandez, A. F., Gammon, S. T., Kaye, J., Lebleu, V. S., et al., (2015). Glypican-1 identifies cancer exosomes and detects early pancreatic cancer. *Nature, 523*, 177–182.

150. Pedersen, J. W., Gentry-Maharaj, A., Nostdal, A., Fourkala, E. O., Dawnay, A., Burnell, M., Zaikin, A., et al., (2014). Cancer-associated autoantibodies to MUC1 and MUC4-a blinded case-control study of colorectal cancer in UK collaborative trial of ovarian cancer screening. *Int. J. Cancer, 134*, 2180–2188.

151. Borsig, L., Wong, R., Feramisco, J., Nadeau, D. R., Varki, N. M., & Varki, A., (2001). Heparin and cancer revisited: Mechanistic connections involving platelets, P-selectin, carcinoma mucins, and tumor metastasis. *Proc. Natl. Acad. Sci. U.S.A, 98*, 3352–3357.

152. Noujaim, A. A., Schultes, B. C., Baum, R. P., & Madiyalakan, R., (2001). Induction of CA125-specific B and T-cell responses in patients injected with MAb-B43.13-evidence for antibody-mediated antigen-processing and presentation of CA125 *in vivo. Cancer Biother. Radiopharm., 16*, 187–203.

153. Musselli, C., Ragupathi, G., Gilewski, T., Panageas, K. S., Spinat, Y., & Livingston, P. O., (2002). Reevaluation of the cellular immune response in breast cancer patients vaccinated with MUC1. *Int. J. Cancer, 97*, 660–667.

154. Goss, P. E., Baptiste, J., Fernandes, B., Baker, M., & Dennis, J. W., (1994). A phase I study of Swainsonine in patients with advanced malignancies. *Cancer Res., 54*, 1450–1457.

155. Liekens, S., Leali, D., Neyts, J., Esnouf, R., Rusnati, M., Dell'era, P., Maudgal, P. C., et al., (1999). Modulation of fibroblast growth factor-2 receptor binding, signaling, and mitogenic activity by heparin-mimicking polysulfonated compounds. *Mol. Pharmacol., 56*, 204–213.

# CHAPTER 6

# Glycans in the Host-Pathogen Interaction

MUZAFAR JAN[1] and SUNIL K. ARORA[2]

[1]*Assistant Professor, Department of Biochemistry GDC Dooru, University of Kashmir, Srinagar, Jammu and Kashmir (UT) – 192211, India, E-mail: muzijan@gmail.com*

[2]*Professor, Department of Immunopathology, Postgraduate Institute of Medical Education and Research (PGIMER), Chandigarh (UT) – 160012, India*

## ABSTRACT

Living organisms on this earth have evolved to use carbohydrates for most of their biological processes, such as energy source for survival, structural framework, and biological functions. Glycans have many biochemical, structural, and functional features that make them ideal to be universally used by living cells in energy metabolism, cellular interactions, and the formation of protective physical barriers against the outside environment. The glycans being ubiquitous constituents of all the cell surfaces and the surrounding cellular environments, are not only important for the biological processes of the host but also for the binding of pathogens to them. Pathogens are in constant contact with their host, and overtime have co-evolved traits to explore the host glycans for their survival as well as pathogenesis. Pathogens belonging to diverse dimensions of life like bacteria, fungi, helminths, protozoans, and viruses use different types of glycans like negatively charged sialic acid, oligomannose glycans, and glycosaminoglycans (GAGs) to establish host-pathogen relations. The enormous diversity, abundance, and density of the host glycans are subjected to tremendous immune selection pressure to evade the constantly targeting and more rapidly evolving pathogens that infect them. This immense selection pressure leads to development of diverse glycan expression patterns on the host cells without compromising their own survival. This may attribute to the significant structural variation of glycans in nature, which add to biological diversity and to speciation as well. Furthermore, the glycans also play a critical role in antigenicity and virulence in these host-pathogen interactions and are therefore considered as potential drug targets.

## 6.1   INTRODUCTION

Among the four most abundant biologically important biomolecules constituting living organism, carbohydrates occupy a prominent position as a major component found in all life forms. The organic matter on earth is mostly composed of carbohydrates which find their places in almost all the biological processes needed by living organisms for their survival such as fuels, structural framework, stores, and metabolic intermediates. Carbohydrates are omnipresent within the biological sphere and form structural framework of nucleic acids as deoxyribose and ribose sugars, are principle structural components of bacterial and plant cell walls. Carbohydrates also combine with other biomolecules like proteins and lipids to form glycoproteins and glycolipids predominant entities in the extracellular matrix. In nature, the carbohydrates exist as polymers of diverse molecular weights composed of smaller constituents called monosaccharides and their derivatives. The carbohydrate polymers may exist as homo-polysaccharides (i.e., of only one kind of monosaccharide units) or hetero-polysaccharides (i.e., of more than one kind of monosaccharide units). Besides the central role of carbohydrates in energy metabolism, they also form building blocks of the organism and are mainly distributed on the cell surface and cell-matrix interface in the form of conjugates with proteins and lipids, which include glycoproteins, proteoglycans, and glycolipids. During the post-translational and co-translational modifications of the proteins within the endoplasmic reticulum (ER) and Golgi, the proteins are modified with the attachment of glycans to generate glycoproteins, which influences their biological properties with enhanced functional diversity. Glycolipids, carbohydrates attached to lipids, also play an important role in cellular recognition. As mentioned above, the close association in function and structure with other cellular macromolecules, some of the concepts involved in glycosylation, are also common to lipidomics, metabolomics, and proteomics [1–3]. However, as a field, glycomics is the least explored and lags behind genomics and proteomics, owing to the intrinsic complications in isolation of glycans, the subsequent determination of their structure and function and the availability of a restricted number of analytical methods for use in their study.

Glycans are the most diverse of the proteins, nucleic acids, glycans, and lipids, the fundamental building blocks of cellular life. Although there are many examples where cells function without nucleus, but there is not a single instance of living cells described to function without surface glycans. Anything approaching the living cell, being it a microorganism, a protein, or another cell, in order to have access to the cell has to interact with the glycan coat of the cell [4]. Linear or branched glycans are formed by combining monosaccharide units via α- or β-glycosidic bonds in which hydroxyl groups of adjacent monosaccharides undergo condensation reaction with loss of water molecule. The entire repertoire of glycans within mammals is assembled from a group of ten precursors which combine to generate enormous diversity of glycans. The stereochemistry of the anomeric carbon formed during the glycosidic bond (α or β) also contributes to the complexity of carbohydrates. For example, the possible oligosaccharide isomers of a hexasaccharide are estimated to be approximately around $10^{12}$ structures [5]. In addition to diversity in monosaccharide precursors and types of linkage, there are other mechanisms of generation of carbohydrate diversity within living cells, including covalent modifications of the monosaccharide units such as sulfation, methylation, phosphorylation, and in some cases, the addition of amino acids [6, 7]. Together these mechanisms generate an enormously diverse and complex repertoire of glycans in living organisms. The glycans may remain attached to proteins (proteoglycans and glycoproteins), lipids (glycolipids), or remain free (milk oligosaccharides and hyaluronan). They together form complex arrays of glycans, which shroud the surface of all living cells in the form of membrane-bound glycolipids and glycoproteins, while the extracellular matrix surrounding the cells predominantly has glycosaminoglycans (GAGs), mucins, and other forms of proteoglycans. All the cellular interactions with other cells or microorganisms take place in this highly complex glycosylated environment of the living cells.

Briefly, the present chapter will summarize the important glycan types present and discuss the involvement of different glycans in the realm of host-pathogen interactions. According to the chemical structure of glycan core, the type of glycan linkage and the type of molecule a glycan is linked to glycans and glycoconjugates can be categorized in different classes. As mentioned already that oligosaccharides are covalently attached to proteins by a process known as protein glycosylation to form glycoproteins and there are two types of linkages through which glycans are attached to the resulting glycoproteins: N- and O-linked glycans (Figure 6.1). Glycans are covalently linked to the asparagine (N) amino acid via β-N-glycosidic bond within the sequence N-X-S/T (Asparagine-any amino acid (X)-serine/threonine) to form N-linked glycans (NLGs). The N-X-S/T is encoded by the genome of the organism and all NLGs have a conserved Man3GlcNAc2β-NXS/T core structure [8–10]. The O-linked glycans form O-glycosidic bond with the oxygen atom of serine (S) or threonine (T) amino acids. The most common of the O-linked glycans is mucin type or GalNAcα-S/T type. Non-mucin type O-linked glycans include GlcNAcβ-S/T (N-acetylglucosamine), α-linked O-mannose, β-linked O-xylose, α-linked O-fucose, α-/β-linked O-galactose, and α-/β-linked O-glucose glycans [11–14]. In addition to N- and O-linked glycoproteins, proteins are also modified with GAGs. The repeating disaccharide unit of GAGs is linear acid/galactose residue and an amino sugar. Proteins modified with GAGs are commonly referred to as proteoglycans (Figure 6.2) [4, 14, 15]. Similarly, as for protein-linked glycosylation, there are also different types of lipid-linked glycosylation modifications, the glycosphingolipids [4, 16] and the glycophospholipid anchors or glycosylphosphatidylinositol (GPI) anchors [4, 17] represent important glycolipid classes (Figure 6.3). Despite differences in their linkages to carrier molecules and core structures, most glycan types have conserved structural characteristics as they follow partially overlapping biosynthetic pathways. Despite the fact that few glycan features may be entirely found in one particular glycan class, numerous (sub) terminal glycan modifications can be observed

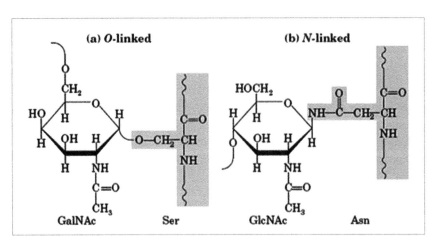

**FIGURE 6.1** Types of linkage in glycoproteins O- and N-linked glycans [N-glycans are added to conserve Asn-X-Ser/Thr amino acid sequences and the final maturation products of glycosylation modification enzymes are if three types: oligomannose, complex, and hybrid glycans with a common core in all of Man₃GlcNAc₂Asn. O-linked glycoproteins are formed by covalent α-glycosidic linkage of glycans with the -OH group of serine or threonine via N-acetylgalactosamine (GalNAc)].

in different glycan classes. Common (sub) terminal modifications involve poly-N-acetyl-lactosamine chains, ABH, and Lewis histo-blood group antigens (HBGA), and sialic acids in various linkages [4, 15]. Therefore, the glycan components of glycoproteins and glycolipids frequently have much in common than one would anticipate based on their core structure.

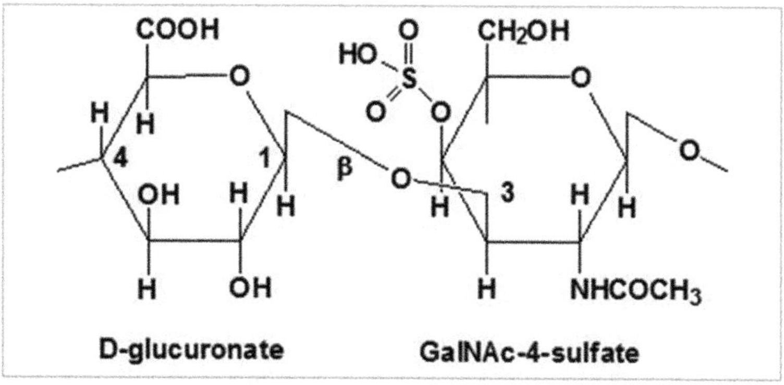

**FIGURE 6.2** Basic disaccharide unit of glycosaminoglycans [GAGs consisting of uronic acid and amino sugar. Many proteins are conjugated to glycosaminoglycans called proteoglycans. Glycosaminoglycans are linear polymers of repeating disaccharide units of uronic acid/galactose residue and an amino sugar. They are usually negatively charged polymers of carbohydrates].

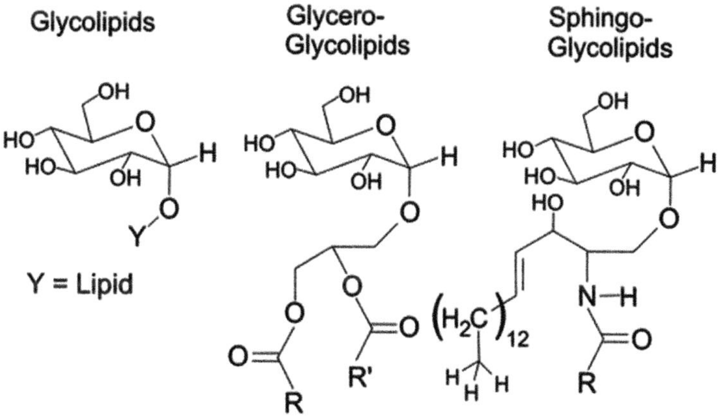

**FIGURE 6.3** Basic glycolipids core structure [Glycolipids comprise a glycan moiety (monosaccharide or oligosaccharide) glycosidically attached to the primary -OH group of either glycerol or sphingosine backbone giving rise to two main categories of glycolipids, glyceroglycolipids, and sphingolipids].

The glycan biosynthetic machinery which synthesizes, matures, and diversifies glycans both on secreted molecules as well as on cell surfaces, operate within the ER and Golgi compartments, the cellular secretory pathway. The glycosylation biosynthetic process is a highly ordered assembly, template-independent, and is subjected to multiple sequential and competitive enzymatic pathways. The final products of glycosylation biosynthetic machinery depend on the presence and availability of the biosynthetic pathway enzymes, the types of cells, and the structure of the protein. The glycan patterns on cells are highly difficult to predict from the gene expression patterns alone due to non-template nature of synthesis of glycans. As a response to small variations in the extracellular and intra-cellular environment and events there is a dynamic change in the glycan composition of the cells. Regardless of this, there is a distinct expression pattern of glycans on each cell type in each organism and these expression patterns tend to remain conserved within a species. Furthermore, these cell-type-specific glycan expression patterns are susceptible to dramatic and stereotypic species-specific changes in the course of development suggesting that the expression patterns of glycans are under rigorous regulatory control. For example, the evolutionarily conserved switch from peanut-agglutinin negative to positive during thymic development of T-cells.

Glycans with their universal presence on the cell surfaces and cell surroundings are not only involved to perform some of the important biological functions of the host but also for the binding of pathogens to them. The glycans are key partners in the establishment of protective physical barriers against the outside environment, mediating cell-cell and cell-matrix interactions, or regulating intracellular signaling through organization of membrane receptors. Throughout evolution, opportunistic pathogens have developed both the ability to target glycan structures on host cells to facilitate infection as well as to adopt the host glycosylation machinery to acquire stealth, enabling them to evade immune surveil-lance. The host glycans due to their abundance and diversity remain under constant immune selection pressure in order to sustain the constantly invading pathogen infections. The pathogens evolve more rapidly than their host does by changing their pattern of glycan expression without compromising their own survival. This may attribute to the enormous diversity of the glycans in nature and the host-pathogen interactions which may also contribute even to speciation. Glycans evolve like all other biological molecules by neutral processes, natural, and/or sexual selection. Glycan evolution is rapidly driven by infectious agents that constantly invade the host and rapid evolutionary changes have been also suggested to mediate speciation. As this field of research is growing day by day, new concepts and knowledge about the novel roles of glycans is emerging. The role of glycans present on cell surfaces and in extracellular matrix to mediate host-pathogen interactions, cell-matrix, and cell-cell interactions are well-known, intracellular glycans also perform important functions, for example, can serve as dynamic regulatory switches, to compete with protein modifications such as phosphorylation. At molecular level, there is a universal requirement for survival of all organisms is the capacity to differentiate between self and non-self, a fundamental tenet of immune recognition. Tolerance against self and rejection of the non-self-structures is a common phenomenon and at the molecular level, appears to rely on the interaction of surface components of the organisms that come in contact with the host cells. The host and its pathogens interact in a constant co-evolution process influencing each other. In terms of glycans, just as the host develops highly diverse arrays of glycans on their outermost cell surface, so are oligosaccharides and polysaccharides found on the surface of all bacteria, viruses, and other pathogens. Hence, almost all interactions of pathogens with their hosts are controlled to a crucial level by the arrangement of glycans and glycan-binding receptors that each expresses [1].

## 6.2   GLYCANS IN HOST-PATHOGEN INTERACTIONS

The cellular surface and extracellular matrix primarily contain glycans and glycoconjugates as main constituents. The universal distribution of these glycans decorating the cell surfaces of eukaryotes as well as prokaryotes makes them ideal targets of interaction for organisms approaching the cells, including pathogenic organisms. Besides being utilized by pathogens for interaction with host cells, the glycans influence the host-pathogen interactions in many ways; such as being utilized as ligands or receptors to access cellular entry, molecular mimicry by pathogens to evade host defense mechanisms, mediation of lateral movements of pathogens to enhance pathogenesis; and in being used as decoys as a prevention mechanism of infection by host cells (Figures 6.4 and 6.5).

**A.  Nontypeable *Haemophilus influenzae***

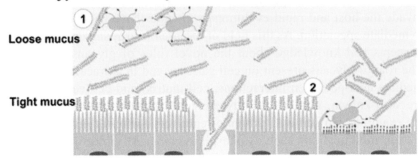

**B.  Influenza A virus**

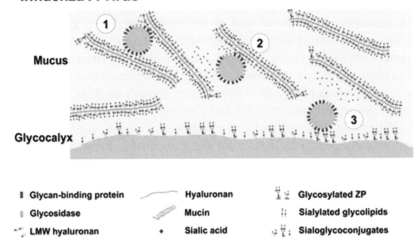

**FIGURE 6.4**  (A-1) Nontypeable *Haemophilus influenzae* (NTHi) binds glycosylated mucus and remains in the loose mucus layer; (**A-2**) Impaired mucus clearance due to ciliary, or tissue damage enables NTHi binding to sialylated glycolipids on host cells; (**B-1**) Influenza A virus bearing hemagglutinin (HA) and neuraminidase (NA) binds to host mucins; (**B-2**) The virus can free itself from mucins by cleaving sialic acids; (**B-3**) and binds to sialoglycoconjugates on the host cells. The yellow shaded area represents the highly hydrated cumulus and mucus layers.

*Source*: Reprinted from Ref. [18]. (http://creativecommons.org/licenses/by/4.0/).

## 6.2.1  GLYCANS AS LIGANDS FOR CELL-MICROBE INTERACTIONS

The plasma membrane of eukaryotic cells has its constituent lipids and proteins tethered to glycans to form glycolipids and glycoproteins performing diverse functions. Although glycolipids and glycoproteins form predominant

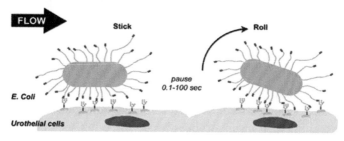

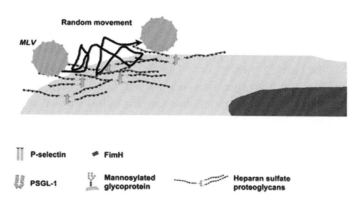

**FIGURE 6.5** Glycan-mediated lateral movement on host tissues. (**A**) FimH proteins localized to the tip of *E. coli* pili interact with oligomannose on urothelial cells. The interactions persist for 0.1 to 100 s, then bacteria detaches and rolls on the cells. Switching between detaching and rolling leads to stick and roll adhesion; (**B**) Weak interactions of murine leukemia virus (MLV) with glycosaminoglycans on the cell surface mediate multidirectional movement with frequent jumps on the cell surface. This surfacing movement does not depend on the cytoskeleton.

*Source*: Reprinted from Ref. [18]. (http://creativecommons.org/licenses/by/4.0/).

entities of glycans on cellular membranes, the extracellular matrix has glycosaminoglycans (GAGs) and mucins as well as the secreted glycoproteins. Together this contributes to the extracellular matrix mesh enriched with glycosylated proteins wherein all cellular interactions take place. Due to this ubiquitous presence of glycans in the milieu of living cells, numerous cellular interactions with self and non-self (pathogens) components involve glycans. Different microbial pathogens such as viruses, bacteria, and parasites exploit cellular glycans as specific binding sites and as recognition targets for many plant and bacterial toxins. These microbial pathogens express surface proteins containing specific carbohydrate binding domains (CBDs) called glycan binding proteins (GBPs) which mediate initial recognition and binding to glycans on target cells [18]. The GBPs recognize and interact with the glycans in a highly precise and specific manner and different types of GBPs differ in their recognition of the types of carbohydrate structures. Glycans being highly abundant on cell and tissue surfaces increase the chances of interaction between host glycans and pathogens expressing GBP. The biological interactions involving glycans although specific typically are of weak avidity (mM-μM *Kd* values for single glycan-protein interaction). The avidity of this interaction is enhanced by using multivalent binding with their interaction involving more than one pair of partners in close proximity. The reversibility and specificity of glycan interactions is mainly contributed by engaging multivalent binding of glycans [19, 20]. This simple multivalency of either glycan binding protein or the glycan is sufficient to promote binding in some cases; for example, influenza virus hemagglutinin (HA) has very low affinity but high specificity for its sialylated glycan ligands. High avidity of binding results due to its abundance on viral surface and multivalency of HA trimerization [21]. The interactions of GBPs with glycans can also be enhanced by multivalent aggregation of GBPs; for example, binding of cholera toxin B subunit recognizes and binds to ganglioside GM1 with very high avidity by forming pentamer (*Kd 40*nM). The multivalent binding to glycan ligands increases the binding affinity of an order of magnitude compared to typical carbohydrate-protein interaction (Figure 6.4(A and B)) [21, 22].

The glycans, due to enormous structural diversity, can be spatially organized into clustered saccharide patches (CSP) at the cellular membranes or capsular polysaccharides of pathogens. In CSPs, several closely spaced saccharides or structurally similar glycans aggregate to form a specific recognition site, thereby generates a distinct and unique topology for each protein. There is ample evidence that formation of CSPs provide a potential way to enhance affinity as well as specificity of GBP-glycan interaction. In addition, cells may employ a strategy to assemble and organize unique CSPs so that they can be differentially recognized by different GBPs [18–20]. Similarly, clustering of glycan-binding proteins on the membrane can form docking sites for the glycans, the best example for this is employed by C-type lectin microdomain (100–200 nm) formation which provides a recognition and binding site for several viral glycans [2, 23].

## 6.2.2    GLYCANS AS RECEPTORS DURING PATHOGEN INVASION

As described above, many pathogens including bacteria, viruses, and protozoa exploit the host glycans and glycoconjugates as their receptors to enter into the cells [24]. This interaction needs the pathogen to display close proximity to the cellular membrane of the host. Many pathogens which target the respiratory, circulatory, gastrointestinal, or reproductive tracts being covered with thick layers of mucus and secretions rich in glycans require traversing or penetrating through these defensive layers. The surface of mucosal tissue of the respiratory, reproductive, and gastrointestinal tract represents the route of access or primary site of infection for many pathogens and adhesion, recognition, and binding to this surface via carbohydrates is a prerequisite for the initiation of infection [25]. The mucin producing cells in the epithelium or sub-mucosal glands and in the glycocalyx underneath are coated with impenetrable mucus layers utilized as main targets by these pathogens [26]. Most of bacteria have on their surface adhesins (GBPs) and express glycosidases. Some bacteria recognize these mucins and bind them to colonize it while the nutrients are being obtained from hexose sugars by cleaving mucin glycoproteins or from the host diet using glycosidases. However, the mucus and its secretions form the first layer of defense against most pathogens by forming a protective barrier [3].

Opportunistic commensals such as nontypeable *Haemophilus influenzae (NTHi)* infect damaged epithelium areas resulting due to impaired mucus clearance. Other human specific gram-negative bacteria express several adhesins including hemagglutinating pili, P2 and P5 colonize the respiratory tract by binding to sialylated-linked glycans on the mucins [27–29]. The HMWA protein of NTHi acts to facilitate the infection by binding to sialylated lacto/neolacto glycolipids and gangliosides on the host cells. Furthermore, NTHi can bind to sulfated GAGs on cells (Figure 6.4(A)).

The cell surface is negatively charged due to the presence of proteoglycans such as GAGs and as the components of glycocalyx has many sulfated saccharides, impart negative charge to the cellular membrane. Like utilization of mucins by bacterial invasion, proteoglycans are also exploited as entry receptors by many pathogens such as herpes simplex virus (HSV) [30]. The virus binds to host cell via glycosaminoglycan heparin sulfate and is the first step in a cascade of interactions between pathogen and host (i.e., HSV, and host cell) [31]. The brain and neural cells express glycolipids such as gangliosides, the major cell surface determinants of these cells. These glycolipids also act as receptors for various pathogens to initiate the interaction with host cells. Blood group antigen oligosaccharides also form another major group of glycans that are involved in various host pathogen interactions. These HBGA (ABH and Lewis) expressed in abundance in gastrointestinal epithelium serve as receptors for many pathogens including *norovirus* and *H. pylori* [4, 32, 33].

Similarly, glycan and their conjugates are utilized as receptors by many protozoa and helminths and form an important part of their pathogenesis, for example, the most critical factor for *Toxoplasma gondii* attachment and invasion into host epithelial cells is its interaction with sialic acid containing receptors [34].

### 6.2.3   GLYCANS AS DECOYS DURING PATHOGEN INVASION

Glycans besides being utilized as receptors to allow host-pathogen interactions are also employed as barriers or decoys by the host to prevent infection. Mucus layers and secreted mucin within the respiratory, gastrointestinal, and reproductive tracts are the targets of many pathogens, as discussed above. However, the principle ingredient of these mucus layers and secretions are various glycans and glycoconjugate components of glycocalyx with immense diversity in their chemical designs. Similar to recognition and invasion components utilized by certain pathogens as discussed under Sections 6.2.1 and 6.2.2, the binding with these glycans and glycoconjugates also directly prevents the microbes from accessing entrance to the cellular membranes beneath the impenetrable mucus layers. The glycan receptor interaction with microbes captures them at mucus layers to prevent their access to cells. The mucus secretions may also trap the microbes and wash them off along with the shedding off secreted mucus [35, 36]. The best example of utilization of glycans as both invasion receptors and decoys is an enveloped RNA virus of *Orthomyxoviridae* family, *Influenza A* virus causing respiratory infections. The virus envelope expresses two major glycoproteins: Sialic acid-cleaving neuraminidase (NA) and Sialic acid-binding HA. When *Influenza A* virus encounters the host, sialylated mucins on host mucus layers act as decoy ligands by interacting with HA and prevents the virus infection [3, 37, 38].

The secreted mucins, on the other hand, function as decoys against pathogens expressing adhesins. The secreted mucins besides containing anti-microbial agents also contain many oligosaccharide structures and are constantly secreted in large amounts thereby washing the mucosal surfaces along with the trapped microbes [24]. Among the mucosal cells within the follicle-associated epithelium covering the gut associated lymphoid tissue (GALT) follicles such as Peyer's patches are scattered a specialized antigen transporting cells called M cells. These M cells are known to express on their surface enormous diversity of glycoconjugates to efficiently adhere, attach the microbes and can efficiently engulf them [39, 40]. Glycoconjugate GP2 expressed by M cells on their membrane is utilized for bacterial attachment by these cells [41, 42]. Thus, the glycans and glycoconjugates expressed on mucosal tissue and other cells are efficient in preventing the contact of pathogens and other microbes trapped or targeting outer mucus layers to contact the underlying epithelial cells.

## 6.2.4   GLYCAN MEDIATED LATERAL MOVEMENT ON HOST TISSUE

The initial contact of a pathogen with the host cell may result into non-simultaneous association and dissociation of low affinity single protein-glycan interaction between them. This process of association and dissociation often paves the way for a very strong protein-protein interaction followed by signaling and internalization (Figure 6.5(A-B)). The microbes targeting the host are more often under flow induced shear force due to flow of accessible body fluids like urine, blood, and tears or ciliary beating. The shear force prevents adhesive interactions of the pathogen and attachment of the virus with the host cells eventually washing off bound antigen [18, 43]. Certain microbes have evolved mechanisms to thrive under this shear force to enhance their protein-glycan interactions with the host like regulation of receptor-ligand bond by tensile mechanical force, formation of catch bonds, which mediate shear enhanced adhesion [44]. *Escherichia coli* switch from rolling adhesion to stationary adhesion at low to high shear stress respectively by forming a catch bond using a catch bond-forming protein FimH [45]. The fimbrae covering the surface of *E. coli* is composed of FimA subunits and a single adhesive protein FimH at the distal tip with two domains: a pilin domain and a lectin domain. Pilin domain functions to anchor the FimH to its fimbrae while lectin domain recognizes and binds terminal mannose residues present on host cells [4, 43]. The interaction of FimH proteins and host glycoconjugates is of low affinity but multiple FimH proteins interact simultaneously and affinity is further enhanced due to conformational change in the FimH under shear force [44]. The presence of shear force cause conformational change so that pilin and lectin domains separate, leading to increase in FimH affinity for mannose (200 fold), while in the absence of shear force the two domains bind tightly to each other resulting in low mannose affinity [46–48].

The binding of bacteria immobilizes it for variable time periods (100 ms–100 s), resulting into bacterial switching between rolling and adhesion, hence the term stick and roll (Figure 6.5(A)). Bacteria do not require any other ligand for its firm adhesion to host cells and can easily revert from stationary back to rolling. Under the circumstances when bacteria has to face shear stress above 20 dynes cm$^{-2}$, the FimH mediated adhesion rolling of bacteria stops while much higher levels of shear stress causes bacteria to roll slowly without detaching. This stick and roll adhesion allows bacteria to rapidly colonize new surfaces due to weak rolling adhesion [45, 49]. This causes the bacteria to preferentially bind high shear microenvironments with higher nutrients and also reduces their susceptibility to competitive inhibitors under such conditions [49, 50].

Viruses utilize host GAGs for their lateral movements on host cells. There are examples of viruses which utilize HSPGs to bind cells such as retroviruses including human immunodeficiency virus (HIV) and murine leukemia virus (MLV) [5, 51, 52]. The interactions between the virus and these HSPGs are relatively weak and can occur even in the absence of a specific viral receptor. Such weak interactions mediate surfacing of the virus on the host cellular membrane, which is a multidirectional movement and causes frequent jumps of virus, and also requires multivalent interactions for maintenance (Figure 6.5(B)) [53].

### 6.2.5   MOLECULAR MIMICRY OF HOST GLYCANS BY PATHOGEN

We have been discussing above the involvement of microbial glycans in recognition and attachment with the host cells which confer pathogenicity to these pathogens to invade the host. The microbial glycoconjugates may also serve as immunodominant antigens to activate and mount immune responses which may prove detrimental. However, the host-pathogen interaction over time has enabled these pathogens to evolve mechanisms to evade host immune responses. These pathogens decorate their surfaces with host-derived polysaccharides to thwart the host immune system, a phenomenon termed "molecular mimicry." In order to evade the host immune responses, the microbes coat their surface with a coating of glycans identical or nearly identical to the invading host [39]. Group A streptococcus hyaluronic acid capsule provides an elegant example of host glycan mimicking by bacterial glycoconjugate hyaluronic acid, an important constituent of many extracellular matrix and critical component of many cellular processes. Another example is being provided by mimicking of human fetal brain glycoproteins (FBGs) by bacterium *Neisseria meningitides* group B ($\alpha2\rightarrow8$) sialic acid capsule which is structurally similar to FBGs, unlike group C ($\alpha2\rightarrow9$) polysaccharide, does not evoke any immune response. There are also certain examples of parasites known to evolve in a manner to thwart the host immune responses via molecular mimicry [54]. Exposure of molecules similar in structure to the host as in case of molecular mimicry leads to a pathophysiological condition that compromises the tolerance, an important attribute of immune system. This process ultimately causes autoimmune diseases after infection of such microbes as the immune responses against these pathogens cross-react with host antigens [55, 56].

In addition to the mechanisms mentioned, there are ample evidences of pathogens containing host glycans. These pathogens with intimate associations with their hosts derive glycans from the host and incorporate these glycans in their structures. The best example of such pathogens is being provided by nontypeable *H. Influenza* (NTHi) which causes middle ear infection by manipulating the host immune system by acquiring sialic acids from the host and uses this to decorate its lipopolysaccharide (LPS) [57, 58]. Perhaps, not surprisingly microbes appear to have evolved to accomplish this state of mimicry for thwarting host immune responses and may explore every possible way, including convergent evolution toward similar biosynthetic pathways, direct or indirect appropriation of host glycans and even lateral gene transfer [15].

## 6.3  GLYCANS IN HOST-BACTERIAL INTERACTIONS

Glycosylation of proteins was once considered to be a characteristic feature of eukaryotes only; the perception however was revised overtime as the studies on surface layer glycoprotein in addition to the existence of various glycosylated components including enzymes proved the capability of prokaryotes to modify their proteins by tethering to carbohydrates. Various other improved molecular and analytical techniques in combination with the genome sequencing further validated the presence of glycosylated proteins in prokaryotes and archaea as well. The relationship shared by hosts with components of their microbes span a continuum from mutually beneficial (symbiotic) to benefiting one partner without necessarily being detrimental to other (commensal) to benefiting one partner while producing significant loss of fitness in other (pathogenic). During such interactions between a host and the bacteria, glycans form the first interface as they are the constituents of both outermost surfaces of bacteria as well as host tissue cells. Glycans serve as receptors as well as co-receptors for most bacterial pathogens while adhering to the host tissue cells. An indispensable virulence factor for bacterial pathogenesis is the specific adherence often mediated through bacterial adhesins, carbohydrate-binding proteins, such as fimbriae (Table 6.1). Apart from acting as receptors and co-receptors, glycans determine the host and tissue specificity, likewise pathogenic profile, tissue tropism, and enormity of virulence of a bacterium are determined by various adhesion factors released by them [59].

Role of bacterial glycans such as LPSs in host-bacterial interactions include recognition as PAMPs by host cells resulting in elicitation of host immune response and as shield to host immune recognition as well as unfavorable environment, when modified. In addition bacteria manifest and augment their virulence using glycans via many other processes including biofilm formation, ectodomain shedding of heparin sulfate proteoglycans (HSPGs) [60]. Many bacterial toxins have also been known to bind and act via glycans, for example, toxin of *Vibrio cholera* acts via binding to GM1 ganglioside receptor. *Vibrio cholera* toxin is composed of two subunits, A, and B in AB5 ratio and requires Galβ1–3GalNAc glycan moiety for binding via CRDs located on the base of its B subunit (Figure 6.6).

**TABLE 6.1**   Bacterial Adhesins That Interact with Host Glycans*

| Adhesin Protein | Bacterial Species | Receptor | Target Tissue |
|---|---|---|---|
| FimH (Type 1 pilus) | *Escherichia coli* | Mannose-oligosaccharides | G.I Tract |
| PapG (P pilus) | *Escherichia coli* | Galα-4Galβ-in glycolipids | Urinary Tract |
| BabA | *Helicobacter pylori* | Lewis b (Leb) blood group antigen and H-1 antigen | Stomach |
| SabA | *Helicobacter pylori* | Sialyl-Le$^x$ blood group antigen | Stomach |
| FHA | *Bordetella pertussis* | Sulfated glycolipids, heparin | Respiratory |
| HS antigen | *Streptococcus gonococci* | (α2-3)-linked sialic acid containing glycans. | Respiratory |
| OpA adhesion | *Neisseria meningitides* | Heparan sulfate Proteoglycans | Respiratory |
| HMW1 | *Haemophilus influenza* | (α2-3)-linked sialic acid containing glycans. | Respiratory |

*Some of the known bacterial adhesion protein utilizing host glycoprotein receptors.

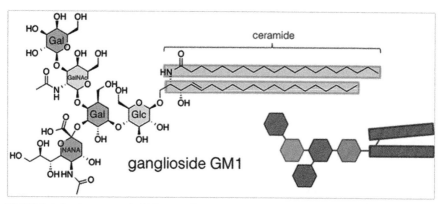

**FIGURE 6.6** Structure of GM1 ganglioside [GM1 (monosialotetrahexosylganglioside) is the prototype ganglioside with one sialic acid (N-acetylnueraminic acid, NeuNAc) containing oligosaccharide covalently attached to a ceramide lipid. GM1 acts as a binding receptor for cholera toxin and *E. coli* enterotoxin. The cholera-secreted toxin binds to host mucosal cells by interacting with GM1 gangliosides. A1 subunit of cholera toxin enters to intestinal epithelial cells with the help of B subunit of toxin using GM1 receptor].

### 6.3.1   GLYCANS AS RECEPTORS AND CO-RECEPTORS IN BACTERIAL PATHOGENESIS

The prerequisite for successful colonization and pathogenesis of bacteria is their adherence to the host tissue cells which can occur at various sites within the body and is also typical of a normal human microflora. Bacterial adherence to the host tissue cell requires involvement of two factors: a receptor and a ligand and the interaction between the two can be specific as well as non-specific. Specific interactions generally involve interactions of bacteria with matrix glycoproteins (fibronectin, laminin, etc.), or mucosal surface tissue mucins, mediated via several adhesion molecules secreted by bacteria [61]. Bacteria can have numerous of such adhesins including pili or fimbriae, outer surface proteins and cell wall constituents (LPS), each with different carbohydrate specificities [33]. Non-specific interactions on the other hand entail various charged molecules on host cell surfaces such as negatively charged sialic acid molecules [62].

Various hair like threads protruding from bacterial surfaces known as fimbriae or pili are often primarily composed of two subunits, a recurring structural subunit that provides extension and a distinct subunit known as tip adhesin that mediates interaction between bacteria and host and controls adherence at initial step of infection. Bacterial membranes are provided with fastened binding effect to epithelial surfaces by such adhesins, *Salmonella* pathogenic strains for example, adhere to human intestinal cell mucosa via various fimbriae resulting in food poisoning and infectious diarrhea. Among various types of fimbriae present in most bacterial species, Type 1 fimbriae are composed of a major and a minor component, FimA, and FimH respectively. FimH lying at the tip recognizes mannose-containing oligosaccharides in most bacterial species with substantially high affinity for cell surface glyco-proteins mainly consisting of Manα1-3, such as Manα1-3Manβ1-4GlcNAc and Manα1-6(Manα1-3) Manα1-6(Manα1-4) Man residues (Figure 6.7). The number of bacteria binding to host cell via FimH-Mannose interaction is in cognate with the expression of Type 1 fimbriae on bacterial surface [63, 64]. Type 4 fimbriae which include long polar fimbriae (LPF) and plasmid-code fimbriae (PEF) are thin and flexible and like Type 1 fimbriae they are also known to act via specific glycan receptors. The role of Type 4 fimbriae in bacterial virulence has been demonstrated by the decrease in bacterial viru-lence after their deletion, albeit their glycan receptors are yet to be identified [65, 66].

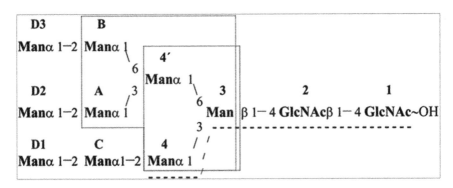

**FIGURE 6.7**   High mannose structure of N-linked glycans [The high mannose structure is formed during *N*-linked glycosylation on Asn residues of Asn-X-Ser/Thr sequences on glycoproteins. The $Glc_3Man_9GlcNAc_2$ is synthesized on the endoplasmic reticulum and is transferred to newly synthesized polypeptides on Asp residues via formation of covalent linkage with the GlcNAc (*N*-acetylglucosamine) moiety].

The glycan-lectin interactions between a bacteria and a host cell play a pivotal role in bacterial invasion and colonization of host tissues by enabling bacteria to infiltrate and invade epithelial barriers, followed by propagation through blood stream and produce deeper infections [59]. The bacterial adhesion via lectin like adhesins followed by invasion and colonization is best exemplified by *Helicobacter pylori,* a gram-negative bacterium causing gastritis, peptic ulcers, as well as gastric cancers. Various fucosylated blood group antigens have been known to act as mediators in adhesion of *H. pylori* to human gastric epithelial cells by binding to different lectin-like adhesins expressed by *H. pylori* [33, 67].

Human blood group Lewis antigens are of two types: Type 1 and Type 2 each composed of different backbones, with Type 1 chain composed of Gal and GlcNAc as Gal-$\beta$(1,3)-GlcNAc, and Type 2 chain has Gal and GlcNAc as Gal-$\beta$(1,4)-GlcNAc. Fucozylation of Type 1 and Type 2 chains at different sites on the backbone produces different Lewis antigens such as Lewis a (Le$^a$), Lewis b (Le$^b$), sialyl-Le$^a$ antigens and Lewis x (Le$^x$), Lewis y (Le$^y$), sialyl-Le$^x$ antigens, respectively [68]. Blood group-binding adhesin (BabA) expressed by *H pylori* binds type 1 antigen such as Le$^a$ and Le$^b$, expressed on surface of gastric epithelium resulting in severe gastric injury and high *H pylori* density. Type 2 antigens such as Le$^x$ and sialyl-Le$^x$ situated deeper in the glands interact with SabA adhesin expressed by *H pylori,* causing gastric degeneration, intestinal metaplasia, as well as gastric cancer [69]. Neutrophil-activating protein and 25KDa protein are the other *H pylori* adhesins that mediate pathogenesis via other glycans such as sulfated carbohydrate structures and laminin glycoprotein in the extracellular matrix, respectively [70, 71]. Other adhesins include adherence-associated lipoprotein A and B (AlpA/B) and HorB, however their gastric receptors are yet to be identified [72, 73].

Besides using glycans as receptors, certain pathogenic bacteria exploit glycans and their conjugates as co-receptors for invasion and pathogenesis of host cells. Bacterial pathogen *Neisseria gonorrheae* is the causative agent of human specific sexually transmitted disease, gonorrhea, and is known to exhibit pathogenesis via binding to cell surface proteoglycans, specifically to the representatives of syndecan family (syndecan 1-4) [73, 74]. *Neisseria gonorrheae* binds via its OpaA surface protein to heparin sulfate glycoproteins, utilizing it as co-receptor as binding to the glycoproteins (syndecans) is insufficient for invading host epithelial tissue [74, 75]. Binding of OpaA expressing bacteria to heparin sulfate glycoproteins (HSPGs) on epithelial cell surfaces result in the temporary aggregation of nearby actin, followed by the uptake of bacteria. The entry of bacteria could occur either through the integrin receptors where both vitronectin and fibronectin may act as linking molecules between the bacteria and its HSPG co-receptor, or through direct interaction of OpaA and HSPG as in Chang conjunctiva epithelial cells

[75–77]. In case of integrin mediated interaction, triggering of a signaling cascade occurs within the target cell which is dependent upon the activation of protein kinase C, while as in case of direct interaction a signal transduction cascade is triggered that leads to phosphatidylcholine-dependent phospholipase C-mediated formation of diacylglycerol, which successively triggers acidic spingomyelinase and results in the production of ceramide from spingomyelin. Such processes inflect cytoskeletal rearrangements for endocytosis of pathogens [78, 79].

## 6.3.2   BACTERIAL GLYCANS AS VIRULENCE FACTORS IN PATHOGENESIS

Various studies involving bacterial infections have clearly validated bacterial surface glycans such as polysaccharide capsule and O-antigen as virulence factors. These bacterial glycans have been known to exhibit virulence through several mechanisms such as immune evasion, immune modulation, biofilm production, and antiphagocytic as well as antibacteriolytic activity [80]. Being highly immunogenic they evoke strong immune responses which frequently lead to the prevention of subsequent infections. An effective immune response against these bacterial polysaccharide antigens entails the opsonization by phagocytes such as neutrophils and macrophages which have receptors for activated complement proteins or antibody Fc domains, which consequently result in the engulfment and killing of bacteria by host immune cells. Various anionic sugars expressed on the bacterial capsule such as sialic acids however, have the ability to debilitate the activity of alternative complement pathway by binding to host regulatory protein factor H, which is a soluble complement regulator necessary for controlling the alternative pathway. In addition surface polysaccharides can be utilized by bacteria in shrouding the protein structures present on their surface which might elicit protective immune responses [81]. Although humans are generally apt in eliciting a good immune response against bacterial polysaccharide capsules which however declines with age and is reduced in infants as well.

Diversity in the capsular polysaccharides and O-antigens form the basis of the classification of bacterial species into different serological groups which ultimately results in intraspecies antigenic variability. *E coli*, for example, have numerous serotypes which are differentiated on the basis of types of O antigens and capsular polysaccharides synthesized. Different strains of a same bacterial species produce capsular polysaccharides with different compositions and linkages of repeating sugar units and the genetic basis for such diversity in these polysaccharides has been attributed to the clustering of their genes into a single operon. The heterogenous products of polysaccharide biosynthesis loci in bacteria involving precursor nucleotide-charged monosaccharides, glycosyltransferases that create the linkages between monosaccharides, and products involved in transport, assembly, and regulation of polysaccharide expression have considerable genetic differences or single gene alterations. Therefore, immune response elicited against one serotype strain polysaccharide does not necessarily provide protection against a different polysaccharide type of the same bacterial species [54].

A characteristic feature of gram-negative bacterium outer membrane is the LPS component which serves as a pathogen associated molecular pattern (PAMP) recognized by innate immune system and has a role in triggering various inflammatory responses against bacteria which may eventually result in fatal sepsis syndrome. Bacterial LPS comprises of a lipid moiety, Lipid A and two carbohydrates moieties, a core oligosaccharide and an O-antigen (Figure 6.8). CD14 which is a GPI-linked protein expressed on the surface of many TLR4 expressing cells along with Toll-like receptor 4 (TLR4) a membrane protein, interact with soluble LPS released during the invasion process of bacteria. Interaction with TLR4 results in a signaling cascade which ultimately activates transcription factor NF-κB, which in turn controls the expression of various proinflammatory cytokine genes. Perhaps, many Gram-negative bacteria evade these immune responses by various mechanisms such as manipulating overall LPS structure by incorporating such components which reduce the overall negative charge and repel the cationic peptides present in host antimicrobial agents such as defensins [82]. Studies on *Salmonella enterica* LPS modifications in model organism *S. enterica serovar Typhimurium* have shown that the antimicrobial agents of the host can be confronted by manipulating the LPS structure by the addition of 4-aminoarabinose to the phosphate group of the lipid A backbone [6, 83]. Similarly, LPS modification in *Pseudomonas aeruginosa* involves synthesis of a unique hexa-acylated lipid A containing palmitate and 4-aminoarabinose conferring resistance against antimicrobial peptides [84].

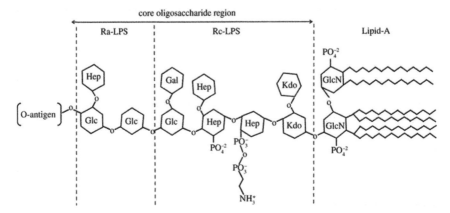

**FIGURE 6.8** Structure of lipopolysaccharid [Lipopolysaccharide (LPS) present in the outer membrane of gram-negative bacteria consists of three molecules, lipid A, a core oligosaccharide, and a highly variable O-antigen. The membrane anchoring part of LPS is lipid A that contains fatty acids linked to glucosamine with a variable number of phosphate groups and 1–4 units of ketodeoxyoctulosonic acid (Kdo). The core domain part of LPS always contains an oligosaccharide component that attaches to lipid A and contains sugars such as heptose (Hep) and Kdo].

In contrast to LPS that is ubiquitous among Gram-negative bacteria, many pathogens such as *Neisseria gonorrheae* possess lipooligosaccharides (LOS) in their outer membranes, which lack the repeating O-antigen sugar [85]. *N gonorrheae* has a capability of switching between several antigenically different types of LOS by modifications with oligosaccharide substitutions, which results in both inter and intra strain variability and can change the manner of association with host, hence the course of infection. It efficiently transforms a serum-sensitive organism to a serum-resistant organism by acquiring a sialic acid moiety from the traces of CMP-sialic acid present in host body fluids [87]. In addition, the sugar moieties of host glycosphingolipids can be mimicked by LOS epitopes present at the end of oligosaccharide chains, providing a strategy for immune evasion as well as utilizing host-derived molecules. Triggering the release of various immune factors by the gonococcal LOS including, proteases, phospholipases, and TNF promotes pathogenesis besides conferring resistance to serum bactericidal activity [88].

### 6.3.3 BIOFILM PRODUCTION IN BACTERIAL PATHOGENESIS

Biofilm formation is an important property of bacteria for its survival, dissemination, and is critical for the bacterial pathogenesis. It is a multistep process which involves the irreversible immobilization of bacteria by adhering to a given surface utilizing polysaccharides to interact with the host cell surfaces. The adhered bacteria produce extracellular polymers that facilitate attachment and matrix formation, and cell-cell interaction leading to formation of polymicrobial community or microcolony with altered growth properties and gene expression [89]. The biofilm matrix or extracellular polymeric substances (EPSs) is composed of various molecules; however, the focus of this chapter is on the extracellular polysaccharides important for biofilm formation. These EPSs consist primarily of polysaccharides and provide matrix or structure for the biofilm and have also been known to serve itself as the primary carbon reserve for biofilm microorganisms during substrate deprivation. The EPSs produced by biofilm forming microbes are highly variable in their composition and in their chemical and physical properties. Majority of EPSs are anionic, of either uronic acids (D-glucuronic, D-galacturonic, or D-mannuronic acids) or ketal-linked pyruvate and inorganic residues such as phosphate or sulfate, may also contribute to the negative charge. However, neutral or cationic EPS are also reported. In most natural and experimental environments, EPSs are found in ordered compositions, with long, thin molecular chains, ranging in mass from 0.5 to $2.0 \times 10^6$ Da. To further add to the diversity of EPSs, environmental factors and association with other biomolecules such as lectins, lipids, proteins, and bacterial and host extracellular DNA influence the composition of biofilm. Furthermore, biofilms can be composed of heterogeneous population of organism of multiple bacterial or even fungal species, whereby a range of EPSs may interact to generate further permutations of unique architectures. Biofilms are usually highly hydrated (98%) and heterogeneous with "water channels" that allow transport of oxygen and essential nutrients to the growing organisms and have a propensity to act as molecular filters entrapping various particles including minerals and host components. For example, dental plaque represents an oral biofilm in which thick mushroom-like clumps of bacteria protrude from the surface of the tooth enamel, interspersed with bacteria-free channels occupied with extracellular polysaccharides (EPS) that function as diffusion channels. Bacteria within these biofilms often communicate with each other through soluble signaling molecules by a mechanism known as "quorum sensing" to optimize gene expression for survival [7, 90].

Formation of biofilm provides microbes with several advantages under the protective immune response of host and other environmental factors. Biofilm production confers on bacteria a sufficient decrease in antibacterial susceptibility, which could be intrinsic, inherent in the biofilm mode of growth or acquired, due to acquisition of resistance plasmids [90]. For example, *Staphylococcus aureus* biofilms have been shown to require >10 times the minimal bactericidal concentration (MBC) of vancomycin to provide a 3-log reduction [91]. The antimicrobial resistance provided by biofilms results from various factors such as the retarded diffusion of antimicrobial agents by EPS, which chemically reacts with them and limit their transport rate, reduced growth rate of biofilm bacteria which minimizes antimicrobial agent uptake hence inactivation kinetics, and environmental conditions surrounding the biofilm which may provide conditions to protect bacteria [89, 92]. Bacteria from these biofilms may sometimes detach either as a result of removal of biofilm aggregates or cell growth and division which may eventually lead to systemic infections, depending upon various factors including host immune response. In addition biofilms sometimes arise on catheters and other medical devices as well further baffling the treatment of biofilm infections [89].

### 6.3.4   GLYCAN ECTODOMAIN SHEDDING IN BACTERIAL PATHOGENESIS

Besides utilizing direct interactions with host glycan and glycoconjugates as a mechanism of pathogenesis, shedding of cell surface glycans can also be used as an alternate indirect mechanism to promote pathogenesis. Proteoglycan heparin sulfate shedding from the surface of cells is utilized by the bacteria to promote their pathogenesis [93]. A family of metalloprotein enzymes collectively known as secretases or sheddases trims the surface molecules from the bacterial cells and releases the ectodomains from these cell surface molecules, known as ectodomain shedding [94]. Many proteins including cytokines, enzymes, growth factors and cell adhesion molecules, require for their activation and secretion to undergo ectodomain shedding. Furthermore, during septic shock, host defense and cell proliferation and many other pathophysiological conditions, the shed ectodomains perform critical functions. During host response to tissue injury there is generation of intact soluble HSPG ectodomains due to shedding of cell surface HSPGs and causes a rapid loss of the amount of cell surface heparin sulfate [95].

The secretion of various virulence factors by many microbes like *Bacillus anthracis* and *Pseudomonas aeruginosa* are known to accelerate the shedding of surface HS proteoglycans (syndecan-1). The LasA virulence factor secreted by *Pseudomonas aeruginosa* causes syndecan-1 shedding *in vitro* and has also been shown to enhance shedding *in vivo*. The shedding of syndecan-1 ectodomain has been shown to cause enhanced bacterial infection in the newborn mice. Newborn mice lacking syndecan-1 are resistant to lung infection by *P. aeruginosa*, but when given heparin or purified syndecan-1 ectodomain they become susceptible to infection. However, newborn mice did not get susceptible to infection when given with only core protein of ectodomain indicating that the effectors are the heparin sulfates of ectodomain. Hence, the accelerated shedding of syndecan-1 by the virulence factors of bacteria efficiently enables these bacteria to colonize and invade the host cells [96, 97].

## 6.4 GLYCANS IN HOST-VIRAL INTERACTIONS

In addition to human and animal hosts, their pathogens can also gain from the glycosylation biosynthetic pathways. Glycans being the most important part of the surface of many obligatory intracellular pathogens such as viruses, they have evolved multiple strategies to exploit the host glycosylation machinery. The viral components derive their glycosylation from the host cellular glycosylation machinery hence resemble to the host cells [98]. Although similar in composition to the host there are certain important differences in glycosylation pattern between viruses and the host. The architecture of respective glycan chains can deviate between host and virus, and viral proteins are often glycosylated to a great extent than their host cellular counterparts [99]. In line with the resemblance between host and viral glycosylation, various primary functions covered by glycans in humans and other animal physiology and in viral infection biology are essentially associated to the biology of the host and are essentially the same. Glycans have important structural functions and are directly involved in protein structure and function such as resistance of proteins to proteases, folding, and solubility and controls the antigenicity of the proteins ('glycan shielding') [15, 38]. In addition, glycans also perform nonstructural roles such as in molecular recognition, being recognized with precise specificity by complementary GBPs called lectins [38, 100].

Viruses employ diverse mechanisms to target the cells and require specific binding of virus to receptors expressed on the host cell surface. It is not surprising that viruses have evolved mechanisms to explore the host glycoconjugates as entry receptors due to high abundance and enormous diversity of glycoconjugates [99, 101]. Most often viruses utilize the carbohydrate parts of these glycoprotein receptors to gain access into the host cells. Structurally the glycans on the cells have constituent components covalently attached to proteins (glycoproteins), plasma membrane lipids (glycolipids), and macromolecules polymeric glycans (proteoglycans) and each class of glycans or glycoconjugates are used by different viruses to establish host-pathogen interaction. Viruses usually use different mechanisms to enter into the cells and each is used for a different purpose [102, 103].

The first interaction between a virus and the cell is to get adsorbed to the surface of cell, so most of the viruses that interact with the cell (Table 6.2) interact via the cellular glycoconjugates and more commonly bind to carbohydrates with a terminal sialic acid or sulfated oligosaccharide motifs of GAGs. The presence of negatively charged residues helps in establishment of interaction between the virus and the cell and these charges are imparted by sulfate groups of GAGs and carboxyl group of sialic acids. Although the electrostatic interactions are highly non-specific but they help to bring the virus closer to the cell. The formation of hydrogen bonds coupled with other non-covalent interactions with GAG and sialic acid containing glycoconjugates leads more often to a high degree of specificity in the binding [104–106].

**TABLE 6.2**    Classification of Viruses Using Glycoepitopes as Receptors*

| Virus Family (Subfamily/ Genus) | Virus Type | Receptor |
|---|---|---|
| Adenoviridae | Adenovirus | ($\alpha$2-3)-linked sialic acid Heparan sulfate |
| Arenaviridae | Lassa virus | Dystroglycan glycans |
| Caliciviridae Noroviruses | Norwalk and others | Histo-blood group glycoepitopes in secretor-positive individuals |
| Coronaviridae | Coronavirus OC43 | 9-O-acetyl-sialic acid |
| Flaviviradae Hepaciviruses Flavivirus | Hepatitis C virus Denguevirus Japanese encephalitis virus. West Nile virus | Heparan sulfate Heparan sulfate Heparan sulfate |

**TABLE 6.2** *(Continued)*

| Virus Family (Subfamily/ Genus) | Virus Type | Receptor |
|---|---|---|
| Herpesvirdae<br>αherpsviruses<br>βherpsviruses<br>γ herpsviruses | Herpes simplex virus type 1 and 2<br>Varicella-zoster virus<br>Cytomegalovirus, Human herpesvirus types 6 and 7<br>Human herepesvirus type 8 | Heparin sulfate (chondrotin sulfate)<br>Heparan sulfate<br>Heparan sulfate<br>Heparan sulfate |
| Ortomyxoviridae | Influenza A virus<br>Influenza B virus<br>Influenza C virus | ($\alpha$2-3)-linked sialic acid: Bird virus<br>($\alpha$2-6)-linked sialic acid: Human virus<br>($\alpha$2-6)-linked sialic acid<br>($\alpha$2-3)-linked sialic acid<br>9-O-acetylsialic acid |
| Papillomaviridae<br>Papillomavirus | Human papillomavirus types | Heparan sulfate |
| Paramyxoviridae<br>Respirovirus<br>Pnemovirus<br>Metapneumov | Paramyxovirus 1–3<br>Respiratory syncytial virus<br>Human metapneumovirus | Sialic acid<br>Heparan sulfate<br>(chondrotin sulfate)<br>Heparan sulfate |
| Parvoviridae<br>Erythrovirus<br>Dependovirus | B 19<br>Adenoassociated virus (AAV) types 4 and 5<br>AAV type 2 | Globosid/histo-blood group P substance<br>Sialic acid;<br>Glycosaminoglycan |
| Picornavirus<br>Enterovirus<br>Rhinovirus | Enterovirus<br>Rhinovirus | Sialic acid<br>Sialic acid |
| Polyomaviridae<br>Polyomavirus | JC and BK virus | Sialic acid |
| Poxviridae<br>Ortopoxvirus | Vaccinia virus | Heparin sulfate, chondroitin sulfate |
| Retroviridae<br>Lentivirus | HIV-1 | Sulfaatide; galactosylceramide, Heparan sulfate (chondrotin sulfate) |

*Most of the viruses have evolved to exploit the host glycans as means of entry into the host and establishment of infection. The glycan receptors include negatively charged sialic acid containing receptors, glycosaminoglycans, and other carbohydrate containing receptors.

## 6.4.1   SIALIC ACID CONTAINING GLYCOCONJUGATES AS VIRAL ATTACHMENT AND ENTRY RECEPTORS

Sialic acids, a family of monosaccharides derived from a nine-carbon neuraminic acid, as a part of glycoconjugates are utilized by various viruses as shown in Table 6.1 as their receptors in pathogenesis. The substitutions may be at any of the carbon hydroxyl groups of neuraminic acid however only a few of the substitutions have been explained for human cells, most frequent being the substitution at C5 with an N-acetyl group forming N-acetylneuraminic acid, also referred to as sialic acid. Most of the sialic acid containing viral receptors are characterized by the presence of terminal sialic acid attached to penultimate galactose either in (α2-6)- or (α2-3)-linkage. Their distribution varies differentially in different tissues, thereby influencing tropism of sialic acid-dependent viruses [8, 107, 108].

Adenoviridae, a well-defined virus family with about 50 genotypes (Ad1-51), exploits sialic acid containing glycans as sole target receptors for their entry into host cells. Human adenovirus infections primarily result in respiratory, gastrointestinal or eye infections depending upon the infecting serotypes. Moreover, it is the relative distribution of sialic acids in the tissue which govern the tissue tropism of these viruses. For example, adenovirus 37 (Ad37) preferentially binds (α2-3)-linked sialic acid which is in fact the most frequent type of sialic acid linkage in the corneal and conjuctival cells and causes epidemic keratoconjuctivitis (EKC). Such (α2-3)-linked sialic acid containing glycoconjugates are utilized as receptors by other viruses as well [108, 109].

Influenza viruses is another group which utilize (α2-6)- and (α2-3)-linked sialic acid as receptors for entry into the host cell and its receptor structure was the first to be described (Figure 6.9). Influenza virus has three main types, Influenza A, B, and C each exposing two surface glycoproteins: hemagglutinin (H) and neuraminidase (N), which mediate binding to the host cell surface sialic acid and release of virus from the host cell by cleaving sialic acid group from the glycoprotein, respectively. The specificity of the human influenza A virus (HA) is predominantly directed towards (α2-6)-linked sialic acid however for some HA subtypes elements of cross reactivity with (α2-3)-linked sialic acid may occur. The avian influenza HA, on the other hand, binds exclusively to a (α2-3)-linked sialic acid and the frequent epidemic of avian influenza or flu occurs due to the cross reactivity of the receptors [110, 111].

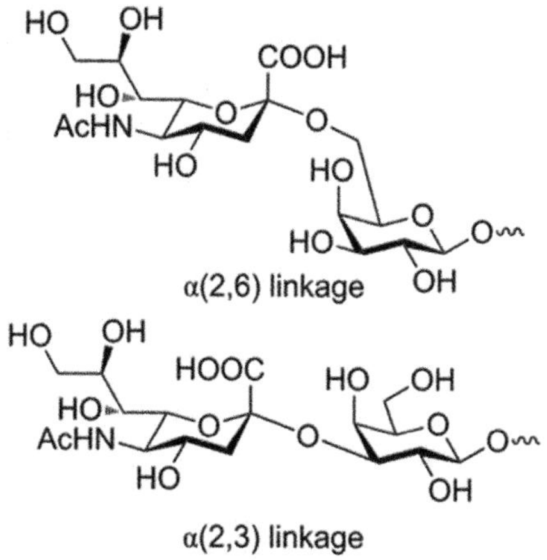

**FIGURE 6.9** Sialic acid linkage in influenza viruses [Sialic acid is linked to the sugar galactose by either α2-3 or α2-6 linkage. Avian influenza virus preferentially binds to α2-3 linked sialic acids expressed abundantly on avian epithelial cells. Human influenza virus strains display specificity to α2-6 linked sialic acids preferentially expressed on human respiratory epithelial cells].

## 6.4.2 GLYCOSAMINOGLYCAN (GAG) UTILIZATION OF VIRUSES FOR ATTACHMENT AND ENTRY INTO HOST CELL

GAG receptors are a second class of glycan receptors which significantly mediate attachment and entry of virus into the host cell. GAG molecules constitute one of the biggest groups of carbohydrate receptors for human viruses, representing members from various leading virus families (Table 6.1). The different members of this receptor family include HS, chondroitin sulfate (CS), dermatan sulfate (DS), keratan sulfate (KS) and hyaluronic acid (HA) (Figure 6.10) [112–114].

The GAG chains are synthesized of long (20–70 monosaccharides), unbranched, and negatively charged polysaccharides which are connected to serine residues on proteoglycans. Syndecans and glypicans represent the two major families of proteoglycans that have been characterized. Syndecans are transmembrane proteins and carry 3–5 GAG chains of HS and/or CS. Internalization through endocytosis and binding to cytoskeletal proteins are the two main properties of these proteoglycans that are of interest for viral entry. Glypicans, which is the other family of proteoglycans, is comprised mainly of cysteine rich globular proteins and are attached to glycosylphosphatidylinositol (GPI) molecules which are successively hooked to the outer membrane layer. Glypicans, on the other hand, carry 2–3 HS chains located in the of the GPI anchor, and may also take part in endocytosis [112, 115, 116].

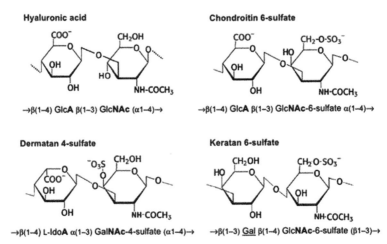

**Typical disaccharide units comprising glycosaminoglycans:**

**Hyaluronic acid**

→β(1–4) GlcA β(1–3) GlcNAc (α1–4)→

**Chondroitin 6-sulfate**

→β(1–4) GlcA β(1–3) GlcNAc-6-sulfate α(1–4)→

**Dermatan 4-sulfate**

→β(1–4) L-IdoA α(1–3) GalNAc-4-sulfate (α1–4)→

**Keratan 6-sulfate**

→β(1–3) Gal β(1–4) GlcNAc-6-sulfate (β1–3)→

**FIGURE 6.10** Structure of representative glycosamioglycans [Glycosaminoglycans (GAGs) based on core disaccharide structure are of four types mostly of negatively charged polymers of sulfated disaccharide units (except hyaluronic acid which is not sulfated) performing besides other functions have structural roles in organisms. GAGs are polymers consisting of an amino sugar (N-acetylglucosamine or N-acetylgalactosamine) along with an uronic acid sugar (glucuronic acid or iduronic acid) or galactose. Chondroitin sulfate, Keratin sulfate, Dermatin sulfate, and Hyaluronic acid are the representative glycosaminoglycans utilized by pathogens for entry into the host].

Many viruses have evolved to exploit these GAG-containing glycoconjugates as receptors for their particular biological usage in addition to tropism to target tissues and organs. For example, human herpesviruses, i.e., HSV-1, HSV-2, varicella-zoster virus (VZV), cytomegalovirus (CMV), human herpesvirus (HHV)-7 and HHV-8 may utilize cell surface HS for binding. HSV infection like all other viral infections requires attachment of the virus to the cell-surface membrane before the viral transfer of genetic material, after penetration and entry into the cytoplasm. The two glycoproteins expressed by HSV include gB and gC are involved in the initial attachment through interaction with negatively charged heparin sulfate (HS) of cell surface proteoglycan. After this initial attachment, a second high affinity attachment is provided by the gD glycoprotein through interaction with the tumor necrosis factor-nerve growth factor (TNF/NGF) receptor family. Subsequent to the initial attachment, the virus perforates the cell by fusion of the virus envelope with the cell plasma membrane. GAG containing glyco-conjugates are exploited by many other viruses as receptors for entry and attachment into the host cell, hence pathogenesis and may include viruses like respiratory syncytial virus (RSV) and viruses in the flaviviridae (dengue virus, yellow fever virus, Japanese encephalitis virus, zika virus, tick-borne encephalitis virus (TBE) and hepatitis C virus (HCV)) [9, 117–120].

### 6.4.3   HIV-1 USES ITS SURFACE GLYCOPROTEIN FOR VIRAL ENTRY

Human immunodeficiency virus 1 (HIV-1) is the causative agent of global pandemic of acquired immunodeficiency syndrome (AIDS). The envelope (env) glycoprotein like other enveloped viruses is the only molecular entity visible on the surface of HIV-1 embedded into the host-derived lipid bilayer, mediating the first steps of cell attachment and entry through the primary receptors CD4 and co-receptors CCR5/CXCR4 [121–123]. The HIV-1 env spike theoretically is the exclusive component of virus accessible to anti-HIV-1 neutralizing antibodies; thus, env has been a target of intense research as a vaccine immunogen, and this glycoprotein has formed the basis of extensive phase III clinical trials [124, 125]. The outermost domain of gp120 is extensively covered with hypervariable loops and thick NLGs contributing about 50% of the env molecular mass, thus playing central roles in the determination of env structure, epitope exposure, and consequently, affecting antigenicity, immunogenicity, antibody neutralization, infectivity, and receptor binding [126–128]. This unusual extensive degree of HIV-1 env glycosylation thought for more than a decade as an immunologically silent shield masking the preserved functional sites on the HIV-1, has emerged as an amazing target for recognition by broadly neutralizing antibodies (bnAbs) [129].

Besides protecting the virus by formation of immunologically silent glycan shield, the HIV-1 has evolved to use these glycans to facilitate its infection of host. There is growing evidence that env also plays a major role in the viral capture, transmission, and dissemination during early stages of HIV-1 infection by hijacking the natural functions of the cells of innate immune system. Langerhans Cells (LCs), dendritic cells (DCs) and macrophages, which are strategically located at all the entry sites epidermis, mucosa, sub-mucosa, lymphoid, and circulatory system to facilitate optimal interaction of the virus during sexual transmission, the most common cause of HIV infection. The enigmatic features of these innate cell interactions with env are mediated primarily by the pattern recognition receptors (PRRs), of membrane bound c-type lectin receptor (CLRs) family like Langerin, DCIR, DC-SIGN, Mannose receptor (MR), BDCA2 and SIGLEC-1, the most important CLRs which recognize exclusively the glycans on the surface of HIV-1 env [130]. The innate cells and CLRs they express are spatially and temporally distributed along the sexual transmission pathway to generally form the first line of defense and perform differential functions leading finally to viral internalization and endosomal degradation for efficient antigen presentation, and modulate TLR-induced cytokine expression to enhance the infection of HIV [131].

Lectin receptors recognize specific carbohydrate structures by means of one or more carbohydrate recognition domains (CRDs) and are grouped on the basis of the presence of a conserved structural motif in their CRDs. Various CLRs have distinguishable carbohydrate specificity, which are related to their amino acid sequence in their respective CRDs [132, 133]. The innate cell (DCs and macrophages) CLRs primarily interact with pathogens via the recognition of mannose, fucose, and glucan carbohydrate structures [134]. The HIV-1 env is known to be having highest glycan content and in particular has a very high proportion of high mannose glycans of $Man_{5-9}GlcNAc_2$ type, which makes HIV-1 a susceptible target of these CLRs that recognize mostly high mannose type glycans (Figure 6.10). Several of the CLRs encountered by HIV-1 in the sexual route DC-SIGN, MR, Langerin, and BDCA-2 exclusively recognize the high mannose glycans on gp120 of HIV-1, thus enhancing the recognition and capture of HIV-1 on these cells. These cells are strong antigen presenting cells (APCs) that take of pathogens and process them, and readily migrate to lymph nodes to present the antigen to adaptive immune system. These events are anticipated to play essential roles in the initial events of HIV-1 transmission by transporting the virus from the peripheral mucosa to lymph node, the place with concentrated amount of T-cells, making favorable the interaction of virus with the T-cells, its ultimate targets. There are two ways DCs direct the transmission of HIV-1 to CD4[+] T-cells: DCs after capturing virus by CLRs transfer captured virus in the absence of productive infection which is referred to as *in trans*-infection or it endocytoses virus within proteosome resistant compartments in which the infectious virus is prevented from degradation and are released to infect the cells. There are mounting evidences that this hijacking of natural functions of DCs, degradation, and presentation of antigens by HIV-1 is determined by the glycan composition [2, 135]. As mentioned above the glycan composition on env of different viruses is heterogeneous, as glycan maturation is primarily driven by the relative exposure of polypeptide during folding to ER and Golgi glycosylation modification enzymes. The specific constitution of env between various HIV-1 strains may play a decisive role in a well-studied CLR, DC-SIGN binding and transmission efficiency. A recent study showed the importance of the composition of HIV-1 glycans for DC-SIGN-mediated transmission. Virions carrying gp120 with higher numbers of oligomannose-type glycans are more efficiently endocytosed through DC-SIGN and more proficiently processed for antigen presentation than HIV-1 containing gp120 with heterogeneous glycans. The transmission of oligomannose-enriched HIV-1 was relatively inefficient. Thus, the expression of oligomannose by HIV-1 enhances capture of DC-SIGN and transmission, but too much oligomannose negatively affects transmission by enhancing viral degradation.

## 6.5   CONCLUSION

The role of glycans in the establishment of host-pathogen interaction seems overwhelming. Although, the field of glycomics has not been explored similar to genomics and proteomics, there is reason to believe that great progress will be made in understanding their inter-relationship in the near future. There have been tremendous technological advances in the field of glycomics in combination with success in functional genomics, microarray technology during the past two decades which have produced an unprecedented opportunity to examine the impact of glycans in host-pathogen interactions. The development of these integrated approaches to study the structure-function relationships of glycans and glycoconjugates have greatly improved our understanding of their role in mediating host-pathogen interactions. Given their widespread role in mediating host-pathogen interactions, there is great potential for the development of glycan based therapeutics, vaccines, and diagnostics for infectious disease. The present chapter summarized the exploration of host glycans by pathogens, such as bacteria, viruses, protozoans that involve carbohydrates such as sialic acids, heparin sulfate, oligomannose glycans, extracellular matrix glycans, and other glycoconjugates. As the role of glycans in the interactions of host-pathogen interaction unfurls, we find that most of the pathogens have evolved to use the host glycans both as an entry mechanism and shield themselves from the host immune system, as carbohydrates are relatively resistant to the development of immune responses. However, further advances in the integration of the diverse technologies to study structure-function relationships of glycans are critical to uncovering novel roles for glycans in host-pathogen interactions. Such approaches would not only shed light on the molecular mechanisms of microbial pathogenesis but also provide a basis for the development of additional strategies for targeted antimicrobial therapies.

## KEYWORDS

- **acquired immunodeficiency syndrome**
- **antigenicity**
- **carbohydrate binding domains**
- **extracellular polymeric substances**
- **microbe**
- **pathogenicity**

## REFERENCES

1. Roseman, S., (2001). Reflections on glycobiology. *J. Biol. Chem., 276*, 41527–41542.
2. German, J. B., Gillies, L. A., Smilowitz, J. T., Zivkovic, A. M., & Watkins, S. M., (2007). Lipidomics and lipid profiling in metabolomics. *Curr. Opin. Lipidol., 18*, 66–71.
3. Dettmer, K., Aronov, P. A., & Hammock, B. D., (2007). Mass spectrometry-based metabolomics. *Mass Spectrom. Rev., 26*, 51–78.
4. Varki, A., (2011). Evolutionary forces shaping the Golgi glycosylation machinery: Why cell surface glycans are universal to living cells. *Cold Spring Harb. Perspect. Biol., 3*.
5. Laine, R. A., (1994). A calculation of all possible oligosaccharide isomers both branched and linear yields $1.05 \times 10(12)$ structures for a reducing hexasaccharide: The isomer barrier to development of single-method saccharide sequencing or synthesis systems. *Glycobiology, 4*, 759–767.
6. Twine, S. M., Paul, C. J., Vinogradov, E., McNally, D. J., Brisson, J. R., Mullen, J. A., McMullin, D. R., et al., (2008). Flagellar glycosylation in clostridium botulinum. *FEBS J., 275*, 4428–4444.
7. Voisin, S., Houliston, R. S., Kelly, J., Brisson, J. R., Watson, D., Bardy, S. L., Jarrell, K. F., & Logan, S. M., (2005). Identification and characterization of the unique N-linked glycan common to the flagellins and S-layer glycoprotein of methanococcus voltae. *J. Biol. Chem., 280*, 16586–16593.
8. Weerapana, E., & Imperiali, B., (2006). Asparagine-linked protein glycosylation: From eukaryotic to prokaryotic systems. *Glycobiology, 16*, 91R–101R.
9. Larkin, A., & Imperiali, B., (2011). The expanding horizons of asparagine-linked glycosylation. *Biochemistry, 50*, 4411–4426.

10. Schwarz, F., & Aebi, M., (2011). Mechanisms and principles of N-linked protein glycosylation. *Curr. Opin. Struct. Biol., 21*, 576–582.
11. Van, D. S. P., Rudd, P. M., Dwek, R. A., & Opdenakker, G., (1998). Concepts and principles of O-linked glycosylation. *Crit. Rev. Biochem. Mol. Biol., 33*, 151–208.
12. Peter-Katalinic, J., (2005). Methods in enzymology: O-glycosylation of proteins. *Methods Enzymol., 405*, 139–171.
13. Jensen, P. H., Kolarich, D., & Packer, N. H., (2010). Mucin-type O-glycosylation-putting the pieces together. *FEBS J., 277*, 81–94.
14. Gill, D. J., Clausen, H., & Bard, F., (2011). Location, location, location: New insights into O-GalNAc protein glycosylation. *Trends Cell Biol., 21*, 149–158.
15. Anonymous, (2009). In: Varki, A., Cummings, R. D., Esko, J. D., Freeze, H. H., Stanley, P., Bertozzi, C. R., Hart, G. W., & Etzler, M. E., (eds.), *Essentials of Glycobiology* (2nd edn.). Cold Spring Harbor (NY).
16. Yu, R. K., Tsai, Y. T., Ariga, T., & Yanagisawa, M., (2011). Structures, biosynthesis, and functions of gangliosides: An overview. *J. Oleo. Sci., 60*, 537–544.
17. Paulick, M. G., & Bertozzi, C. R., (2008). The glycosylphosphatidylinositol anchor: A complex membrane-anchoring structure for proteins. *Biochemistry, 47*, 6991–7000.
18. Cohen, M., (2015). Notable aspects of glycan-protein interactions. *Biomolecules, 5*, 2056–2072.
19. Cohen, M., & Varki, A., (2010). The sialome-far more than the sum of its parts. *OMICS, 14*, 455–464.
20. Cummings, R. D., & Esko, J. D., (2009). Principles of glycan recognition. In Varki, A., Cummings, R. D., Esko, J. D., Freeze, H. H., Stanley, P., Bertozzi, C. R., Hart, G. W., & Etzler, M. E., (eds.), *Essentials of Glycobiology* (2nd edn.). Cold Spring Harbor (NY).
21. Cohen, M., & Varki, A., (2014). Modulation of glycan recognition by clustered saccharide patches. *Int. Rev. Cell Mol. Biol., 308*, 75–125.
22. Kuziemko, G. M., Stroh, M., & Stevens, R. C., (1996). Cholera toxin binding affinity and specificity for gangliosides determined by surface plasmon resonance. *Biochemistry, 35*, 6375–6384.
23. Cambi, A., Koopman, M., & Figdor, C. G., (2005). How C-type lectins detect pathogens. *Cell Microbiol., 7*, 481–488.
24. Kato, K., & Ishiwa, A., (2015). The role of carbohydrates in infection strategies of enteric pathogens. *Trop. Med. Health, 43*, 41–52.
25. Linden, S. K., Sutton, P., Karlsson, N. G., Korolik, V., & McGuckin, M. A., (2008). Mucins in the mucosal barrier to infection. *Mucosal Immunol., 1*, 183–197.
26. McGuckin, M. A., Linden, S. K., Sutton, P., & Florin, T. H., (2011). Mucin dynamics and enteric pathogens. *Nat. Rev. Microbiol., 9*, 265–278.
27. Foxwell, A. R., Kyd, J. M., & Cripps, A. W., (1998). Nontypeable Haemophilus influenzae: Pathogenesis and prevention. *Microbiol. Mol. Biol. Rev., 62*, 294–308.
28. Davies, J., Carlstedt, I., Nilsson, A. K., Hakansson, A., Sabharwal, H., Van, A. L., Van, H. M., & Svanborg, C., (1995). Binding of hemophilus influenzae to purified mucins from the human respiratory tract. *Infect. Immun., 63*, 2485–2492.
29. Bernstein, J. M., & Reddy, M., (2000). Bacteria-mucin interaction in the upper aerodigestive tract shows striking heterogeneity: Implications in otitis media, rhinosinusitis, and pneumonia. *Otolaryngol. Head Neck Surg., 122*, 514–520.

30. Bernfield, M., Kokenyesi, R., Kato, M., Hinkes, M. T., Spring, J., Gallo, R. L., & Lose, E. J., (1992). Biology of the syndecans: A family of transmembrane heparan sulfate proteoglycans. *Annu. Rev. Cell Biol., 8*, 365–393.

31. Shukla, D., & Spear, P. G., (2001). Herpesviruses and heparan sulfate: An intimate relationship in aid of viral entry. *J. Clin. Invest., 108*, 503–510.

32. Ravn, V., & Dabelsteen, E., (2000). Tissue distribution of histo-blood group antigens. *APMIS, 108*, 1–28.

33. Moran, A. P., Gupta, A., & Joshi, L., (2011). Sweet-talk: Role of host glycosylation in bacterial pathogenesis of the gastrointestinal tract. *Gut., 60*, 1412–1425.

34. Monteiro, V. G., Soares, C. P., & De Souza, W., (1998). Host cell surface sialic acid residues are involved on the process of penetration of Toxoplasma gondii into mammalian cells. *FEMS Microbiol. Lett., 164*, 323–327.

35. Cone, R. A., (2009). Barrier properties of mucus. *Adv. Drug Deliv. Rev., 61*, 75–85.

36. Fahy, J. V., & Dickey, B. F., (2010). Airway mucus function and dysfunction. *N. Engl. J. Med., 363*, 2233–2247.

37. Suzuki, Y., (2005). Sialobiology of influenza: Molecular mechanism of host range variation of influenza viruses. *Biol. Pharm. Bull., 28*, 399–408.

38. Taylor, H. P., Armstrong, S. J., & Dimmock, N. J., (1987). Quantitative relationships between an influenza virus and neutralizing antibody. *Virology, 159*, 288–298.

39. Kraehenbuhl, J. P., & Neutra, M. R., (2000). Epithelial M cells: Differentiation and function. *Annu. Rev. Cell Dev. Biol., 16*, 301–332.

40. Owen, R. L., (1999). Uptake and transport of intestinal macromolecules and microorganisms by M cells in Peyer's patches: A personal and historical perspective. *Semin. Immunol., 11*, 157–163.

41. Ohno, H., & Hase, K., (2010). Glycoprotein 2 (GP2): Grabbing the FimH bacteria into M cells for mucosal immunity. *Gut. Microbes, 1*, 407–410.

42. Hase, K., Kawano, K., Nochi, T., Pontes, G. S., Fukuda, S., Ebisawa, M., Kadokura, K., et al., (2009). Uptake through glycoprotein 2 of FimH(+) bacteria by M cells initiates mucosal immune response. *Nature, 462*, 226–230.

43. Thomas, W., (2008). Catch bonds in adhesion. *Annu. Rev. Biomed. Eng., 10*, 39–57.

44. Marshall, B. T., Long, M., Piper, J. W., Yago, T., McEver, R. P., & Zhu, C., (2003). Direct observation of catch bonds involving cell-adhesion molecules. *Nature, 423*, 190–193.

45. Anderson, B. N., Ding, A. M., Nilsson, L. M., Kusuma, K., Tchesnokova, V., Vogel, V., Sokurenko, E. V., & Thomas, W. E., (2007). Weak rolling adhesion enhances bacterial surface colonization. *J. Bacteriol., 189*, 1794–1802.

46. Thomas, W. E., Nilsson, L. M., Forero, M., Sokurenko, E. V., & Vogel, V., (2004). Shear-dependent 'stick-and-roll' adhesion of type 1 fimbriated *Escherichia coli. Mol. Microbiol., 53*, 1545–1557.

47. Sokurenko, E. V., Vogel, V., & Thomas, W. E., (2008). Catch-bond mechanism of force-enhanced adhesion: Counterintuitive, elusive, but widespread? *Cell Host. Microbe., 4*, 314–323.

48. Aprikian, P., Interlandi, G., Kidd, B. A., Le Trong, I., Tchesnokova, V., Yakovenko, O., Whitfield, M. J., et al., (2011). The bacterial fimbrial tip acts as a mechanical force sensor. *PLoS Biol., 9*, e1000617.

49. Thomas, W. E., Trintchina, E., Forero, M., Vogel, V., & Sokurenko, E. V., (2002). Bacterial adhesion to target cells enhanced by shear force. *Cell, 109*, 913–923.

50. Nilsson, L. M., Thomas, W. E., Sokurenko, E. V., & Vogel, V., (2006). Elevated shear stress protects *Escherichia coli* cells adhering to surfaces via catch bonds from detachment by soluble inhibitors. *Appl. Environ. Microbiol., 72*, 3005–3010.

51. Walker, S. J., Pizzato, M., Takeuchi, Y., & Devereux, S., (2002). Heparin binds to murine leukemia virus and inhibits Env-independent attachment and infection. *J. Virol., 76*, 6909–6918.

52. Connell, B. J., & Lortat-Jacob, H., (2013). Human immunodeficiency virus and heparan sulfate: From attachment to entry inhibition. *Front Immunol., 4*, 385.

53. Sherer, N. M., Jin, J., & Mothes, W., (2010). Directional spread of surface-associated retroviruses regulated by differential virus-cell interactions. *J. Virol., 84*, 3248–3258.

54. Comstock, L. E., & Kasper, D. L., (2006). Bacterial glycans: Key mediators of diverse host immune responses. *Cell, 126*, 847–850.

55. Acharya, S., Shukla, S., Mahajan, S. N., & Diwan, S. K., (2010). Molecular mimicry in human diseases-phenomena or epiphenomena? *J. Assoc. Physicians India, 58*, 163–168.

56. Behar, S. M., & Porcelli, S. A., (1995). Mechanisms of autoimmune disease induction. The role of the immune response to microbial pathogens. *Arthritis Rheum., 38*, 458–476.

57. Severi, E., Hood, D. W., & Thomas, G. H., (2007). Sialic acid utilization by bacterial pathogens. *Microbiology, 153*, 2817–2822.

58. Varki, A., & Gagneux, P., (2012). Multifarious roles of sialic acids in immunity. *Ann. N. Y. Acad. Sci., 1253*, 16–36.

59. Sharon, N., (2006). Carbohydrates as future anti-adhesion drugs for infectious diseases. *Biochim. Biophys. Acta, 1760*, 527–537.

60. Bernfield, M., Gotte, M., Park, P. W., Reizes, O., Fitzgerald, M. L., Lincecum, J., & Zako, M., (1999). Functions of cell surface heparan sulfate proteoglycans. *Annu. Rev. Biochem., 68*, 729–777.

61. Baumler, A. J., Tsolis, R. M., & Heffron, F., (1996). Contribution of fimbrial operons to attachment to and invasion of epithelial cell lines by *Salmonella typhimurium. Infect Immun., 64*, 1862–1865.

62. Sakarya, S., Gokturk, C., Ozturk, T., & Ertugrul, M. B., (2010). Sialic acid is required for nonspecific adherence of *Salmonella enterica* ssp. enterica serovar Typhi on Caco-2 cells. *FEMS Immunol. Med. Microbiol., 58*, 330–335.

63. Misselwitz, B., Kreibich, S. K., Rout, S., Stecher, B., Periaswamy, B., & Hardt, W. D., (2011). *Salmonella enterica serovar Typhimurium* binds to HeLa cells via Fim-mediated reversible adhesion and irreversible type three secretion system 1-mediated docking. *Infect. Immun., 79*, 330–341.

64. Krogfelt, K. A., Bergmans, H., & Klemm, P., (1990). Direct evidence that the FimH protein is the mannose-specific adhesin of *Escherichia coli* type 1 fimbriae. *Infect. Immun., 58*, 1995–1998.

65. Van, D. V. A. W., Baumler, A. J., Tsolis, R. M., & Heffron, F., (1998). Multiple fimbrial adhesins are required for full virulence of *Salmonella typhimurium* in mice. *Infect. Immun., 66*, 2803–2808.

66. Hicks, S., Frankel, G., Kaper, J. B., Dougan, G., & Phillips, A. D., (1998). Role of intimin and bundle-forming pili in enteropathogenic *Escherichia coli* adhesion to pediatric intestinal tissue *in vitro. Infect. Immun., 66*, 1570–1578.

67. Ilver, D., Arnqvist, A., Ogren, J., Frick, I. M., Kersulyte, D., Incecik, E. T., Berg, D. E., et al., (1998). Helicobacter pylori adhesin binding fucosylated histo-blood group antigens revealed by retagging. *Science, 279*, 373–377.

68. Green, C., (1989). The ABO, Lewis, and related blood group antigens: A review of structure and biosynthesis. *FEMS Microbiol. Immunol., 1*, 321–330.

69. Yamaoka, Y., (2008). Roles of helicobacter pylori BabA in gastroduodenal pathogenesis. *World J. Gastroenterol., 14*, 4265–4272.

70. Teneberg, S., Miller-Podraza, H., Lampert, H. C., Evans, D. J. Jr., Evans, D. G., Danielsson, D., & Karlsson, K. A., (1997). Carbohydrate binding specificity of the neutrophil-activating protein of Helicobacter pylori. *J. Biol. Chem., 272*, 19067–19071.

71. Valkonen, K. H., Wadstrom, T., & Moran, A. P., (1997). Identification of the N-acetylneuraminyllactose-specific laminin-binding protein of helicobacter pylori. *Infect. Immun., 65*, 916–923.

72. Odenbreit, S., Till, M., Hofreuter, D., Faller, G., & Haas, R., (1999). Genetic and functional characterization of the alpAB gene locus essential for the adhesion of helicobacter pylori to human gastric tissue. *Mol. Microbiol., 31*, 1537–1548.

73. Snelling, W. J., Moran, A. P., Ryan, K. A., Scully, P., McGourty, K., Cooney, J. C., Annuk, H., & O'Toole, P. W., (2007). HorB (HP0127) is a gastric epithelial cell adhesin. *Helicobacter, 12*, 200–209.

74. Edwards, J. L., & Apicella, M. A., (2004). The molecular mechanisms used by Neisseria gonorrhoeae to initiate infection differ between men and women. *Clin. Microbiol. Rev., 17*, 965–981.

75. Van, P. J. P., Duensing, T. D., & Cole, R. L., (1998). Entry of OpaA+ gonococci into HEp-2 cells requires concerted action of glycosaminoglycans, fibronectin and integrin receptors. *Mol. Microbiol., 29*, 369–379.

76. Duensing, T. D., & Putten, J. P., (1998). Vitronectin binds to the gonococcal adhesin OpaA through a glycosaminoglycan molecular bridge. *Biochem. J., 334*(Pt. 1), 133–139.

77. Duensing, T. D., & Van, P. J. P., (1997). Vitronectin mediates internalization of Neisseria gonorrhoeae by Chinese hamster ovary cells. *Infect. Immun., 65*, 964–970.

78. Dehio, M., Gomez-Duarte, O. G., Dehio, C., & Meyer, T. F., (1998). Vitronectin-dependent invasion of epithelial cells by *Neisseria gonorrhea* involves alpha(v) integrin receptors. *FEBS Lett., 424*, 84–88.

79. Grassme, H., Gulbins, E., Brenner, B., Ferlinz, K., Sandhoff, K., Harzer, K., Lang, F., & Meyer, T. F., (1997). Acidic sphingomyelinase mediates entry of *N. gonorrhea* into nonphagocytic cells. *Cell, 91*, 605–615.

80. Schmidt, M. A., Riley, L. W., & Benz, I., (2003). Sweet new world: Glycoproteins in bacterial pathogens. *Trends Microbiol., 11*, 554–561.

81. Paoletti, L. C., Ross, R. A., & Johnson, K. D., (1996). Cell growth rate regulates expression of group B streptococcus type III capsular polysaccharide. *Infect. Immun., 64*, 1220–1226.

82. Hornef, M. W., Wick, M. J., Rhen, M., & Normark, S., (2002). Bacterial strategies for overcoming host innate and adaptive immune responses. *Nat. Immunol., 3*, 1033–1040.

83. Gunn, J. S., Lim, K. B., Krueger, J., Kim, K., Guo, L., Hackett, M., & Miller, S. I., (1998). PmrA-PmrB-regulated genes necessary for 4-aminoarabinose lipid A modification and polymyxin resistance. *Mol. Microbiol., 27*, 1171–1182.

84. Ernst, R. K., Yi, E. C., Guo, L., Lim, K. B., Burns, J. L., Hackett, M., & Miller, S. I., (1999). Specific lipopolysaccharide found in cystic fibrosis airway *Pseudomonas aeruginosa*. *Science, 286*, 1561–1565.

85. Griffiss, J. M., O'Brien, J. P., Yamasaki, R., Williams, G. D., Rice, P. A., & Schneider, H., (1987). Physical heterogeneity of neisserial lipooligosaccharides reflects

oligosaccharides that differ in apparent molecular weight, chemical composition, and antigenic expression. *Infect. Immun., 55,* 1792–1800.

86. Smith, H., Parsons, N. J., & Cole, J. A., (1995). Sialylation of neisserial lipopolysaccharide: A major influence on pathogenicity. *Microb. Pathog., 19,* 365–377.

87. Van, P. J. P., (1993). Phase variation of lipopolysaccharide directs interconversion of invasive and immuno-resistant phenotypes of *Neisseria gonorrhoeae. EMBO J., 12,* 4043–4051.

88. Mandrell, R. E., (1992). Further antigenic similarities of *Neisseria gonorrhea* lipooligosaccharides and human glycosphingolipids. *Infect. Immun., 60,* 3017–3020.

89. Donlan, R. M., (2001). Biofilm formation: A clinically relevant microbiological process. *Clin. Infect. Dis., 33,* 1387–1392.

90. Donlan, R. M., (2000). Role of biofilms in antimicrobial resistance. *ASAIO J., 46,* S47–52.

91. Williams, I., Venables, W. A., Lloyd, D., Paul, F., & Critchley, I., (1997). The effects of adherence to silicone surfaces on antibiotic susceptibility in *Staphylococcus aureus. Microbiology, 143*(Pt. 7), 2407–2413.

92. Hoyle, B. D., Wong, C. K., & Costerton, J. W., (1992). Disparate efficacy of tobramycin on $Ca^{(2+)}$, $Mg^{(2+)}$, and HEPES-treated Pseudomonas aeruginosa biofilms. *Can J. Microbiol., 38,* 1214–1218.

93. Park, P. W., Reizes, O., & Bernfield, M., (2000). Cell surface heparan sulfate proteoglycans: Selective regulators of ligand-receptor encounters. *J. Biol. Chem., 275,* 29923–29926.

94. Arribas, J., Coodly, L., Vollmer, P., Kishimoto, T. K., Rose-John, S., & Massague, J., (1996). Diverse cell surface protein ectodomains are shed by a system sensitive to metalloprotease inhibitors. *J. Biol. Chem., 271,* 11376–11382.

95. Kainulainen, V., Wang, H., Schick, C., & Bernfield, M., (1998). Syndecans, heparan sulfate proteoglycans, maintain the proteolytic balance of acute wound fluids. *J. Biol. Chem., 273,* 11563–11569.

96. Park, P. W., Pier, G. B., Hinkes, M. T., & Bernfield, M., (2001). Exploitation of syndecan-1 shedding by *Pseudomonas aeruginosa* enhances virulence. *Nature, 411,* 98–102.

97. Park, P. W., Pier, G. B., Preston, M. J., Goldberger, O., Fitzgerald, M. L., & Bernfield, M., (2000). Syndecan-1 shedding is enhanced by LasA, a secreted virulence factor of Pseudomonas aeruginosa. *J. Biol. Chem., 275,* 3057–3064.

98. Hooper, L. V., & Gordon, J. I., (2001). Glycans as legislators of host-microbial interactions: Spanning the spectrum from symbiosis to pathogenicity. *Glycobiology, 11,* 1R–10R.

99. Olofsson, S., & Bergstrom, T., (2005). Glycoconjugate glycans as viral receptors. *Ann. Med., 37,* 154–172.

100. Ji, X., Chen, Y., Faro, J., Gewurz, H., Bremer, J., & Spear, G. T., (2006). Interaction of human immunodeficiency virus (HIV) glycans with lectins of the human immune system. *Curr. Protein Pept. Sci., 7,* 317–324.

101. Spear, P. G., (2004). Herpes simplex virus: Receptors and ligands for cell entry. *Cell Microbiol., 6,* 401–410.

102. Balzarini, J., (2007). Carbohydrate-binding agents: A potential future cornerstone for the chemotherapy of enveloped viruses? *Antivir. Chem. Chemother., 18,* 1–11.

103. Scanlan, C. N., Offer, J., Zitzmann, N., & Dwek, R. A., (2007). Exploiting the defensive sugars of HIV-1 for drug and vaccine design. *Nature, 446*, 1038–1045.

104. Ghazal, P., Gonzalez, A. J. C., Garcia-Ramirez, J. J., Kurz, S., & Angulo, A., (2000). Viruses: Hostages to the cell. *Virology, 275*, 233–237.

105. Holland, J., & Domingo, E., (1998). Origin and evolution of viruses. *Virus Genes., 16*, 13–21.

106. Domingo, E., (2003). Host-microbe interactions: Viruses. Complexities of virus-cell interactions. *Curr. Opin. Microbiol., 6*, 383–385.

107. Haywood, A. M., (1994). Virus receptors: Binding, adhesion strengthening, and changes in viral structure. *J. Virol., 68*, 1–5.

108. Matrosovich, M. N., Matrosovich, T. Y., Gray, T., Roberts, N. A., & Klenk, H. D., (2004). Neuraminidase is important for the initiation of influenza virus infection in human airway epithelium. *J. Virol., 78*, 12665–12667.

109. Rogers, G. N., & D'Souza, B. L., (1989). Receptor binding properties of human and animal H1 influenza virus isolates. *Virology, 173*, 317–322.

110. Connor, R. J., Kawaoka, Y., Webster, R. G., & Paulson, J. C., (1994). Receptor specificity in human, avian, and equine H2 and H3 influenza virus isolates. *Virology, 205*, 17–23.

111. Matrosovich, M. N., Gambaryan, A. S., Teneberg, S., Piskarev, V. E., Yamnikova, S. S., Lvov, D. K., Robertson, J. S., & Karlsson, K. A., (1997). Avian influenza A viruses differ from human viruses by recognition of sialyloligosaccharides and gangliosides and by a higher conservation of the HA receptor-binding site. *Virology, 233*, 224–234.

112. Weigel, P. H., & DeAngelis, P. L., (2007). Hyaluronan synthases: A decade-plus of novel glycosyltransferases. *J. Biol. Chem., 282*, 36777–36781.

113. Esko, J. D., & Selleck, S. B., (2002). Order out of chaos: Assembly of ligand binding sites in heparan sulfate. *Annu. Rev. Biochem., 71*, 435–471.

114. Trowbridge, J. M., & Gallo, R. L., (2002). Dermatan sulfate: New functions from an old glycosaminoglycan. *Glycobiology, 12*, 117R–125R.

115. Lehel, F., & Hadhazy, G., (1966). Effect of heparin on herpes simplex virus infection in the rabbit. *Acta Microbiol. Acad. Sci. Hung., 13*, 197–203.

116. Nahmias, A. J., & Kibrick, S., (1964). Inhibitory effect of heparin on herpes simplex virus. *J. Bacteriol., 87*, 1060–1066.

117. Feyzi, E., Trybala, E., Bergstrom, T., Lindahl, U., & Spillmann, D., (1997). Structural requirement of heparan sulfate for interaction with herpes simplex virus type-1 virions and isolated glycoprotein C. *J. Biol. Chem., 272*, 24850–24857.

118. Laquerre, S., Argnani, R., Anderson, D. B., Zucchini, S., Manservigi, R., & Glorioso, J. C., (1998). Heparan sulfate proteoglycan binding by herpes simplex virus type 1 glycoproteins B and C, which differ in their contributions to virus attachment, penetration, and cell-to-cell spread. *J. Virol., 72*, 6119–6130.

119. Spear, P. G., Shieh, M. T., Herold, B. C., WuDunn, D., & Koshy, T. I., (1992). Heparan sulfate glycosaminoglycans as primary cell surface receptors for herpes simplex virus. *Adv. Exp. Med. Biol., 313*, 341–353.

120. Hadigal, S. R., Agelidis, A. M., Karasneh, G. A., Antoine, T. E., Yakoub, A. M., Ramani, V. C., Djalilian, A. R., et al., (2015). Heparanase is a host enzyme required for herpes simplex virus-1 release from cells. *Nat. Commun., 6*, 6985.

121. Alkhatib, G., Combadiere, C., Broder, C. C., Feng, Y., Kennedy, P. E., Murphy, P. M., & Berger, E. A., (1996). CC CKR5: Arantes, MIP-1alpha, MIP-1beta receptor as a fusion cofactor for macrophage-tropic HIV-1. *Science, 272*, 1955–1958.

122. Deng, H., Liu, R., Ellmeier, W., Choe, S., Unutmaz, D., Burkhart, M., Di Marzio, P., et al., (1996). Identification of a major co-receptor for primary isolates of HIV-1. *Nature, 381*, 661–666.

123. Dragic, T., Litwin, V., Allaway, G. P., Martin, S. R., Huang, Y., Nagashima, K. A., Cayanan, C., et al., (1996). HIV-1 entry into CD4+ cells is mediated by the chemokine receptor CC-CKR-5. *Nature, 381*, 667–673.

124. Wei, X., Decker, J. M., Wang, S., Hui, H., Kappes, J. C., Wu, X., Salazar-Gonzalez, J. F., et al., (2003). Antibody neutralization and escape by HIV-1. *Nature, 422*, 307–312.

125. Pejchal, R., Doores, K. J., Walker, L. M., Khayat, R., Huang, P. S., Wang, S. K., Stanfield, R. L., et al., (2011). A potent and broad neutralizing antibody recognizes and penetrates the HIV glycan shield. *Science, 334*, 1097–1103.

126. Leonard, C. K., Spellman, M. W., Riddle, L., Harris, R. J., Thomas, J. N., & Gregory, T. J., (1990). Assignment of intrachain disulfide bonds and characterization of potential glycosylation sites of the type 1 recombinant human immunodeficiency virus envelope glycoprotein (gp120) expressed in Chinese hamster ovary cells. *J. Biol. Chem. 265*, 10373–10382.

127. Reitter, J. N., Means, R. E., & Desrosiers, R. C., (1998). A role for carbohydrates in immune evasion in AIDS. *Nat. Med., 4*, 679–684.

128. Binley, J. M., Ban, Y. E., Crooks, E. T., Eggink, D., Osawa, K., Schief, W. R., & Sanders, R. W., (2010). Role of complex carbohydrates in human immunodeficiency virus type 1 infection and resistance to antibody neutralization. *J. Virol., 84*, 5637–5655.

129. Crispin, M., & Doores, K. J., (2015). Targeting host-derived glycans on enveloped viruses for antibody-based vaccine design. *Curr. Opin. Virol., 11*, 63–69.

130. Lore, K., Sonnerborg, A., Brostrom, C., Goh, L. E., Perrin, L., McDade, H., Stellbrink, H. J., et al., (2002). Accumulation of DC-SIGN+CD40+ dendritic cells with reduced CD80 and CD86 expression in lymphoid tissue during acute HIV-1 infection. *AIDS, 16*, 683–692.

131. Iwasaki, A., (2016). Exploiting mucosal immunity for antiviral vaccines. *Annu. Rev. Immunol., 34*, 575–608.

132. Loris, R., (2002). Principles of structures of animal and plant lectins. *Biochim. Biophys. Acta, 1572*, 198–208.

133. Saifuddin, M., Hart, M. L., Gewurz, H., Zhang, Y., & Spear, G. T., (2000). Interaction of mannose-binding lectin with primary isolates of human immunodeficiency virus type 1. *J. Gen. Virol., 81*, 949–955.

134. Izquierdo-Useros, N., Naranjo-Gomez, M., Archer, J., Hatch, S. C., Erkizia, I., Blanco, J., Borras, F. E., et al., (2009). Capture and transfer of HIV-1 particles by mature dendritic cells converges with the exosome-dissemination pathway. *Blood, 113*, 2732–2741.

135. Van, M. T., Eggink, D., Boot, M., Tuen, M., Hioe, C. E., Berkhout, B., & Sanders, R. W., (2011). HIV-1 N-glycan composition governs a balance between dendritic cell-mediated viral transmission and antigen presentation. *J. Immunol., 187*, 4676–4685.

**CHAPTER 7**

# Glycome in Microbial Infections and Immune Evasion

ABID QURESHI

*Biomedical Informatics Centre, Sher-i-Kashmir Institute of Medical Sciences (SKIMS), Srinagar, Jammu and Kashmir, India, E-mail: abidbioinf@gmail.com*

## ABSTRACT

Carbohydrate-mediated host-pathogen interaction has been implicated not only in a pathogenic attachment to cells but also in eliciting immune responses against pathogens. These interactions are facilitated by the highly specific complementarity between the carbohydrates and their protein-binding partners, the lectins. The latter interact with carbohydrates via hydrogen bonds, hydrophobic, and electrostatic interactions as well as metal ion coordination complexes. The carbohydrates on the lipopolysaccharide (LPS) capsule of pathogenic bacteria have been identified to mediate the attachment to the host cell. Carbohydrates are also involved in binding of certain bacterial toxins to host cells through gangliosides. Likewise, viruses such as influenza possess sialic acid binding surface glycoproteins (hemagglutinins (HAs)) that are responsible for initiating the viral infection. Many protozoans possess proteins that attach to heparan sulfate proteoglycans on host cells to initiate the infection. Furthermore, lectins such as mannose binding proteins (MBPs) on immune cells help in the identification of pathogenic organisms via the sugars present on them. Since sugar mediated interactions play a critical role in microbial pathogenesis, researchers have devised pertinent glycomimetics to thwart the invasion. This chapter tries to provide mechanistic insights of role of glycans in mediating host-pathogen interaction and their immune responses.

## 7.1 INTRODUCTION

Carbohydrates play a pivotal role in numerous cell adhesion events including attachment of pathogens such as bacteria, viruses, as well as microbial toxins to the host cells [1]. They are also involved in immune response and identification of foreign cells [2]. Carbohydrates have the ability to form specificity determinants which can be identified by complementary molecular structures on other biomolecules. The specificity needed for the binding is particularly afforded via complementary stereochemistry between the multifarious carbohydrates and proteins capable of binding carbohydrates [3]. Sugar-protein interactions are responsible for pathogen specificity of different cell types [4]. Complex sugars present on the surface of cells interact with carbohydrate binding protein receptors such as lectins, the sugar binding proteins [5]. The sugar binding proteins are very selective in differentiating diverse carbohydrates including their conjugates such as glycoproteins and glycolipids and hence play a major role in cell identification [6].

The sugar-lectin interactions have been implicated in multiple cellular activities including the attachment of pathogens like bacteria and viruses and immune responses such as leukocyte recruitment (Figure 7.1) [7]. Consequently, the sugar gamut on the host cell determines its vulnerability to microbial pathogens [8]. Bacteria use their surface lectins to bind cellular surface oligosaccharides during their attachment phase [9]. Macrophages also phagocytose bacteria via sugar-lectin binding [10]. Microbes having surface mannose glycans attach to the mannose specific lectins on the macrophages [11]. The capricious lipopolysaccharide (LPS) sugars on the surface of bacterial cells are also involved in the ensuing infection and immune retort [12]. Certain bacteria produce disease-causing toxins that bind sugar elements of molecules like gangliosides on the cell surface [13].

Similarly, viruses extensively make use of glycans as receptors. Some viruses also use chemokine receptors, glycolipids, and integrins as co-receptors [14]. For example, influenza virus binds to N-acetylneuraminic acid on host cell surface for attachment [15]. However, at the same time complex glycans such as glycocalyx and mucins can act as barricades to prevent virus entry into the cell [16]. In pathogenic protozoans, glyco-conjugates, and glycan-binding proteins play an important role in infection initiation and immune evasion [17]. Various lectins on the immune cell membrane distinguish self and non-self or pathogenic glycans. For instance, interaction of C-type lectins with glycans causes antigen internalization and further signaling reactions [18]. It is clear that sugar mediated interactions play a pivotal role in cellular as well as host pathogen relationship since a variety of microbes also make use of use glycoproteins and glycolipids on host cells for initiation of infection (Figure 7.2) [19]. As a result, scientists have devised apposite saccharides and glyco-mimetics to hinder this interaction [20]. The present chapter focuses on the molecular basis and role of different carbohydrates in mediating host-pathogen interactions in bacteria, viruses, protozoa as well as the immune responses against these pathogens.

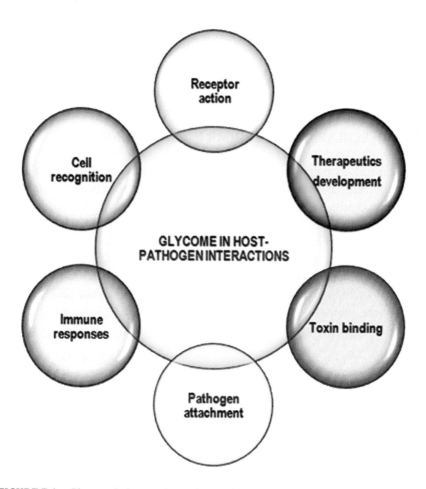

**FIGURE 7.1**  Glycome in host-pathogen interactions.

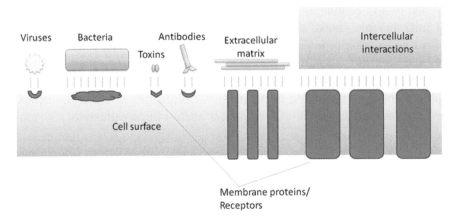

**FIGURE 7.2** Sugar mediated interactions implicated in host-pathogen and cellular interactions.

## 7.2  MOLECULAR BASIS OF SUGAR INTERACTIONS

Interaction of carbohydrates with other biomolecules is responsible for a variety of cellular roles such as cell adhesion, specificity determination, cell-cell recognition, etc. In addition, the sugars are also involved in host-pathogen interactions like attachment of viruses and bacteria [21]. The selective binding is usually facilitated by the stereochemical complementarity amid the sugar and the sugar binding proteins [22]. The sugars are usually present on the cell surface and interact with protein receptors like lectins (Table 7.1) which are very specific for carbohydrate moiety [23]. Lectins interact with their sugar partners using hydrogen bonds, hydrophobic, and electrostatic interactions and metal ion coordination complexes (Figure 7.3) [24].

**TABLE 7.1**  Commonly Known Lectins and Their Carbohydrate Ligands Implicated in Host Pathogen Interactions

| SL. No. | Lectin | Ligand | Pathogen |
|---------|--------|--------|----------|
| 1. | Botulinum neurotoxin B | NeuAcα2,3Galβ4Glc | *Clostridium botulinum* |
| 2. | FimH | Man | *Escherichia coli* |
| 3. | PapG | GalNAcβ3Galα4-Galβ4-GlcβCer | |
| 4. | Hemagglutinin | NeuAcα2,3βGalβ3-GlcNAcβ3Galβ4Glc | *Influenza virus* |

**TABLE 7.1** *(Continued)*

| SL. No. | Lectin | Ligand | Pathogen |
|---------|--------|--------|----------|
| 5. | PA-IL | Gal | *Pseudomonas* |
| 6. | PA-IIL | L-Fuc | *aeruginosa* |
| 7. | RS-IIL | Man | *Ralstonia solanacearum* |
| 8. | Enterotoxin B | Galβ4Glc | *Staphylococcus aureus* |
| 9. | Cholera toxin | NPαGal | *Vibrio cholerae* |

*Abbreviations:* NeuAc: N-acetylglucosamine, Gal: Galactose, Glc: Glucose, Man: Mannose, Cer: Ceramide, Fuc: Fucose, NP: Nitrophenyl, α, β denote the orientation of the OH group at the anomeric carbon.

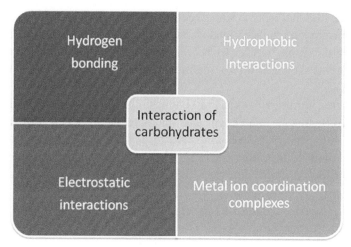

**FIGURE 7.3** Molecular basis of sugar-protein interactions. Represents different ways of interaction of carbohydrates with proteins.

Hydrogen bonds contribute greatly to the specificity of sugar-protein interactions. The hydrogen bonds are formed by linking of hydroxyl groups on the sugar with certain amino acids like serine, arginine, glutamine, asparagine, glutamic acid, aspartic acid, etc., [24a, 25]. Hydrophobic interactions such as the stacking of sugars on side chains of aromatic amino acids have been described in certain bacterial toxins and galectins. The methyl group of acetamido saccharides also interacts with aromatic amino acids of lectins via hydrophobic interactions [26]. Other types of ionic interactions have been described in sialic acids and glycoaminoglycans. Certain C-type lectins possess protein-bound $Ca^{2+}$ which is able to interact with saccharide hydroxyls [27]. The sugars as well as proteins are also bound to water molecules via hydrogen bonding with certain water molecules being conserved in such structures. During interaction or complex formation, hydrogen bonds with water molecules are swapped by glycan-protein bonds [28].

## 7.3   SUGARS INVOLVED IN BACTERIAL PATHOGENESIS

Carbohydrates play vital role in the binding of bacteria to the host cell [1c]. The adhesin proteins on bacterial surface or pilli have been identified to mediate the attachment [29]. The sugars on host as well as bacterial surface are also responsible in the infection and immune reactions [30]. The carbohydrates on the pathogenic bacteria are components of LPS or capsular polysaccharides having variant structures in the serotypes [12]. In *Homo sapiens*, urinary tract infections (UTIs) have been correlated with *E. coli* generating the Type-1, P, S or F1C fimbriae that possess adhesion proteins. These proteins show affinity for mannose, galabiose, and sialylated galactose moieties [31].

The type 1 fimbriae having FimH adhesin have been widely studied. The adhesin possesses an attachment site for mannose moieties [32]. Likewise the PapG adhesins of *E. coli* that bind to galabiose glycolipids have been correlated with kidney disease [33]. The bacteria bind to the host tissue surface through multiple fimbriae [34]. Other disease causing bacteria like *P. aeruginosa, S. pneumoniae, H influenza,* etc., need GalNAcb1-4Gal for adhesion [35]. *P. aeruginosa* further generates virulent lectins, PA-IL (*P. aeruginosa* Lectin-I) and PA-IIL (*P. aeruginosa* Lectin-II) having affinity for galactose and fucose, that regulate quorum sensing and biofilm formation in lung diseases [36]. *H. pylori*, the causative agent of stomach lesions and cancer, has been well researched for its adhesion affinity for the diverse gangliosides [37]. *H. pylori* recognizes sugars such as Le$^b$, neolacto, ganglio, and sulfated structures using the adhesins like BabA, SabA, and HP0721, etc., [38]. Researchers are working to develop antibiotics to prevent the bacterial infection at the very initial step by exploiting the adhesion system of the pathogens [39]. Design of inhibitors against the adhesion process that can be used as therapeutics against the pathogens [40].

Further numerous pathogenic bacteria like *V. cholerae* and *S. dyzenteriae* generate toxins that bind to tissue surface via defined sugar elements for instance the GM1 ganglioside or Gb3 globo-triaosylceramide [13]. The toxins per se cause millions of deaths per year globally [41]. For example, the AB$_5$

cholera toxins possess the subunit A (disease causing) enclosed by five sugar attaching B subunits. The B subunits are bound to the tissue surfaces and are responsible for delivering the disease causing subunit during the initial phase of infection. Similar toxins produced by other pathogenic bacteria include Shiga toxins (*S. dyzenteriae*) which bind to Gb3 and enterotoxin (*E. coli*) that attaches via GM1 [13]. Synthetic ligands (usually oligosaccharide derivatives) are being designed to bind these toxins and hence hold back the disease at an early stage [42].

## 7.4  SUGARS INVOLVED IN VIRAL PATHOGENESIS

The first step in the interaction between virus and host often occurs via terminal sugars of surface glycoproteins [1a, 17]. The virus also uses co-receptors like integrins or chemokine receptors for attachment and fusion with the cell membrane [43]. The influenza virus initiates contact to the host cell sialic acid receptors using its sialic acid binding surface hemagglutinin (HA) glycoprotein [44]. On the other hand, the neuraminidase (NA) enzyme of influenza cleaves sialic acid residues from glycosides to facilitate virus release and spread infection [45]. NA inhibitors such as oseltamivir (tamiflu) are valuable for fighting influenza infection [46].

The human parainfluenza virus surface glycoprotein HA-NA also attaches to host cells using sialic acid receptors [47]. Another fusion (F) glycoprotein on the surface of this virus helps in the fusion of viral and cellular membranes for virus internalization [48]. Similarly, the porcine transmissible gastroenteritis virus S fusion protein also binds to sialic acids as well as aminopeptidase N protein receptor [49]. Likewise, polyomaviruses attach to sialic acid receptors to enter the host cell [50].

Noroviruses employ blood group antigen oligosaccharides as receptors and as a result diverse noroviruses display dissimilar receptor affinities [51]. Certain parvoviruses bind to sialic acid as well as transferrin receptor for initiating infection [52]. Adeno-associated viruses attach to glycans having terminal sialic acids while others bind to heparan sulfate (Figure 7.4) [53]. Similarly, the herpes simplex virus (HSV) glycoproteins gB and gC bind to heparan sulfate on host cell surface for penetration [54]. The human immunodeficiency virus (HIV) envelope glycoprotein gp120 attaches to cell surfaces using the T-cell CD4 receptor and CCR5 chemokine co-receptor [55]. The glycoprotein gp41 brings the virus and host cell membranes in close contact for fusion and ensuing entry of the virus [56]. Both gp120 and gp41 are heavily glycosylated which to evade the immune response [57].

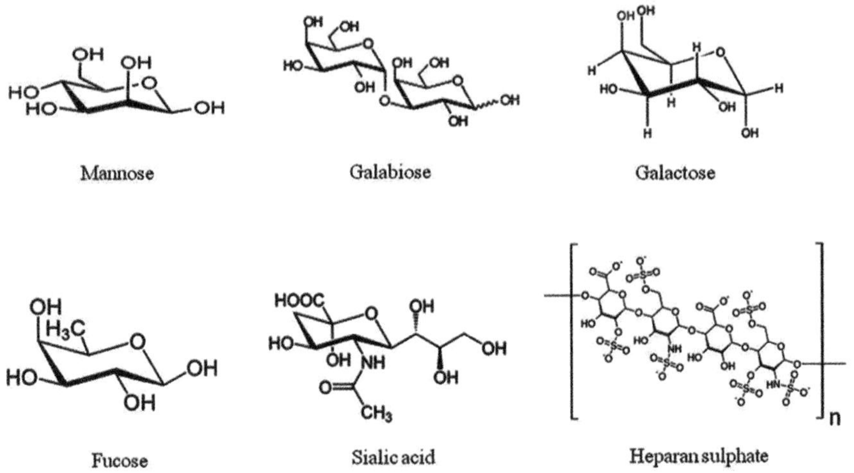

**FIGURE 7.4**    List of some common sugars frequently involved in host-pathogen interactions.

## 7.5   SUGARS INVOLVED IN PROTOZOAN PATHOGENESIS

Parasitic protozoans possess specialized carbohydrates and glycan-binding proteins that aid them to invade the host [58]. The glyco-conjugates also form a shielding barrier to swindle the immune response [59]. The malaria causing *P. falciparum* circumsporozoite protein attaches to heparan sulfate proteoglycans on hepatocytes to initiate the infection [60]. The *Plasmodium* merozoites also cause the discharge of glycosylphosphatidylinositols (GPIs) virulence factors which greatly contributes to malaria pathogenesis [61]. Sleeping sickness causing trypanosomes escape the host defenses by varying their surface GPI-anchored glycoproteins [62]. For instance, *T. cruzi* has a layer of mucins and glycosylinositolphospholipids (GIPLs) that not only protects the pathogen but also aids it in binding to macrophages [63].

In *Leishmania*, surface lipophosphoglycan (LPG) mediates the attachment of the parasite in the mid-gut of the sandfly [64]. In the mammalian host, this LPG also provides protection to the pathogen against the complement system [65]. In *Entamoeba histolytica*, the GPI-anchored Gal/GlcNAc binds to colonic mucins and hence plays a decisive role in survival of the pathogen [66]. Apicomplexan parasites stimulate the immune system via the sugars present on their surface [67]. *Acanthamoeba keratitis*, the pathogen that causes eye infections, binds to host cells through a mannose-binding protein [68]. Likewise, the severe diarrhea causing *Cryptosporidium parvum* possesses surface glycoproteins GP900 and GP40 having sugar residues that bind to αGalNAc-specific lectins that are implicated in the attachment of pathogens to host tissue by adhering to internal mucus [69]. The study of antigens is quite promising in development of diagnostics and vaccines against these pathogens [70].

## 7.6   SUGARS INVOLVED IN PLANT PATHOGEN INTERACTIONS

In plants, carbohydrates are not only responsible for acting as energy sources but also as signaling molecules in various metabolic pathways [71]. Moreover, they play a significant role in host-pathogen interactions [72]. Carbohydrates have been reported to turn on a variety of pattern recognition and pathogenesis-related genes [73]. Synchronized interaction of carbohydrate and hormonal pathways leads to efficient defense reactions in plants [74]. Phloem mobile carbohydrates such as raffinose and trehalose are known to accumulate isoflavones during plant defense responses [75]. Galactinol and raffinose have been implicated to arouse plant resistance pathways during pathogen confrontation [76]. Likewise studies have revealed that trehalose activates plant defense genes to defy powdery mildew attack in wheat [77]. Plant resistance genes have also been shown to be activated by oligogalacturonides released from the cell wall in retort to fungal infections [78]. In rice, the monosaccharide sugar psicose up-regulates the plant immunity against bacterial blight [79].

Sucrose has also been reported to elicit endogenous signals to advance immune reactions against plant pathogens via accumulation of antimicrobial anthocyanins [80]. Pattern recognition proteins like cell wall invertases that break sucrose to fructose and glucose have also been shown to be essential for the production of reactive oxygen species (ROS). The latter act as triggers in the activation of plant defense signaling [81]. Furthermore, pathogen invertase enzymes and carbohydrate transporters have also been implicated to influence host pathogen interactions in plants [82].

## 7.7   SUGARS INVOLVED IN INFLAMMATORY RESPONSE

The lectin proteins on host defense cells recognize the sugars present on pathogenic organisms [10]. Collectins such as the mannose binding proteins (MBPs) are vital for recognizing pathogens as well as for the commencement of complement system [83]. MBPs attach to sugars like mannose, glucose, and N-acetylglucosamine that are commonly present on bacteria, viruses, and fungi. MBPs that get attached to microbes are ingested by macrophages via their collectin receptors. The MBP-pathogen complex can also instigate the complement system to kill the pathogen [84]. Macrophages also have other lectin receptors such as CD206, CD207, CD209, CD301, etc., that aid in the endocytosis of microbes [85]. Diverse types of siglecs or sialic acid binding lectins reported in hematopoietic and other immune cells also help in recognition of pathogenic microorganisms [86].

Galectins also take part in immune responses by attaching to different microbes. Galectins can further function as cytokines and adhesion molecules. Macrophages secrete galectins 3 and 9 that bind lipo-phosphoglycans of certain bacteria and fungi and also take part in creating oligomers of microbes [87]. DC-SIGN, a C-type-lectin receptor on dendritic cells (DCs) has been associated with HIV infection since the lectin binds high-mannose-type N-linked glycans (NLGs) on gp120 of HIV. The HIV thus bound on DCs is carried from mucosa to lymph nodes where the virus drifts to T-cells via contact with CD4 and other co-receptors for further infection [88]. Galectins 1 and 9 are implicated in programmed cell death during T-cell selection. T-cells with self-reactive receptors are destined for apoptosis via galectin mediated cross-linking of glycoproteins with core-2 O-linked glycans [89]. Similarly, T-cells with siglec-1 binds to sialylated mucins MUC-1 and CD43 on macrophages to stimulate the immune system [90]. Galectin-1 on stromal cells is also implicated in B-cell development via synapse formation between stromal and pre-B cells [91]. Likewise, Siglec-2 regulates B-cell signaling via interaction with sialylated glycans of B-cells [90, 92]. Consequently, carbohydrate moieties and their binding partners play critical roles in host-pathogen relationships and immune responses.

## 7.8  CONCLUSION

Highly specific complementarity between the carbohydrates and their protein binding partners or lectins are responsible for the adhesion of pathogenic microbes and toxins to the host cell surface. Immune cells also make use of lectins to recognize pathogens via their sugars signatures. The sugar-binding proteins are very selective in differentiating diverse carbohydrates, including their conjugates such as glycoproteins and glycolipids. The capability of therapeutic polysaccharides and sugar analogs to restrain the initiation of microbial infection can lead to the development of promising antimicrobial agents.

## 7.9 CHALLENGES AND FUTURE PERSPECTIVES

Despite advances in host-pathogen glycobiology, many questions remain unanswered. For instance, the critical number of receptors/co-receptors required for a pathogen to result in productive infection is unclear. In addition, researchers need to look for alternative mechanisms and pathways that may be used by the microbes to enter the host cell. The role of the relative affinity of carbohydrate ligands for cellular receptors in productive attachment is also unclear. In addition, the signaling networks that activate pathogen entry and transport need to be further explored. To design potent carbohydrate-based microbial binding inhibitors, we further need to have high-resolution structural information about the binding domains of the receptor proteins. This approach can lead to the discovery of novel antimicrobial agents.

## KEYWORDS

- glycans
- glycomimetics
- herpes simplex virus
- immune response
- lectins
- microbial pathogenesis

## REFERENCES

1. (a) Brandley, B. K., & Schnaar, R. L., (1986). Cell-surface carbohydrates in cell recognition and response. *J. Leukoc. Biol., 40*(1), 97–111. (b) Bucior, I., Scheuring, S., Engel, A., & Burger, M. M., (2004). Carbohydrate-carbohydrate interaction provides adhesion force and specificity for cellular recognition. *J. Cell Biol., 165*(4), 529–537. (c) Pieters, R. J., (2011). Carbohydrate mediated bacterial adhesion. *Adv. Exp. Med. Biol., 715*, 227–240.
2. Cobb, B. A., & Kasper, D. L., (2005). Coming of age: Carbohydrates and immunity. *Eur. J. Immunol., 35*(2), 352–356.
3. Haseley, S. R., Talaga, P., Kamerling, J. P., & Vliegenthart, J. F., (1999). Characterization of the carbohydrate binding specificity and kinetic parameters of lectins by using surface plasmon resonance. *Anal. Biochem., 274*(2), 203–210.
4. Zopf, D., & Roth, S., (1996). Oligosaccharide anti-infective agents. *Lancet, 347*(9007), 1017–1021.

5.  Weis, W. I., & Drickamer, K., (1996). Structural basis of lectin-carbohydrate recognition. *Annu. Rev. Biochem.*, *65*, 441–473.
6.  Yamauchi, K., Tomita, M., Giehl, T. J., & Ellison, R. T., (1993). Antibacterial activity of lactoferrin and a pepsin-derived lactoferrin peptide fragment (3rd edn.). *Infect. Immun.*, *61*(2), 719–728.
7.  (a) Ni, Y., & Tizard, I., (1996). Lectin-carbohydrate interaction in the immune system. *Vet. Immunol. Immunopathol.*, *55*(1–3), 205–223. (b) Sequeira, L., (1978). Lectins and their role in host-pathogen specificity. *Annu. Rev. Phytopathol.*, *16*, 453–481.
8.  Cohen, M., (2015). Notable aspects of glycan-protein interactions. *Biomolecules*, *5*(3), 2056–2072.
9.  Sharon, N., (1987). Bacterial lectins, cell-cell recognition, and infectious disease. *FEBS Lett.*, *217*(2), 145–157.
10. Linehan, S. A., Martinez-Pomares, L., & Gordon, S., (2000). Macrophage lectins in host defense. *Microbes. Infect.*, *2*(3), 279–288.
11. Stahl, P. D., (1992). The mannose receptor and other macrophage lectins. *Curr. Opin. Immunol.*, *4*(1), 49–52.
12. Caroff, M., & Karibian, D., (2003). Structure of bacterial lipopolysaccharides. *Carbohydr. Res.*, *338*(23), 2431–2447.
13. Beddoe, T., Paton, A. W., Le, N. J., Rossjohn, J., & Paton, J. C., (2010). Structure, biological functions and applications of the AB5 toxins. *Trends Biochem. Sci.*, *35*(7), 411–418.
14. Flint, S., Enquist, L., Racaniello, V., & Skalka, A., (2009). *Principles of Virology*. ASM Press, Washington, DC.
15. Rogers, G. N., Herrler, G., Paulson, J. C., & Klenk, H. D., (1986). Influenza C virus uses 9-O-acetyl-N-acetylneuraminic acid as a high affinity receptor determinant for attachment to cells. *J. Biol. Chem.*, *261*(13), 5947–5951.
16. Pickles, R. J., Fahrner, J. A., Petrella, J. M., Boucher, R. C., & Bergelson, J. M., (2000). Retargeting the coxsackievirus and adenovirus receptor to the apical surface of polarized epithelial cells reveals the glycocalyx as a barrier to adenovirus-mediated gene transfer. *J. Virol.*, *74*(13), 6050–6057.
17. Kato, K., & Ishiwa, A., (2015). The role of carbohydrates in infection strategies of enteric pathogens. *Trop. Med. Health*, *43*(1), 41–52.
18. Van, V. S. J., Den, D. J., Gringhuis, S. I., Geijtenbeek, T. B., & Van, K. Y., (2007). Innate signaling and regulation of dendritic cell immunity. *Curr. Opin. Immunol.*, *19*(4), 435–440.
19. Karlsson, K. A., (2001). Pathogen-host protein-carbohydrate interactions as the basis of important infections. *Adv. Exp. Med. Biol.*, *491*, 431–443.
20. (a) Sharon, N., (2006). Carbohydrates as future anti-adhesion drugs for infectious diseases. *Biochim. Biophys. Acta*, *1760*(4), 527–537. (b) Sattin, S., & Bernardi, A., (2016). Glycoconjugates and glycomimetics as microbial anti-adhesives. *Trends Biotechnol.*, *34*(6), 483–495.
21. (a) Karlsson, K. A., (1999). Bacterium-host protein-carbohydrate interactions and pathogenicity. *Biochem. Soc. Trans.*, *27*(4), 471–474. (b) Olofsson, S., & Bergstrom, T., (2005). Glycoconjugate glycans as viral receptors. *Ann. Med.*, *37*(3), 154–172.
22. Hooper, L. V., & Gordon, J. I., (2001). Glycans as legislators of host-microbial interactions: Spanning the spectrum from symbiosis to pathogenicity. *Glycobiology*, *11*(2), 1R–10R.

23. Sharon, N., & Lis, H., (2004). History of lectins: From hemagglutinins to biological recognition molecules. *Glycobiology, 14*(11), 53R–62R.

24. (a) Bhattacharyya, L., & Brewer, C. F., (1988). Lectin-carbohydrate interactions. Studies of the nature of hydrogen bonding between D-galactose and certain D-galactose-specific lectins, and between D-mannose and concanavalin A. *Eur. J. Biochem., 176*(1), 207–212. (b) Elgavish, S., & Shaanan, B., (1997). Lectin-carbohydrate interactions: Different folds, common recognition principles. *Trends Biochem. Sci., 22*(12), 462–467.

25. Allison, S. D., Chang, B., Randolph, T. W., & Carpenter, J. F., (1999). Hydrogen bonding between sugar and protein is responsible for inhibition of dehydration-induced protein unfolding. *Arch Biochem. Biophys., 365*(2), 289–298.

26. Basu, S., Ghosh, S., Basu, M., Hawes, J. W., Das, K. K., Zhang, B. J., Li, Z. X., et al., (1990). Carbohydrate and hydrophobic-carbohydrate recognition sites (CARS and HY-CARS) in solubilized glycosyltransferases. *Indian J. Biochem. Biophys., 27*(6), 386–395.

27. Weis, W. I., Drickamer, K., & Hendrickson, W. A., (1992). Structure of a C-type mannose-binding protein complexed with an oligosaccharide. *Nature, 360*(6400), 127–134.

28. Kadirvelraj, R., Foley, B. L., Dyekjaer, J. D., & Woods, R. J., (2008). Involvement of water in carbohydrate-protein binding: Concanavalin A revisited. *J. Am. Chem. Soc., 130*(50), 16933–16942.

29. Pizarro-Cerda, J., & Cossart, P., (2006). Bacterial adhesion and entry into host cells. *Cell, 124*(4), 715–727.

30. Ohlsen, K., Oelschlaeger, T. A., Hacker, J., & Khan, A. S., (2009). Carbohydrate receptors of bacterial adhesins: Implications and reflections. *Top Curr. Chem., 288,* 17–65.

31. (a) Bower, J. M., Eto, D. S., & Mulvey, M. A., (2005). Covert operations of uropathogenic *Escherichia coli* within the urinary tract. *Traffic, 6*(1), 18–31. (b) Bjornham, O., Nilsson, H., Andersson, M., & Schedin, S., (2009). Physical properties of the specific PapG-galabiose binding in *E. coli* P pili-mediated adhesion. *Eur. Biophys. J., 38*(2), 245–254.

32. Martinez, J. J., Mulvey, M. A., Schilling, J. D., Pinkner, J. S., & Hultgren, S. J., (2000). Type 1 pilus-mediated bacterial invasion of bladder epithelial cells. *EMBO J., 19*(12), 2803–2812.

33. Siliano, P. R., Rocha, L. A., Medina-Pestana, J. O., & Heilberg, I. P., (2010). The role of host factors and bacterial virulence genes in the development of pyelonephritis caused by *Escherichia coli* in renal transplant recipients. *Clin. J. Am. Soc. Nephrol., 5*(7), 1290–1297.

34. Krachler, A. M., & Orth, K., (2013). Targeting the bacteria-host interface: Strategies in anti-adhesion therapy. *Virulence, 4*(4), 284–2894.

35. Chang, Y. C., Uchiyama, S., Varki, A., & Nizet, V., (2012). Leukocyte inflammatory responses provoked by pneumococcal sialidase. *MBio, 3*(1).

36. Chemani, C., Imberty, A., De Bentzmann, S., Pierre, M., Wimmerova, M., Guery, B. P., & Faure, K., (2009). Role of LecA and LecB lectins in Pseudomonas aeruginosa-induced lung injury and effect of carbohydrate ligands. *Infect. Immun., 77*(5), 2065–2075.

37. Sgouros, S. N., & Bergele, C., (2006). Clinical outcome of patients with Helicobacter pylori infection: The bug, the host, or the environment? *Postgrad. Med. J., 82*(967), 338–342.

38. (a) Pang, S. S., Nguyen, S. T., Perry, A. J., Day, C. J., Panjikar, S., Tiralongo, J., Whisstock, J. C., & Kwok, T., (2014). The three-dimensional structure of the extracellular adhesion domain of the sialic acid-binding adhesin SabA from Helicobacter pylori. *J. Biol. Chem., 289*(10), 6332–6340; (b) Day, C. J., Semchenko, E. A., & Korolik, V., (2012). Glycoconjugates play a key role in campylobacter jejuni infection: Interactions between host and pathogen. *Front Cell Infect. Microbiol., 2*, 9.

39. Hartmann, M., Papavlassopoulos, H., Chandrasekaran, V., Grabosch, C., Beiroth, F., Lindhorst, T. K., & Rohl, C., (2012). Inhibition of bacterial adhesion to live human cells: Activity and cytotoxicity of synthetic mannosides. *FEBS Lett., 586*(10), 1459–1465.

40. Cozens, D., & Read, R. C., (2012). Anti-adhesion methods as novel therapeutics for bacterial infections. *Expert. Rev. Anti. Infect. Ther., 10*(12), 1457–1468.

41. (a) Barton, B. C., Jones, T. F., Vugia, D. J., Long, C., Marcus, R., Smith, K., Thomas, S., et al., (2011). Deaths associated with bacterial pathogens transmitted commonly through food: Foodborne diseases active surveillance network (FoodNet), 1996–2005. *J. Infect. Dis., 204*(2), 263–7; (b) Henkel, J. S., Baldwin, M. R., & Barbieri, J. T., (2010). Toxins from bacteria. *EXS, 100*, 1–29.

42. Sisu, C., Baron, A. J., Branderhorst, H. M., Connell, S. D., Weijers, C. A., De Vries, R., Hayes, E. D., et al., (2009). The influence of ligand valency on aggregation mechanisms for inhibiting bacterial toxins. *Chembiochem, 10*(2), 329–337.

43. Falasca, L., Agrati, C., Petrosillo, N., Di Caro, A., Capobianchi, M. R., Ippolito, G., & Piacentini, M., (2015). Molecular mechanisms of Ebola virus pathogenesis: Focus on cell death. *Cell Death Differ., 22*(8), 1250–1259.

44. Leung, H. S., Li, O. T., Chan, R. W., Chan, M. C., Nicholls, J. M., & Poon, L. L., (2012). Entry of influenza A virus with a alpha2, 6-linked sialic acid binding preference requires host fibronectin. *J. Virol., 86*(19), 10704–10713.

45. Shtyrya, Y. A., Mochalova, L. V., & Bovin, N. V., (2009). Influenza virus neuraminidase: Structure and function. *Acta Naturae, 1*(2), 26–32.

46. McKimm-Breschkin, J. L., (2013). Influenza neuraminidase inhibitors: Antiviral action and mechanisms of resistance. *Influenza Other Respir. Viruses, 7*(1), 25–36.

47. Winger, M., & Von, I. M., (2012). Exposing the flexibility of human parainfluenza virus hemagglutinin-neuraminidase. *J. Am. Chem. Soc., 134*(44), 18447–18452.

48. (a) Earp, L. J., Delos, S. E., Park, H. E., & White, J. M., (2005). The many mechanisms of viral membrane fusion proteins. *Curr. Top Microbiol. Immunol., 285*, 25–66. (b) Yin, H. S., Wen, X., Paterson, R. G., Lamb, R. A., & Jardetzky, T. S., (2006). Structure of the parainfluenza virus 5 F protein in its metastable, prefusion conformation. *Nature, 439*(7072), 38–44.

49. Krempl, C., & Herrler, G., (2001). Sialic acid binding activity of transmissible gastroenteritis coronavirus affects sedimentation behavior of virions and solubilized glycoproteins. *J. Virol., 75*(2), 844–849.

50. Gee, G. V., Dugan, A. S., Tsomaia, N., Mierke, D. F., & Atwood, W. J., (2006). The role of sialic acid in human polyomavirus infections. *Glycoconj. J., 23*(1/2), 19–26.

51. Marionneau, S., Ruvoen, N., Le Moullac-Vaidye, B., Clement, M., Cailleau-Thomas, A., Ruiz-Palacois, G., Huang, P., et al., (2002). Norwalk virus binds to histo-blood group antigens present on gastroduodenal epithelial cells of secretor individuals. *Gastroenterology, 122*(7), 1967–1977.

52. Vihinen-Ranta, M., Suikkanen, S., & Parrish, C. R., (2004). Pathways of cell infection by parvoviruses and adeno-associated viruses. *J. Virol., 78*(13), 6709–6714.

53. Mietzsch, M., Broecker, F., Reinhardt, A., Seeberger, P. H., & Heilbronn, R., (2014). Differential adeno-associated virus serotype-specific interaction patterns with synthetic heparins and other glycans. *J. Virol., 88*(5), 2991–3003.

54. Spear, P. G., (2004). Herpes simplex virus: Receptors and ligands for cell entry. *Cell Microbiol., 6*(5), 401–410.

55. Myszka, D. G., Sweet, R. W., Hensley, P., Brigham-Burke, M., Kwong, P. D., Hendrickson, W. A., Wyatt, R., et al., (2000). Energetics of the HIV gp120-CD4 binding reaction. *Proc. Natl. Acad. Sci. U.S.A, 97*(16), 9026–9031.

56. Bar, S., & Alizon, M., (2004). Role of the ectodomain of the gp41 transmembrane envelope protein of human immunodeficiency virus type 1 in late steps of the membrane fusion process. *J. Virol., 78*(2), 811–820.

57. (a) Go, E. P., Chang, Q., Liao, H. X., Sutherland, L. L., Alam, S. M., Haynes, B. F., & Desaire, H., (2009). Glycosylation site-specific analysis of clade C HIV-1 envelope proteins. *J. Proteome Res., 8*(9), 4231–4242. (b) Wilen, C. B., Tilton, J. C., & Doms, R. W., (2012). HIV: Cell binding and entry. *Cold Spring Harb. Perspect. Med., 2*(8).

58. Cummings, R., & Turco, S., (2009). Parasitic infections. In: *Essentials of Glycobiology* (2nd edn.). Chapter 40. Cold Spring Harbor Laboratory Press.

59. Rodrigues, J. A., Acosta-Serrano, A., Aebi, M., Ferguson, M. A., Routier, F. H., Schiller, I., Soares, S., et al., (2015). Parasite glycobiology: A bittersweet symphony. *PLoS Pathog., 11*(11), e1005169.

60. Pinzon-Ortiz, C., Friedman, J., Esko, J., & Sinnis, P., (2001). The binding of the circumsporozoite protein to cell surface heparan sulfate proteoglycans is required for plasmodium sporozoite attachment to target cells. *J. Biol. Chem., 276*(29), 26784–26791.

61. Mbengue, B., Niang, B., Niang, M. S., Varela, M. L., Fall, B., Fall, M. M., Diallo, R. N., et al., (2016). Inflammatory cytokine and humoral responses to Plasmodium falciparum glycosylphosphatidylinositols correlates with malaria immunity and pathogenesis. *Immun. Inflamm. Dis., 4*(1), 24–34.

62. La Greca, F., & Magez, S., (2011). Vaccination against trypanosomiasis: Can it be done or is the trypanosome truly the ultimate immune destroyer and escape artist? *Hum. Vaccin., 7*(11), 1225–1233.

63. (a) Melo, R. C., & Machado, C. R., (2001). Trypanosoma cruzi: Peripheral blood monocytes and heart macrophages in the resistance to acute experimental infection in rats. *Exp. Parasitol., 97*(1), 15–23. (b) Brodskyn, C., Patricio, J., Oliveira, R., Lobo, L., Arnholdt, A., Mendonca-Previato, L., Barral, A., & Barral-Netto, M., (2002). Glycoinositolphospholipids from *Trypanosoma cruzi* interfere with macrophages and dendritic cell responses. *Infect. Immun., 70*(7), 3736–3743.

64. Jecna, L., Dostalova, A., Wilson, R., Seblova, V., Chang, K. P., Bates, P. A., & Volf, P., (2013). The role of surface glycoconjugates in Leishmania midgut attachment examined by competitive binding assays and experimental development in sand flies. *Parasitology, 140*(8), 1026–1032.

65. Gupta, G., Oghumu, S., & Satoskar, A. R., (2013). Mechanisms of immune evasion in leishmaniasis. *Adv. Appl. Microbiol., 82*, 155–184.

66. Frederick, J. R., & Petri, W. A. Jr., (2005). Roles for the galactose-/N-acetylgalactosamine-binding lectin of Entamoeba in parasite virulence and differentiation. *Glycobiology, 15*(12), 53R–59R.

67. Sibley, L. D., (2011). Invasion and intracellular survival by protozoan parasites. *Immunol. Rev., 240*(1), 72–91.

68. Panjwani, N., (2010). Pathogenesis of acanthamoeba keratitis. *Ocul. Surf., 8*(2), 70–79.

69. Cevallos, A. M., Bhat, N., Verdon, R., Hamer, D. H., Stein, B., Tzipori, S., Pereira, M. E., et al., (2000). Mediation of cryptosporidium parvum infection *in vitro* by mucin-like glycoproteins defined by a neutralizing monoclonal antibody. *Infect. Immun., 68*(9), 5167–5175.

70. Nyame, A. K., Kawar, Z. S., & Cummings, R. D., (2004). Antigenic glycans in parasitic infections: Implications for vaccines and diagnostics. *Arch Biochem. Biophys., 426*(2), 182–200.

71. Bolouri-Moghaddam, M. R., Le Roy, K., Xiang, L., Rolland, F., & Van, D. E. W., (2010). Sugar signaling and antioxidant network connections in plant cells. *FEBS J., 277*(9), 2022–2037.

72. Morkunas, I., Marczak, L., Stachowiak, J., & Stobiecki, M., (2005). Sucrose-induced lupine defense against *Fusarium oxysporum*. Sucrose-stimulated accumulation of isoflavonoids as a defense response of lupine to *Fusarium oxysporum. Plant Physiol. Biochem., 43*(4), 363–373.

73. Jones, J. D., & Dangl, J. L., (2006). The plant immune system. *Nature, 444*(7117), 323–329.

74. Leon, P., & Sheen, J., (2003). Sugar and hormone connections. *Trends Plant Sci., 8*(3), 110–116.

75. Hofmann, J., El Ashry, A. N., Anwar, S., Erban, A., Kopka, J., & Grundler, F., (2010). Metabolic profiling reveals local and systemic responses of host plants to nematode parasitism. *Plant J., 62*(6), 1058–1071.

76. Kim, M. S., Cho, S. M., Kang, E. Y., Im, Y. J., Hwangbo, H., Kim, Y. C., Ryu, C. M., et al., (2008). Galactinol is a signaling component of the induced systemic resistance caused by *Pseudomonas chlororaphis* O6 root colonization. *Mol. Plant Microbe. Interact., 21*(12), 1643–1653.

77. Tayeh, C., Randoux, B., Vincent, D., Bourdon, N., & Reignault, P., (2014). Exogenous trehalose induces defenses in wheat before and during a biotic stress caused by powdery mildew. *Phytopathology, 104*(3), 293–305.

78. Denoux, C., Galletti, R., Mammarella, N., Gopalan, S., Werck, D., De Lorenzo, G., Ferrari, S., et al., (2008). Activation of defense response pathways by OGs and Flg22 elicitors in *Arabidopsis* seedlings. *Mol. Plant, 1*(3), 423–445.

79. Kano, A., Hosotani, K., Gomi, K., Yamasaki-Kokudo, Y., Shirakawa, C., Fukumoto, T., Ohtani, K., et al., (2011). D-Psicose induces upregulation of defense-related genes and resistance in rice against bacterial blight. *J. Plant Physiol., 168*(15), 1852–1857.

80. Solfanelli, C., Poggi, A., Loreti, E., Alpi, A., & Perata, P., (2006). Sucrose-specific induction of the anthocyanin biosynthetic pathway in Arabidopsis. *Plant Physiol., 140*(2), 637–646.

81. (a) Proels, R. K., & Roitsch, T., (2009). Extracellular invertase LIN6 of tomato: A pivotal enzyme for integration of metabolic, hormonal, and stress signals is regulated by a diurnal rhythm. *J. Exp. Bot., 60*(6), 1555–1567. (b) Xiang, L., Le, R. K., Bolouri-Moghaddam, M. R., Vanhaecke, M., Lammens, W., Rolland, F., & Van, D. E. W., (2011). Exploring the neutral invertase-oxidative stress defense connection in *Arabidopsis thaliana. J. Exp. Bot., 62*(11), 3849–3862.

82. Bolouri, M. M. R., & Van, D. E. W., (2012). Sugars and plant innate immunity. *J. Exp. Bot., 63*(11), 3989–3998.

83. Holmskov, U. L., (2000). Collectins and collectin receptors in innate immunity. *APMIS Suppl, 100,* 1–59.

84. Turner, M. W., (2003). The role of mannose-binding lectin in health and disease. *Mol. Immunol., 40*(7), 423–429.

85. (a) Geijtenbeek, T. B., & Gringhuis, S. I., (2009). Signaling through C-type lectin receptors: Shaping immune responses. *Nat. Rev. Immunol., 9*(7), 465–479. (b) Hollmig, S. T., Ariizumi, K., Cruz, P. D. Jr., (2009). Recognition of non-self-polysaccharides by C-type lectin receptors dectin-1 and dectin-2. *Glycobiology, 19*(6), 568–575.

86. Crocker, P. R., Paulson, J. C., & Varki, A., (2007). Siglecs and their roles in the immune system. *Nat. Rev. Immunol., 7*(4), 2552–2566.

87. Thiemann, S., & Baum, L. G., (2016). Galectins and immune responses-just how do they do those things they do? *Annu. Rev. Immunol., 34,* 243–264.

88. Geijtenbeek, T. B., & Van, K. Y., (2003). DC-SIGN: A novel HIV receptor on DCs that mediates HIV-1 transmission. *Curr. Top Microbiol. Immunol., 276,* 31–54.

89. Bi, S., Earl, L. A., Jacobs, L., & Baum, L. G., (2008). Structural features of galectin-9 and galectin-1 that determine distinct T-cell death pathways. *J. Biol. Chem., 283*(18), 12248–12258.

90. Von, G. S., & Bochner, B. S., (2008). Basic and clinical immunology of siglecs. *Ann. N.Y. Acad. Sci., 1143,* 61–82.

91. Mourcin, F., Breton, C., Tellier, J., Narang, P., Chasson, L., Jorquera, A., Coles, M., Schiff, C., & Mancini, S. J., (2011). Galectin-1-expressing stromal cells constitute a specific niche for pre-BII cell development in mouse bone marrow. *Blood, 117*(24), 6552–6561.

92. Jellusova, J., & Nitschke, L., (2011). Regulation of B cell functions by the sialic acid-binding receptors siglec-G and CD22. *Front Immunol., 2,* 96.

# CHAPTER 8

# Acylation as a Vital Post-Glycosylation Modification of Proteins: Insights and Therapeutics Prospects

USMA MANZOOR, SNOBER SHABEER WANI, FASIL ALI, PARVAIZ A. DAR, and TANVEER ALI DAR

*Clinical Biochemistry, University of Kashmir, Hazratbal, Srinagar – 190006, Jammu and Kashmir, India, Phone: 91-9419639396, E-mail: tanveerali@kashmiruniversity.ac.in (T. A. Dar)*

## ABSTRACT

Acetylation is the most frequent mode of post-glycosylation modification of glycans such as glycoproteins, glycolipids, and proteoglycans observed in different animal phyla. This enigmatic modification is carried out by sialate-O-acetyltransferases (SOAT) at different positions of sialic acids attached to proteins in their activated forms. Many physiological processes like the catabolism of glycoconjugates, identification by viral hemagglutinins (HAs) and bacterial sialidases, developmental morphogenesis of tissues, and modulation of the alternative pathway of complement activation are known to be influenced by O-acetylation. Apart from their involvement in various types of diseases, their possible use as therapeutic targets has also been revealed recently. This chapter has tried to summarize the general mechanism and importance of O-acetylation as a post-glycosylation modification in humans, bacteria, and viruses. The role of this modification in different health hazards along with the therapeutic outlook has also been highlighted.

## 8.1 INTRODUCTION

Glycosylation, an intricate posttranslational modification, is generally found on many cellular surfaces and eukaryotic extracellular proteins. In this process, carbohydrate chains are attached to proteins through covalent bonds, which in turn are edited in the endoplasmic reticulum (ER) and Golgi complex by different biochemical reactions catalyzed by the enzymes glycosidases and glycosyltransferases. However, the regulation of glycan synthesis is not well understood due to the fact that glycan synthesis is neither a template driven process nor coded by any genome. In spite of this, the process generates a huge variety of glycans with multiple functions like cell-cell interaction, intra-, and inter-cellular transfer, folding of proteins, and their stability. However, these post-translationally bound glycans, especially glycoproteins/proteoglycans, can, in turn, be modified by various other functional groups like methylation, sulfation, phosphorylation, acylation, etc., via their hydroxyl/amino groups [1]. Modification of these glycans on biomolecules, in turn, leads to their modulation of biological function. Among all the modifications, acylation is one of the important biological modifications of the glycans (especially glycoproteins and glycolipids) that involve the acetyl group transfer in the simple or complex form to the hydroxyl group of carbohydrates associated with the glycan. In fact, sialic acid O-acetylation is the most frequent mode of acetylation observed in different phyla [1]. Sialic acids are negatively charged, hydrophilic sugars typically found at the terminal, non-reducing end of glycan structures of glycoconjugates (glycoproteins and glycolipids). Hence cell surfaces, glycoconjugates, and capsular polysaccharides of certain bacteria are chief sites where acetylated sialic acids are found. This chapter will provide an update of the current scenario of acylation as a post-glycosylation modification with its biological significance and therapeutic opportunities.

## 8.2   PHYSIOLOGICAL ROLE OF O-ACETYLATION

Among all kinds of glycans, sialic acid O-acetylation is the most frequently occurring posttranslational modification. Sialic acids are a class of sugars with a skeleton of nine carbons and present as $N$-acetylneuraminic acid and N-glycosylneuraminic acid or their substituted forms. They are naturally found at the non-reducing end of carbohydrates of different glycoconjugates. Many enzymatic reactions involved in the catabolism of glycoconjugates, identification by viral hemagglutinins (HAs) and bacterial sialidases, developmental morphogenesis of tissues, and modulation of the alternative pathway of complement activation are known to be influenced by O-acetylation [2]. In fact, O-acetylation impacts the sialic acid residues on glycans so as to modulate their ligand functions by: (i) decreasing their rate of degradation on glycoconjugates; (ii) generating new ligands; and (iii) veiling existing ligands. These impressions are sufficient enough to elucidate many important trends of O-acetylated sialic acid residues, such as stimulatory role on cell growth [3], upsurge, and diminution of glycoconjugate antigenicity, disposition as differentiation and tumor antigens. Indeed the introduction of increased hydrophobic entities on sialic acid leads to O-acetyl group mediated changes in the charge, size, hydrogen bond pattern, and conformation of glycoconjugates. O-acetylation is therefore considered as a key highly regulated tissue-specific posttranslational modification and not merely a developmentally regulated molecule. More importantly, it is known to be responsible for the pathogenesis of disease and metastasis. Certain properties of glycans like solubility, hydrophobicity, degradation, and molecular recognition are altered by changes in the levels of acylation [1]. From NMR studies, it has been demonstrated that O-acetylation does not alter the inclusive conformation of gangliosides, although a frail interaction takes place between the acetyl group and the carbohydrate beneath, but that does not change the approachability and elasticity of this group towards receptor protein [4]. Sialic acid-O-acetylation is considered as an imperative and ingenious post-glycosylation modification of biomolecules in humans as well as in lower organisms like bacteria and viruses.

### 8.2.1   IN BACTERIA

Certain gram-positive human pathogens like *L. monocytogenes, Staphylococcus aureus* and gram-negative ones such as *Neisseria gonorrhea, Neisseria meningitides*, and *Proteus mirabilis* have shown up to be O-acetylated at C6 hydroxyl group of N-acetylmuramyl residues in the peptidoglycans [5]. In fact, this modification confers resistance to peptidoglycans in the cell wall of bacteria against the hydrolytic activity of hen egg-white lysozyme [6]. Mutants of *Staphylococcus aureus* and *S. pneumonia* that are deficient in peptidoglycan O-acetylation turn out to be more responsive to exogenous lysozyme [7]. However, the O-acetyl groups do not affect the competence of *Chalaropsis* and *Streptomyces globisporus* against lysozyme [8]. Bacteria are well recognized to synthesize acetylated polysaccharides. Virtually all-important pathogens amongst bacteria are O-acetylated, signifying a connection between pathogenicity and O-acetylation. This has been clearly demonstrated in typhoid fever causing *Salmonella typhi*, where the O-acetyl groups stabilize the capsular polysaccharides that act as a virulent factor [9] meningococcal diseases causing pathogen *Neisseria meningitides* group C, exhibited a rather unwavering link between O-acetylation and pathogenicity. The majority of the strains from this group turn out to be acetylated at C7 and C8 of Neu5Ac, and overall about 85% of disease-causing strains among them were recognized as O-acetyl positive [10]. Various findings propose that O-acetylation masks the immunogenic epitopes of capsular polysaccharides in meningococci serogroup C and Y, consequently allowing them to flee the immune scrutiny. This indicates the enzyme responsible for O-acetylation may possibly comprise avirulence or fitness aspect. A similar kind of immuno-dominance was established in *E. coli* K1126 [11]. In addition, 21701 strains where a strong association was found between acetylation and escalating degree of sepsis in patients [12]. Brown seaweeds and a few bacteria, such as mucoid P. *aeruginosa*, produce alginates (polysaccharides) where mannuronate residues are acetylated on one or both $O_2$ and $O_3$ positions.

## 8.2.2   IN VIRUSES

O-acetylation either inhibits viral binding to sialic acid-containing glycoproteins or facilitates their binding by distinguishing between 4- or 9-O-acetylated Sialic acids. The attachment and ensuing assault by influenza C viruses require 9-O-acetylated sialic acids on the surface of the host cell, although this pattern of acetylation inhibits influenza A and B virus [13, 14] and malaria parasite attachment. It has been widely accepted that α, 2–6-linked Neu5Ac, and Neu5Ac9Lt are the excellent ligands for H1N1 and H3N2 influenza viruses. Similarly, 4-O-acetylated sialic acids are specific to mouse hepatitis virus strain S [15, 16].

## 8.2.3   IN HUMANS

Siglecs, sialic acid-dependent adhesion molecules belonging to immunoglobulin superfamily like sialoadhesin and CD22, are allegedly involved in macrophage-lymphocyte interactions, protection, and consistency of myelinated axons in myelin-associated glycoprotein (MAG). O-acetylation of sialic acids on these siglecs prevents them from binding the other glycoconjugates on cell surfaces apart from affecting the hematopoietic development. Acetylated glycans have also been reported to thwart opsonic killing and instigation of alternative complement pathway. Owing to this reason, O-acetyl groups modulate virulence in patients of cystic fibrosis [17]. Another example is the normal intestinal milieu, where O-acetylated sialoglycans defend against enteric bacterial flora to prevent mucin degradation [18]. Various gangliosides are regarded as tumor antigens or differentiation markers, particularly GD3 with acetylated sialic acid as terminal sugar. Likewise, 4-O-acetyl GM3 has been noticed in human colon carcinoma. Additionally, sialic acid O-acetylation of glycoproteins is known to resist the sialidases and concealing the binding sites on mammalian surfaces used by bacterial and viral pathogens.

### 8.2.4   O-ACETYLATION UNDER STRESS CONDITIONS: A PHYSIOLOGICAL RESPONSE

A small number of appropriate and accessible biomarkers are known to screen the physiological response during stress conditions. In fact, a new glycomic methodology has been used to recognize the new stress markers in which the O-acetylation profile of sialic acids in fish serum is monitored over a long-term period. The salmon liver has shown the presence of powerful machinery of sialic acid acetylation [19]. In fact, the majority of the glycans in the serum of salmon (about 83%) are monoacetylated sialic acids. It has been shown that under stress conditions, the cysteine proteinase inhibitors of Atlantic salmon undergo O-acetylation on sialic acids. Parallel findings on proteins like kininogen isolated from spotted wolfish and Atlantic cod showed similar di-O-acetylated sialic acids [20]. O-acetylation on the transferrin is believed to be a physiological response against stress [19]. It has also been reported that long-term stress exposure to fish can be immunosuppressive, whereas a temporary exposure to stress helps it to survive the offense and redeem homeostasis. During stress conditions, the levels of di-o-acetylated sialic acids increased enormously in the fish in comparison to mono-o-acetylated sialic acids that decreased during this period [21]. The sialic acid O-acetylation is associated with an acute response to stress since, the acetylation pattern returned to the one found in normal or control group after some time [21]. So far nothing has been reported on the role played by sialic acid O-acetylation in vertebrates during stress. Although, the above study suggests the role of hyper O-acetylation in salmon defense system, it seems quite reasonable to claim that during stress conditions the degree of O-acetylation might be altered.

## 8.3   MECHANISM OF ACETYLATION PATHWAY

Golgi apparatus (GA) is home to various enzymes involved in O-acetylation where Sialate-*O*-acetyltransferase attaches an acetyl co A group to a particular hydroxyl group of the glycan on the glycoprotein. For diverse sialyltransferases, CMP-activated sialic acids serve as substrates when transported to GA (Figure 8.1). All the hydroxyl positions of biologically occurring sialic acids, i.e., C7, C8 C9, and C4, can be acetylated. Typically, mono O-acetylated forms predominate; however, oligo O-acetylated derivatives are generated by acetylation at two or more hydroxyl groups. The acetyl groups for O-acetylation are derived from Ac CoA, transported through AcCoA-transporter into the Golgi where O-acetyltransferase (OAT) relocate them to glycosidically bound sialic acids probably at C7. An extra acetylation reaction is also catalyzed enzymatically at position C9 after the acetyl group migrates to C9 position [22]. A similar mechanism operates for sialic acid 4-O-acetylation. On the other hand, O-acetylation of C9 of sialic acid can occur directly as well. Three acetyltransferases that exclusively esterify sialic acids at different sites on glycans have been discovered and they include sialate-7-*O*-acetyltransferase, sialate-9-*O*-acetyltransferase, and sialate-4-*O*-acetyltransferase. Initially, acetyl groups are supposed to be added to C7 of sialic acid; yet they can successively transfer to position C9 of same residue. This migration of the ester group to the C9 position of sialic acid from C7 hydroxyl group is due to the activity of a specific enzyme probably mutase and not because of any physical process as claimed earlier. Moreover, the rapid rate of migration (3.75%/min) justifies it to be a non-physical process.

### 8.3.1   ENZYMES FOR O-ACETYLATION

In bacteria, two genes responsible for O-acetylation have been identified as *NeuO and NeuD*. In recent times CASD1 gene has emerged as an essential contender for 9-O-acetylation of sialic acids [23]. *In vitro* experiments using purified CASD1 N terminal luminal domain have revealed that the transfer of AcCoA to CMP-activated sialic acid takes place through the covalent acetyl-enzyme intermediate formation [24]. In fact, the studies confirm about CASD1 being a sialate OAT and a crucial enzyme for 9-O-acetylated glycan biosynthesis. The purification of sialate-O-acetyltransferase has proven to be very complicated. cDNAs that probably encode sialate-O-acetyltransferase have been isolated by cloning of human melanoma library in ganglioside expressing COS cells especially GD3 as substrates. A clone from the same library has shown to induce 9-O-acetylation, instructing a fusion protein of P3 plasmid sequence and tetracycline resistance gene repressor of bacteria. 9-O-acetylation has been induced by a different clone from the rat liver cDNA library in the sialic acid expressing COS cells; however, the encoded sequence was similar to the Vitamin D binding protein. Such results concerning different types of cDNAs involved in the 9-O-acetylation of sialic acid in COS cell lines, raise the chances of actual enzyme to be multimeric and challenging to clone [25]. Furthermore, a highly specific and soluble 9-O-acetyltransferase from *Campylobacter jejuni* has also been cloned. Likewise, human CASD1 gene translates into a transmembrane protein with a molecular weight of about 87.5kDa, and it displays a subtle sialate OAT activity. This protein has a serine-glycine-asparagine-histidine hydrolase domain and a C-terminal domain whose expression results in the formation of multimeric proteins. Apart from it there are 9 histidine and 22 lysine residues in the peptides that connect the transmembrane domains. All these properties comply with the prerequisites of acetyl CoA transporter, and it also lies within the intracellular compartment.

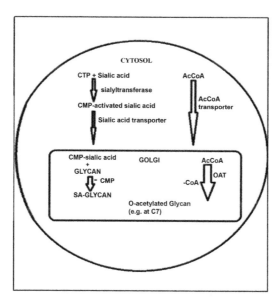

**FIGURE 8.1** Mechanism of acetylation pathway. (*Abbreviations:* CTP: Cytidine triphosphate, CMP: Cytidine monophosphate, SA-GLYCAN: Sialic acid associated glycan, Ac Co A: Acetyl coenzyme A, OAT: O Acetyl transferase.

The 9-O-acetyltransferase activity has been demonstrated in rat liver, bovine submandibular gland, *E. coli*, human colon (fetal and adult), sweat glands and melanomas due to the presence of 9-O-acetylated sialic acids and thus has garnered most of the attention. However, 4-O acetylation being less frequent and the enzyme involved, i.e., 4-O-acetyltransferase has been found in guinea pig liver and equine submandibular gland (Table 8.1). 4-O-acetyltransferase is shown to catalyze methyl group transfer from S-adenosylmethionine to C8 position of sialic acid in starfish *A. rubens*. In general human colon mucosa and bovine submandibular gland display highest concentration of O-acetylated sialic acids in their mucins, indicating high OAT activity. Additionally, increased activity of Sialate-O-acetyltransferase has been found in microsomal fractions and in ALL lymphoblasts that have a high quantity of O-acetylated sialic acids on their cell surfaces. An important class of enzyme namely Sialate O-acetylesterase exists in nature in order to regulate the sialic acid O-acetylation on glycoconjugates. Sialate O-acetylesterase catalyzes the de-O-acetylation of sialic acids and is reported to exist in two forms. The cytosolic SIAE carries out de-O-acetylation of cytosolic 9-O-acetylated sialic acids while as the lysosomal form of SIAE remove 9-O-acetyl groups of sialo-glycoconjugates.

**TABLE 8.1**   Occurrence of Sialate-O-Acetyltransferase Activities

| Cells and Tissues | Position of O-Acetylation | References |
|---|---|---|
| Bovine submandibular gland | C7 of sialic acid | [26, 27] |
| Human colon mucosa | C7 of Sia | [28–30] |
| Acute lymphoblastic leukemia | C7 of Sia | [31] |
| Human skin melanoma | C7 and C9 of Sia | [32] |
| Rat liver | C7 and C9 of Sia | [33, 34] |
| Guinea pig liver | C4 of Sia | [35] |
| *Campylobacter jejuni* | C9 of Sia | [36] |
| *Escherichia coli* | C7 and C9 of Sia | [37, 38] |

## 8.4   ROLE OF ACETYLATED GLYCANS IN DISEASES

### 8.4.1   IN CANCER

#### 8.4.1.1   ROLE OF GLYCOPROTEINS

The sialic acid O-acetylation at C-7, C-8, and C-9 positions, are often found on glycoproteins or glycolipids in all higher vertebrates and in few bacteria.

An increased O-acetylation of sialic acid bearing glycoproteins is a hallmark of oncogenic transformation in pre-B acute lymphoblastic leukemia (ALL). An increased expression of 9-O-acetylated sialoglycoprotein has been acknowledged as a significant determinant on ALL. It mediates the survival of peripheral blood mononuclear cells of ALL [39]. On the other hand, human colon carcinoma exhibit decreased sialic acid O-acetylation which in turn leads to enhanced sialyl Lewis X expression. Sialyl Lewis X, is actually a tumor-associated antigen and along with sialyl-Tn antigen correlates with the metastasis and progression of colorectal cancer [40]. However, some studies report about the unaltered expression of these antigens in normal and cancer cells [41, 42]. The interaction between sialyl Lewis-X antigen of carcinoma cells and E-selectin containing endothelial cells promote extravasation of metastatic cells [43, 44]. These studies infer the significance of sialic acid O-acetylation on the metastatic potential of colorectal cancer cells. In breast carcinoma, 9-O-acetylated sialoglycans were detected at plasma membrane and in the Golgi regions of normal and benign ducts of breast whereas in the carcinoma, they are dispersed disorderly in the cytoplasm.

## 8.4.1.2   ROLE OF GLYCOLIPIDS

The O-acetylated sialic acids especially 9-*O*-Ac-GD3 are reported to likely keep control of cell growth besides affecting microtubule network of neurites. They are recognized to be trademark of many cancers such as leukemia, colorectal cancers and in fact, skin melanoma, basalioma, breast cancer, and childhood. 9-O acetylated sialic acids that are spotted in embryonic development re-emerge as oncofetal antigens in some tumors. Interestingly O-acetyl GD3 endorses the continued existence of ALL due to its anti-apoptotic activity. It has been demonstrated by *in vitro* experiments that GD3 induces apoptosis in lymphoblasts whereas the similar concentration of 9-O-acetyl GD3 proved to be ineffective. Further, they failed in depolarizing mitochondrial membranes, releasing of cytochrome c and activation of caspase-3, unlike the unacetylated form of GD3 [45]. It seems that elevated cellular expression of such acetylated gangliosides prevents apoptosis of tumor cells. However, nothing has been reported on the effects of 7-*O*-acetyl GD3 on tumor cells. In fact, increased expression of acetylated gangliosides takes place during transformation of normal tissue to malignant tissue. Out of the total ganglioside fraction, 12% was found to be 9-O-acetylated in case of basalioma and in normal skin no O-acetylated gangliosides were evident

[46, 47]. Hence, 9-O-acetylated GD3 and GT3 represent a set of unusual gangliosides and hence provide a basis for diagnosis and therapy in human breast carcinoma.

## 8.4.2 IN IMMUNE SYSTEM

### 8.4.2.1 ROLE OF GLYCOPROTEINS

Differences in sialic acid O-acetylation also affect serum protein factor H binding. Classic studies with different strains of mice revealed that the amount of sialic acid on the erythrocytes might limit the degree of control on the alternate complement pathway activation [48]. However, later on, the extent of sialic acid side-chain, O-acetylation was shown to be responsible for this, and therefore, these modified Sias were not considered good targets for factor H binding [49]. Astonishingly sialic acid O-acetylation is actually disadvantageous to the bacterium in case of host-pathogen interaction, either reducing recognition by CD33rSiglecs or augmenting immunogenicity [50]. However, these modifications defend the bacteria against microbial sialidases and bacteriophage-binding proteins and hence survival. O-acetyl group on the sialic acids of glycoproteins obstructs the binding of all Siglecs. Similarly, a surface HA on influenza C virus particularly recognizes 9-O-acetylated sialic acids and is therefore used as a probe against them [51]. Such a particle has been called an InfC hemagglutinin-esterase-fusion protein (InfCHEF) as it holds 9-O-acetyl esterase activity and a fusion activity at acidic pH.

### 8.4.2.2 ROLE OF GLYCOLIPIDS

9-O-acetylation of ganglioside GD3 (CD60) is associated with differentiation and inhibiting the pro-apoptotic pathway in proliferating T-cells [52]. Similarly, it is responsible for the activation and correct development of B-cells. Since it prevents binding of Sia-binding immunoglobulin type lectin CD22 (SIGLEC-2) with the gangliosides on B-cells, which otherwise would inhibit B-cell activation and signaling [53]. On the other hand, CD60b and CD60c positive cells belonging to T-lymphocytes display 9 or 7 O-acetylated sialoglycans that are considered to be a differentiation marker, hence of interest from immunocytological perspective. 7-O-acetyl GD3 expressed on CD60c of T-cells is implicated in signaling as it promotes their proliferation. Furthermore, O-acetyl GD3 was advocated to be more immunogenic than non-acetylated GD3 and the antibodies against O-acetyl GD3 produced in melanoma patients cross-reacted with GD3; thus the antibody response of melanoma patients depends on O-acetylation. Interestingly medullo-blastoma patients exhibited IgM antibodies against-acetyl GD3 in their serum [54]. Therefore, the position of O-acetyl group on outer sialic acid is very important with respect to immunogenicity in mice and humans.

### 8.4.3   IN INFECTIOUS DISEASES

O-acetylated glycans are critically involved in bacterial and viral infections. Patients suffering from cystic fibrosis are believed to produce a thick biofilm in the form of alginates that helps bacteria to evade host defense antibiotic treatment. As a result of bacterial by products or environmental stimulus the O-acetylation in glycans has been suggested to enhance post-natally as it is actually missing before birth in rat colon. Likewise, increased amount of O-acetylation was observed in the colon of naturally aborted fetuses with confirmed bacterial infection as compared to induced aborted fetuses [55]. In fact, O-acetylation has been regarded as a reaction to bacterial occupation. In addition, several viruses such as influenza A virus bind with sialic acids on the host cell surface during initial stages of infection. However, influenza virus C needs 9-O-acetylation on sialic acids for attachment, but at the same time, this pattern of acetylation inhibits influenza A and B viruses from binding to the host cell surface. Thus a significant role is played by O-acetylation in the case of viral adhesion. Certain parainfluenza viruses have been shown to bind only modified sialic acids, particularly acetylated ones found in mammals. Any slight change in 5 or 9 position of sialic acids concludes with different binding affinities and mostly these hPIVs (human parainfluenza viruses) require hydrophobicity at the 5-position and the hydrophilicity at the 9-position of sialic acids. Additionally, other human diseases including type-1 diabetes, rheumatoid arthritis (RA) and some autoimmune disorders result due to any fault in sialate-*O*-acetylesterase activity [56, 57]. In fact a strong genetic link exists between functionally defective SIAE rare and polymorphic variants and relatively common autoimmune disorders [56]. Increased 9-O-acetylated sialic acids was also noted in a patient with Sjogren's syndrome on B-cells, due to heterozygous SIAE (C196F) variant. Similarly, SIAE (R393H) variant was noted in a patient of lupus displaying enhanced 9-O-acetylation of sialic acids. However, the presence of normal SIAE prevents this enhanced 9-O-acetylation and helps set a threshold for B-cell activation. Most probably by thwarting the movement of weakly self-reactive B-cells towards the T-cell zone and accordingly being at danger of somatic mutation. Therefore, SIAE may help to preserve tolerance in germinal centers [56].

## 8.5    ACETYLATION AS A DIAGNOSTIC MARKER AND THERAPEUTIC TARGET

The regulation of sialic acid O-acetylation by stimulating or inhibiting SIAE or SOAT activity, respectively, pledge to develop new drugs in ALL. In that way, sialate-O-acetyltransferases (SOAT) are the most potential and well thought out biomarkers to monitor and screen ALL patients [58]. Lymphoblast O-acetylated sialoglycoprotein exhibit 90, 120, and 135 kDa molecular weights and for this reason, measurement of O-acetylated glycan levels along with SOAT activity should be used to diagnose and follow the disease progression. Probe lectins such as *Cancer antennarius* lectin and achatinin H can also facilitate to detect the O-acetylated sialic acids on glycans but whether they are able to distinguish between 7 or 9 acetylated forms is not understood properly [59]. Similarly, the approach of targeting O-acetylated gangliosides on cell surfaces with specific antibodies is also utilized. Of late a substantial amount of attention has been given to O-acetyl GD2 as a suitable immunotherapeutic target for cancer treatment [60] because it may provide a substitute method to tackle the extreme toxicity associated with therapeutic anti-GD2 antibodies like dinutuximab [61]. This toxicity arises because of the binding of antibody to peripheral nerves that express GD2 and finally following Fc domain mediated CDC activity as advocated in animal studies [62]. Some findings introspected whether O-acetyl GD2 are expressed on normal tissues also, and tried to detect it with the help of mouse monoclonal antibody 8B6, that specifically binds to O-acetyl GD2 [60]. This mouse monoclonal antibody however stained only tumor cells and demonstrated antitumor activities by cell cycle arrest and apoptosis induction even in the absence of NK cells and complement system. Such results were comparable to other antibodies like anti-GD2 14G2a antibody. In fact, Terme and his co-workers developed a chimeric (mouse/human IgG1) version of 8B6 antibody aiming at O-acetyl-GD2 so as to smoothen the progress of clinical development of therapeutic antibodies [63]. Currently using this technique of chimeric monoclonal technology, anti-O-acetylated ganglioside antibodies for passive immunotherapy are further investigated. Similarly, an exclusive IgM response was witnessed against O-acetyl GD3 with an immunogen mix of 9-*O*-GD3, 7-*O*-GD3, and 7,9-di-*O*-GD3. Immunogen generated antibodies produced in humans reacted with hamster melanoma derived 7-O-GD3 containing cells and not with human melanoma derived 9-O-GD3 gangliosides. At the same time, murine antibodies recognized all types of O-acetyl GD3 molecules used for immunization [64]. On the other hand, O-acetylated Ganglioside GD1b became the potent inhibitor of astroblast and astrocytoma proliferation called neurostatin [65, 66].

## 8.6 CONCLUSION AND FUTURE PROSPECTS

The probable effects of O-acetylation on glycans are insightful, influencing many physiological and pathological phenomenons. This subtle chemical modification affects organisms of a diverse range, and the degree of O-acetylation on glycans depends on the relative activity of fascinating enzymes, i.e., SOAT. Several attempts have been made to clone the gene responsible for O-acetylation, but for some unknown reasons that have not been fully accomplished yet. For some instances where the enhanced expressions of these enzymes have been correlated with diseases, specific inhibitors should be explored to regulate their expression. There is a need to further investigate a comparative analysis of role of O-acetylation in other biological processes like in diseased conditions.

## KEYWORDS

- acetyl coenzyme A
- acetylation
- acute lymphoblastic leukemia
- glycans
- glycosylation
- post-glycosylation modification
- sialate-O-acetyltransferases
- sialidases

## REFERENCES

1. Muthana, S. M., Campbell, C. T., & Gildersleeve, J. C., (2012). Modifications of glycans: Biological significance and therapeutic opportunities. *ACS Chem. Biol., 7*(1), 31–43.
2. Varki, A., (1992). Diversity in the sialic acids. *Glycobiology, 2*(1), 25–40.
3. Schauer, R., Schmid, H., Pommerencke, J., Iwersen, M., & Kohla, G., (2001). Metabolism and role of O-acetylated sialic acids. *Adv. Exp. Med. Biol., 491*, 325–342.
4. Siebert, H. C., Von, D. L. C. W., Dong, X., Reuter, G., Schauer, R., Gabius, H. J., et al., (1996). Molecular dynamics-derived conformation and intramolecular interaction analysis of the N-acetyl-9-O-acetylneuraminic acid-containing ganglioside GD1a and NMR-based analysis of its binding to a human polyclonal immunoglobulin G fraction with selectivity for O-acetylated sialic acids. *Glycobiology, 6*(6), 561–572.
5. Vollmer, W., (2008). Structural variation in the glycan strands of bacterial peptidoglycan. *FEMS Microbiol. Rev., 32*(2), 287–306.
6. Brumfitt, W., Wardlaw, A. C., & Park, J. T., (1958). Development of lysozyme-resistance in micrococcus lysodiekticus and its association with an increased O-acetyl content of the cell wall. *Nature, 181*(4626), 1783, 1784.
7. Bera, A., Herbert, S., Jakob, A., Vollmer, W., & Gotz, F., (2005). Why are pathogenic staphylococci so lysozyme resistant? The peptidoglycan O-acetyltransferase OatA is the major determinant for lysozyme resistance of Staphylococcus aureus. *Mol. Microbiol., 55*(3), 778–787.
8. Hamada, S., Torii, M., Kotani, S., Masuda, N., Ooshima, T., Yokogawa, K., et al., (1978). Lysis of Streptococcus mutans cells with mutanolysin, a lytic enzyme prepared from a culture liquor of Streptomyces globisporus 1829. *Arch Oral Biol., 23*(7), 543–549.
9. Bystricky, S., & Szu, S. C., (1994). O-acetylation affects the binding properties of the carboxyl groups on the Vi bacterial polysaccharide. *Biophys. Chem., 51*(1), 1–7.
10. Arakere, G., & Frasch, C. E., (1991). Specificity of antibodies to O-acetyl-positive and O-acetyl-negative group C meningococcal polysaccharides in sera from vaccinees and carriers. *Infect. Immun., 59*(12), 4349–4356.
11. Orskov, F., Orskov, I., Sutton, A., Schneerson, R., Lin, W., Egan, W., et al., (1979). Form variation in *Escherichia coli* K1: Determined by O-acetylation of the capsular polysaccharide. *J. Exp. Med., 149*(3), 669–685.
12. Frasa, H., Procee, J., Torensma, R., Verbruggen, A., Algra, A., Rozenberg-Arska, M., et al., (1993). *Escherichia coli* in bacteremia: O-acetylated K1 strains appear to be more virulent than non-O-acetylated K1 strains. *J. Clin. Microbiol., 31*(12), 3174–3178.
13. Herrler, G., Rott, R., Klenk, H. D., Muller, H. P., Shukla, A. K., & Schauer, R., (1985). The receptor-destroying enzyme of influenza C virus is neuraminate-O-acetylesterase. *EMBO J., 4*(6), 1503–1506.
14. Higa, H. H., Rogers, G. N., & Paulson, J. C., (1985). Influenza virus hemagglutinins differentiate between receptor determinants bearing N-acetyl-, N-glycollyl-, and N,O-diacetylneuraminic acids. *Virology, 144*(1), 279–282.
15. Regl, G., Kaser, A., Iwersen, M., Schmid, H., Kohla, G., Strobl, B., et al., (1999). The hemagglutinin-esterase of mouse hepatitis virus strain S is a sialate-4-O-acetylesterase. *J. Virol., 73*(6), 4721–4727.
16. Schauer, R., (2000). Achievements and challenges of sialic acid research. *Glycoconj. J., 17*(7–9), 485–499.

17. Pier, G. B., Coleman, F., Grout, M., Franklin, M., & Ohman, D. E., (2001). Role of alginate O acetylation in resistance of mucoid Pseudomonas aeruginosa to opsonic phagocytosis. *Infect. Immun., 69*(3), 1895–1901.

18. Corfield, A. P., Wagner, S. A., Clamp, J. R., Kriaris, M. S., & Hoskins, L. C., (1992). Mucin degradation in the human colon: Production of sialidase, sialate O-acetylesterase, N-acetylneuraminate lyase, arylesterase, and glycosulfatase activities by strains of fecal bacteria. *Infect. Immun., 60*(10), 3971–3978.

19. Kvingedal, A. M., Rorvik, K. A., & Alestrom, P., (1993). Cloning and characterization of Atlantic salmon (*Salmo salar*) serum transferrin cDNA. *Mol. Mar. Biol. Biotechnol., 2*(4), 233–238.

20. Ylonen, A., Kalkkinen, N., Saarinen, J., Bogwald, J., & Helin, J., (2001). Glycosylation analysis of two cysteine proteinase inhibitors from Atlantic salmon skin: Di-O-acetylated sialic acids are the major sialic acid species on N-glycans. *Glycobiology, 11*(7), 523–531.

21. Liu, X., Afonso, L., Altman, E., Johnson, S., Brown, L., & Li, J., (2008). O-acetylation of sialic acids in N-glycans of Atlantic salmon (*Salmo salar*) serum is altered by handling stress. *Proteomics, 8*(14), 2849–2857.

22. Vandamme-Feldhaus, V., & Schauer, R., (1998). Characterization of the enzymatic 7-O-acetylation of sialic acids and evidence for enzymatic O-acetyl migration from C-7 to C-9 in bovine submandibular gland. *J. Biochem., 124*(1), 111–121.

23. Arming, S., Wipfler, D., Mayr, J., Merling, A., Vilas, U., Schauer, R., et al., (2011). The human Cas1 protein: A sialic acid-specific O-acetyltransferase? *Glycobiology, 21*(5), 553–564.

24. Baumann, A. M., Bakkers, M. J., Buettner, F. F., Hartmann, M., Grove, M., Langereis, M. A., et al., (2015). 9-O-Acetylation of sialic acids is catalysed by CASD1 via a covalent acetyl-enzyme intermediate. *Nat. Commun., 6*, 7673.

25. Shi, W. X., Chammas, R., & Varki, A., (1998). Induction of sialic acid 9-O-acetylation by diverse gene products: Implications for the expression cloning of sialic acid O-acetyltransferases. *Glycobiology, 8*(2), 199–205.

26. Lrhorfi, L. A., Srinivasan, G. V., & Schauer, R., (2007). Properties and partial purification of sialate-O-acetyltransferase from bovine submandibular glands. *Biol. Chem., 388*(3), 297–306.

27. Srinivasan, G. V., & Schauer, R., (2009). Assays of sialate-O-acetyltransferases and sialate-O-acetylesterases. *Glycoconj. J., 26*(8), 935–944.

28. Corfield, A. P., Myerscough, N., Warren, B. F., Durdey, P., Paraskeva, C., & Schauer, R., (1999). Reduction of sialic acid O-acetylation in human colonic mucins in the adenoma-carcinoma sequence. *Glycoconj. J., 16*(6), 307–317.

29. Shen, Y., Tiralongo, J., Iwersen, M., Sipos, B., Kalthoff, H., & Schauer, R., (2002). Characterization of the sialate-7(9)-O-acetyltransferase from the microsomes of human colonic mucosa. *Biol. Chem., 383*(2), 307–317.

30. Shen, Y., Kohla, G., Lrhorfi, A. L., Sipos, B., Kalthoff, H., Gerwig, G. J., et al., (2004). O-acetylation and de-O-acetylation of sialic acids in human colorectal carcinoma. *Eur. J. Biochem., 271*(2), 281–290.

31. Mandal, C., Srinivasan, G. V., Chowdhury, S., Chandra, S., & Schauer, R., (2009). High level of sialate-O-acetyltransferase activity in lymphoblasts of childhood acute lymphoblastic leukemia (ALL): Enzyme characterization and correlation with disease status. *Glycoconj. J., 26*(1), 57–73.

32. Manzi, A. E., Sjoberg, E. R., Diaz, S., & Varki, A., (1990). Biosynthesis and turnover of O-acetyl and N-acetyl groups in the gangliosides of human melanoma cells. *J. Biol. Chem., 265*(22), 13091–13103.

33. Diaz, S., Higa, H. H., Hayes, B. K., & Varki, A., (1989). O-acetylation and de-O-acetylation of sialic acids. 7- and 9-o-acetylation of alpha 2,6-linked sialic acids on endogenous N-linked glycans in rat liver Golgi vesicles. *J. Biol. Chem., 264*(32), 19416–19426.

34. Higa, H. H., Butor, C., Diaz, S., & Varki, A., (1989). O-acetylation and de-O-acetylation of sialic acids. O-acetylation of sialic acids in the rat liver Golgi apparatus involves an acetyl intermediate and essential histidine and lysine residues-a transmembrane reaction? *J. Biol. Chem., 264*(32), 19427–19434.

35. Iwersen, M., Dora, H., Kohla, G., Gasa, S., & Schauer, R., (2003). Solubilization and properties of the sialate-4-O-acetyltransferase from guinea pig liver. *Biol. Chem., 384*(7), 1035–1047.

36. Houliston, R. S., Endtz, H. P., Yuki, N., Li, J., Jarrell, H. C., Koga, M., et al., (2006). Identification of a sialate O-acetyltransferase from campylobacter jejuni: Demonstration of direct transfer to the C-9 position of terminalalpha-2, 8-linked sialic acid. *J. Biol. Chem., 281*(17), 11480–11486.

37. Bergfeld, A. K., Claus, H., Vogel, U., & Muhlenhoff, M., (2007). Biochemical characterization of the polysialic acid-specific O-acetyltransferase NeuO of *Escherichia coli* K1. *J. Biol. Chem., 282*(30), 22217–22227.

38. Deszo, E. L., Steenbergen, S. M., Freedberg, D. I., & Vimr, E. R., (2005). *Escherichia coli* K1 polysialic acid O-acetyltransferase gene, neuO, and the mechanism of capsule form variation involving a mobile contingency locus. *Proc. Natl. Acad. Sci. U.S.A, 102*(15), 5564–5569.

39. Ghosh, S., Bandyopadhyay, S., Mallick, A., Pal, S., Vlasak, R., Bhattacharya, D. K., et al., (2005). Interferon gamma promotes survival of lymphoblasts overexpressing 9-O-acetylated sialoglycoconjugates in childhood acute lymphoblastic leukemia (ALL). *J. Cell Biochem., 95*(1), 206–216.

40. Mann, B., Klussmann, E., Vandamme-Feldhaus, V., Iwersen, M., Hanski, M. L., Riecken, E. O., et al., (1997). Low O-acetylation of sialyl-Le(x) contributes to its overexpression in colon carcinoma metastases. *Int. J. Cancer, 72*(2), 258–264.

41. Jass, J. R., Allison, L. J., & Edgar, S. G., (1995). Distribution of sialosyl Tn and Tn antigens within normal and malignant colorectal epithelium. *J. Pathol., 176*(2), 143–149.

42. Ogata, S., Ho, I., Chen, A., Dubois, D., Maklansky, J., Singhal, A., et al., (1995). Tumor-associated sialylated antigens are constitutively expressed in normal human colonic mucosa. *Cancer Res., 55*(9), 1869–1874.

43. Fukuda, M., (1996). Possible roles of tumor-associated carbohydrate antigens. *Cancer Res., 56*(10), 2237–2244.

44. Izumi, Y., Taniuchi, Y., Tsuji, T., Smith, C. W., Nakamori, S., Fidler, I. J., et al., (1995). Characterization of human colon carcinoma variant cells selected for sialyl Lex carbohydrate antigen: Liver colonization and adhesion to vascular endothelial cells. *Exp. Cell Res., 216*(1), 215–221.

45. Mukherjee, K., Chava, A. K., Mandal, C., Dey, S. N., Kniep, B., & Chandra, S., (2008). O-acetylation of GD3 prevents its apoptotic effect and promotes survival of lymphoblasts in childhood acute lymphoblastic leukemia. *J. Cell Biochem., 105*(3), 724–734.

46. Paller, A. S., Arnsmeier, S. L., Robinson, J. K., & Bremer, E. G., (1992). Alteration in keratinocyte ganglioside content in basal cell carcinomas. *J. Invest. Dermatol., 98*(2), 226–232.

47. Heidenheim, M., Hansen, E. R., & Baadsgaard, O., (1995). CDW60, which identifies the acetylated form of GD3 gangliosides, is strongly expressed in human basal cell carcinoma. *Br. J. Dermatol., 133*(3), 392–397.

48. Nydegger, U. E., Fearon, D. T., & Austen, K. F., (1978). Autosomal locus regulates inverse relationship between sialic acid content and capacity of mouse erythrocytes to activate human alternative complement pathway. *Proc. Natl. Acad. Sci. U.S.A, 75*(12), 6078–6082.

49. Shi, W. X., Chammas, R., Varki, N. M., Powell, L., & Varki, A., (1996). Sialic acid 9-O-acetylation on murine erythroleukemia cells affects complement activation, binding to I-type lectins, and tissue homing. *J. Biol. Chem., 271*(49), 31526–31532.

50. Weiman, S., Uchiyama, S., Lin, F. Y., Chaffin, D., Varki, A., Nizet, V., et al., (2010). O-Acetylation of sialic acid on Group B Streptococcus inhibits neutrophil suppression and virulence. *Biochem. J., 428*(2), 163–168.

51. Klein, A., Krishna, M., Varki, N. M., & Varki, A., (1994). 9-O-acetylated sialic acids have widespread but selective expression: Analysis using a chimeric dual-function probe derived from influenza C hemagglutinin-esterase. *Proc. Natl. Acad. Sci. U.S.A, 91*(16), 7782–7786.

52. Wipfler, D., Srinivasan, G. V., Sadick, H., Kniep, B., Arming, S., Willhauck-Fleckenstein, M., et al., (2011). Differentially regulated expression of 9-O-acetyl GD3 (CD60b) and 7-O-acetyl-GD3 (CD60c) during differentiation and maturation of human T and B-lymphocytes. *Glycobiology, 21*(9), 1161–1172.

53. Cariappa, A., Takematsu, H., Liu, H., Diaz, S., Haider, K., Boboila, C., et al., (2009). B cell antigen receptor signal strength and peripheral B cell development are regulated by a 9-O-acetyl sialic acid esterase. *J. Exp. Med., 206*(1), 125–138.

54. Ariga, T., Suetake, K., Nakane, M., Kubota, M., Usuki, S., Kawashima, I., et al., (2008). Glycosphingolipid antigens in neural tumor cell lines and anti-glycosphingolipid antibodies in sera of patients with neural tumors. *Neurosignals, 16*(2, 3), 226–234.

55. Muchmore, E. A., Varki, N. M., Fukuda, M., & Varki, A., (1987). Developmental regulation of sialic acid modifications in rat and human colon. *FASEB J., 1*(3), 229–235.

56. Surolia, I., Pirnie, S. P., Chellappa, V., Taylor, K. N., Cariappa, A., Moya, J., et al., (2010). Functionally defective germline variants of sialic acid acetylesterase in autoimmunity. *Nature, 466*(7303), 243–247.

57. Pillai, S., Cariappa, A., & Pirnie, S. P., (2009). Esterases and autoimmunity: The sialic acid acetylesterase pathway and the regulation of peripheral B cell tolerance. *Trends Immunol., 30*(10), 488–493.

58. Pal, S., Chatterjee, M., Bhattacharya, D. K., Bandhyopadhyay, S., & Mandal, C., (2000). Identification and purification of cytolytic antibodies directed against O-acetylated sialic acid in childhood acute lymphoblastic leukemia. *Glycobiology, 10*(6), 539–549.

59. Ravindranaths, M. H., Paulson, J. C., & Irie, R. F., (1988). Human melanoma antigen O-acetylated ganglioside GD3 is recognized by cancer antennarius lectin. *J. Biol. Chem., 263*(4), 2079–2086.

60. Alvarez-Rueda, N., Desselle, A., Cochonneau, D., Chaumette, T., Clemenceau, B., Leprieur, S., et al., (2011). A monoclonal antibody to O-acetyl-GD2 ganglioside and

not to GD2 shows potent anti-tumor activity without peripheral nervous system cross-reactivity. *PLoS One, 6*(9), e25220.

61. Yuki, N., Yamada, M., Tagawa, Y., Takahashi, H., & Handa, S., (1997). Pathogenesis of the neurotoxicity caused by anti-GD2 antibody therapy. *J. Neurol. Sci., 149*(2), 127–130.

62. Slart, R., Yu, A. L., Yaksh, T. L., & Sorkin, L. S., (1997). An animal model of pain produced by systemic administration of an immunotherapeutic anti-ganglioside antibody. *Pain, 69*(1/2), 119–125.

63. Terme, M., Dorvillius, M., Cochonneau, D., Chaumette, T., Xiao, W., Diccianni, M. B., et al., (2014). Chimeric antibody c.8B6 to O-acetyl-GD2 mediates the same efficient anti-neuroblastoma effects as therapeutic ch14.18 antibody to GD2 without antibody-induced allodynia. *PLoS One, 9*(2), e87210.

64. Ritter, G., Ritter-Boosfeld, E., Adluri, R., Calves, M., Ren, S., Yu, R. K., et al., (1995). Analysis of the antibody response to immunization with purified O-acetyl GD3 gangliosides in patients with malignant melanoma. *Int. J. Cancer, 62*(6), 668–672.

65. Abad-Rodriguez, J., Bernabe, M., Romero-Ramirez, L., Vallejo-Cremades, M., Fernandez-Mayoralas, A., & Nieto-Sampedro, M., (2000). Purification and structure of neurostatin, an inhibitor of astrocyte division of mammalian brain. *J. Neurochem., 74*(6), 2547–2556.

66. Abad-Rodriguez, J., Vallejo-Cremades, M., & Nieto-Sampedro, M., (1998). Control of glial number: Purification from mammalian brain extracts of an inhibitor of astrocyte division. *Glia, 23*(2), 156–168.

67. Sharma, V., Chatterjee, M., Mandal, C., Sen, S., & Basu, D., (1998). Rapid diagnosis of Indian visceral leishmaniasis using achatininH, a 9-O-acetylated sialic acid binding lectin. *Am. J. Trop. Med. Hyg., 58*(5), 551–554.

68. Ylonen, A., Helin, J., Bogwald, J., Jaakola, A., Rinne, A., & Kalkkinen, N., (2002). Purification and characterization of novel kininogens from spotted wolffish and Atlantic cod. *Eur. J. Biochem., 269*(11), 2639–2646.

# CHAPTER 9

# Glycome Profiling in Plants: Towards Understanding the Scenario of Carbohydrates

REETIKA MAHAJAN,[1] MUSLIMA NAZIR,[2] JAHANGIR A. DAR,[2]
SHAZIA MUKHTAR,[1] MUNTAZIR MUSHTAQ,[1] SUSHEEL SHARMA,[1]
JASDEEP CHATRATH PADARIA,[3] SHAFIQ A. WANI,[2] and
SAJAD MAJEED ZARGAR[2]

[1]*School of Biotechnology, Sher-e-Kashmir University of Agricultural
Sciences and Technology of Jammu, Chatha, Jammu and Kashmir,
India;* [2]*Division of Plant Biotechnology, Sher-e-Kashmir University of
Agriculture Sciences and Technology of Kashmir, Shalimar, Srinagar,
Jammu and Kashmir, India, E-mail: smzargar@gmail.com (S. M. Zargar);*
[3]*National Research Center on Plant Biotechnology, Indian Agricultural
Research Institute, New Delhi, India*

## ABSTRACT

Glycan (commonly called carbohydrate) is one of the vital macromolecules having an important role in various cellular processes. Term "Glycomics" refers to the study of the structure and function of an organism's entire carbohydrates and their derivatives. Glycans are classified into different groups based on the composition of monomer moieties. Glycans can also be classified based on the attachment of macromolecules like glycoprotein, glycolipids which are commonly known as glycoconjugates. Glycans have an important role in plants; they provide structural stability to the cell wall, cell signaling, cell communication, and interaction between pathogen and plant. Glycans are difficult to study because of their branched or altered monomer structures and genome indirect relation. Moreover, conjugation of other macromolecules to glycan molecule further increases the structural complexity of glycans. Various techniques, like electrophoresis, spectrometry, chromatography, etc., have been frequently used to analyze the glycan of a particular organism. However, there are certain limitations of these techniques which need to be eliminated for the proper analysis of glycans. This chapter has mainly focused on different techniques used in glycome profiling and have highlighted some of the glycome profiling research carried out in different plant species.

## 9.1 INTRODUCTION

Biomolecules like carbohydrates are important for sustaining life as they help in the proper functioning of cellular processes. Glycomics, as an important field, provides knowledge about the structure of sugar and its function in an individual. Glycomics helps in studying sugar's interaction with other macromolecules such as proteins, nucleic acids, and lipids. Analysis of carbohydrates by using high-throughput methods is considered the most challenging task due to template-independent biosynthesis, chemical similarities between the monosaccharides, and inadequate consensus sequences for glycan alteration of cohorts of glycoproteins. Glycome profiling is the first step in detecting various glycan structures present in the samples. Capillary electrophoresis (CE), chromatographic techniques, mass spectrometry (MS), and microarray are important tools for glycomics. This chapter will focus on high throughput methods used for the analysis of whole glycome. Here, a brief overview of different types of glycans, why there is need for glycome analysis, and finally a brief overview of the technologies used for glycome analysis such as electrophoresis, chromatography, MS and microarrays.

## 9.2 GLYCOMICS

The word glycomics is attained from chemical prefix "Glyco meaning sugar" and omic which means comprehension. Glycomics is the subset of glycobiology which is related to systematic analysis of free as well as complex molecules of sugars of an organism. Glycomics consists of pathological, physiological, genetic, and other aspects of glycans [1]. Glycomics helps in the entire glycans identification of a particular organism or cell. It also helps in the identification of glycoproteins encoding genes. Glycomics field includes studying glycoconjugates assembly and its expression in the biological system, which helps to analyze, understand, and relate the interaction of sugar with other macromolecules. Study of processes occurring in cell which involve glycan is termed as glycomics.

## 9.3   GLYCAN

Glycans are the polymers of saccharide molecules linked together by glycosidic bonds. D-glucose is most common constituent of glycans whereas it also includes monosaccharides like D-fructose, D-mannose, D-fucose, D-galactose, D-xylose, L-galactose, and L-arabinose. Glycans also have amino sugars such as D-galactosamine and D-glucosamine and their derivatives like N-acetylmuramic acid and N-acetylneuraminic acid. Moreover, simple sugar acids like iduronic and glucuronic acids are also present in glycans [2]. Glycans are ubiquitous amongst plant kingdom as they range from algae to flowering plants. Glycans exist in plants as free entities or in conjugated form commonly known as glycoconjugates. Glycolipids, glycoproteins, and proteoglycans are different glycocongugates present in an organism [128]. There are two different classes of protein linked glycans namely O-linked and N-linked glycans (NLGs). O-linked glycans are attached to oxygen atom of serine or threonine side chains and N-linked are attached to amide nitrogen atom of asparagine side chain of protein. In addition to linked glycans, plants also have free N-glycans (FNGs) in different parts of plants like seedlings, developing fruits, stems, and culture cells [3–6, 129–131]. Based on reduced terminal structure of FNGs, two different types of glycans are present in plants, GlcNAc1, and GlcNAc2 type. GN1 type free N glycans have one GlcNAc residue whereas GN2 type free N glycans have GlcNAcβ1-4GlcNAc (N-acetyl chitbiosyl unit) [7]. Among these types of glycans, complex N-glycans grafted on plant polypeptide backbone form the core structure of plant glycans [8]. Plants can synthesize different varieties of glycans that can contribute to various processes during their life cycle. Different roles of glycans in plants include: (i) act as structural components in the cell wall; (ii) protect the plants from stress by acting as signaling molecules; (iii) helps in nascent proteins folding [7].

### 9.3.1   TYPES OF GLYCANS

Glycans are categorized on the basis of the following criteria:

1.  **Based on the Composition of Individual Monomer:** On the basis of monosaccharide present, glycans can be classified into two categories, namely homo-polysaccharides and hetero-polysaccharides:

    i.  **Homopolysaccharides:** These are the glycans which on hydrolysis yield single monosaccharide and thus are also referred to as homoglycans. These glycans are having well-defined structures with variable molecular weight. On the basis of branching of side chains, these sugar molecules can be further classified as linear glycans, unbranched glycans, and starch. Major examples that encompass above-mentioned glycans include-cellulose and glycogen, respectively. Cellulose, an unbranched polysaccharide comprises of monomers of D-glucose joined by β-1,4-glycosidic linkage is the most abundant glycan and biomolecule present on the earth. Another important glycan of this kind is glycogen which is a branched homopolysaccharide consisting of monomers of D-glucose joined by α-1,4-glycosidic linkages. Starch lies in between the above-mentioned categories as it consists of two components-amylose and amylopectin. Amylose is a linear polysaccharide having α-1,4-glycosidic linkage similar to that of maltose while amylopectin is a branched polysaccharide having α-1,6-glycosidic linkage similar to that of isomaltose.

    ii. **Heteropolysaccharides:** Glycans that are composed of two different monosaccharides or their derivatives are called heteropolysaccharides or heteroglycans. These glycans on hydrolysis yields mixture of monosaccharides and are abundant in both plant and animal kingdom. Heteropolysaccharides are very difficult to study as they mostly are conjugated glycans linked to either proteins or lipids. Examples of such kind of glycans include glycoproteins and glycolipids. These complex glycans are discussed under the category glycoconjugates.

2.  **Glycoconjugates:** These are biologically indispensable carbohydrate molecules linked covalently with other non-carbohydrate moieties which can be lipids, proteins, and peptides. Such kind of glycans consists of many different categories which include glycolipids, glycoproteins, peptidoglycans, and lipopolysaccharides (LPSss). These glycans are present outside the cell membrane and thus play vital roles in diverse processes of living organisms such as interaction and recognition between two cells, cell, and matrix interactions, signaling cascade, hormone action, immune response, viral infection

and in cell detoxification [9–14]. Glycoproteins and glycolipids are of utmost importance among all the classes of glycoconjugates. Both conjugates are found abundantly in plants as well as in animals:

i. **Glycoproteins:** These are proteins attached covalently with sugar a moiety. Based on structure and mechanism of synthesis glycoproteins are classified into two types namely N-linked and O-linked glycoproteins. Both types of glycoproteins are discussed in subsections.

   a. **N-Linked Glycans (NLGs):** In plants, NLGs are linked with endoplasmic reticulum (ER) via nitrogen (N) group present in the side chain of asparagine in tripeptide sequence (Asn-X-Ser or Asn-X-Thr). In this sequence X can be any amino acid other than proline [132]. Such tripeptide glycan sequence includes various monosaccharides and its derivatives like galactose, mannose, fucose, N-acetylgalactosamine, neuraminic acid, and N-acetylglucosamine. Among these, N-acetylgalactosamine, and N-acetylglucosamine have an important role in proper protein folding. Calreticulin and calnexin are the important chaperone proteins localized in ER and linked to N-linked glycan core via three glucose residues. These proteins assist in protein folding to which the glycans are linked. After proper folding of proteins, three glucose molecules are eliminated, and can be utilized in further reactions. On the other hand, proteins are re-associated to chaperones as glucose residues are attached again to the protein molecule during improper protein folding. Folding process of proteins is activated till protein reaches to proper conformation. Proteins which fail to fold properly are removed from ER and degraded by proteases present in cytoplasm. Steric effects of NLGs play an important role in protein folding process. For instance, cysteine residue may block disulfide bond formation with various cysteine residues temporarily because of nearby glycans present in the peptide. Thus, NLGs control the formation of cysteine residues disulfide bonds present in the cell. All N-glycans have same preliminary structure namely oligmannosidic (Man3GlcNAc2). All plant glycans were initially classified in two broad class's viz.-complex-type N-glycans and the high-mannose-type N-glycans. Based on nuclear magnetic resonance (NMR) and MS-based studies, these glycans were recategorized into four classes, namely, complex type, paucimannosidic type, hybrid-type N-glycans and high-mannose-type N-glycans [15].

b.  **O-Linked Glycans:** In plants, O-linked glycans are formed by the sugar residue incorporation to the OH chain of certain polypeptide chains localized in Golgi apparatus (GA) which comprise of amino acids such as threonine, hydroxylysine, hydroxyproline, and serine in their side chains. In contrast to N-glycoproteins which are N-linked, O-linked glycoproteins do not have conserved sequence and are formed by single sugar addition at a time. For favorable O-linked glycosylation, proline residue should be present at either −1 position or +3 positions relative to threonine or serine [2]. Eukaryotic cell secretes various O-linked glycoproteins which forms part and parcel of extracellular matrix surrounding it. Plant O-linked glycans belongs to hydroxyproline-rich glycoproteins superfamily [16]. Much attention has been gained by two O-glycosylated hydroxyproline-rich glycoproteins which include arabinogalactan proteins and extensins [17–24, 133, 134]. The glycosylation level and type of sugars which is incorporated differs not only between these two families, but also vary between their subfamilies. Thus, plant glycans are usually considered as structurally diverse from the glycans of mammals. In proteoglycans and glycoproteins, human O-glycans are of mucin type having xylose residues which are linked to oxygen at serine or threonine. On the contrary, O-linked plant glycans are formed by the nucleocytoplasmic O-glycosylation during temporary incorporation of only one N-acetylglucosamine residue on serine or threonine residues of the polypeptide [25]. Rice glutelin and plant Golgi membrane proteins have similarity with mucin-type O-glycosylation [26, 27]. Extensins have key roles in plant development and gives resistance against biotic and abiotic stress [22, 24, 28–37, 135].

ii.  **Glycolipids:** Lipids attached to carbohydrates by glycosidic bonds are known as glycolipids. In nature, these molecules are amphipathic comprising of hydrophilic polar sugar backbone and hydrophobic non-polar lipid unit. Owing to its amphipathic nature, glycolipids play key role in stability of cell membrane. Apart from this, glycolipids have a variety of functions in biological processes which include: (i) cell adhesion; (ii) cell signaling; (iii) cell recognition; and (iv) effect the domain formation and fluidity of membranes. Plant glycolipids are composed of glycosphingolipids, glycosyldiacylglycerols, glucose, and sucrose esters and steryl glycosides. Glycolipid comprises about 75% of total membrane lipid in plants [38]. In plants, glycolipids exist in the form of steryl glycosides, free sterols, acylatedsteryl glycosides and acylated sterols. In nature, sterylglucosides are found as 3-β-hydroxyglucosides, where 1$^{st}$ carbon of carbohydrate takes part in the synthesis of glycosidic linkage [39]. Glucose along with mannose, galactose, and gentiobiose are the abundant sugars present in sterylglucoside.

## 9.4   PLANT ORGANELLE GLYCOME

Complement of carbohydrates present in organelles is known as organelle glycome. In plants, glycans are abundant in certain organelles such as cell wall, mitochondria, and chloroplast. Such type of glycome is described as follows.

### 9.4.1   CELL WALL GLYCOME

Plant cell is having a rigid, flexible, semipermeable outermost layer called cell wall which is positioned adjacent to cell membrane. Plant cell wall helps in: (i) proper functioning of tissues and organs by maintaining shapes of tissue and organs; (ii) interaction between macromolecules present internal and external of cells [40]; (iii) interaction between plant and microbes; (iv) defense related aspects of plants [41]. Cellulose, hemicelluloses, and pectin are the important glycans present in the plant cell wall. A detailed description about the various glycans present in plant cell wall is given in this section.

#### 9.4.1.1   CELLULOSE

Cellulose is the most abundant polysaccharide on earth and is the main constituent of primary cell wall. Its composition varies in different organs of plant such as roots, shoots, etc. Cellulose is composed of complex chains of β-(1,4)-linked D-glucose residues interwoven into a network to form crystalline microfibrils. These complex chains are linked by hydrogen bonding with each other [42]. Further, X-ray studies also reported presence of some other polysaccharides which have been predicted as xylans and mannans covalently bonded to the cellulose complex.

## 9.4.1.2  HEMICELLULOSE

Hemicellulose constitutes heteropolymers of pentose and hexose sugar. Arabinose and xylan are the two important pentose sugars present in this heteropolymer whereas glucose, mannose, and galactose are the important hexose sugars in pentose and hexose heteropolymers. Hemicellulose is the second most important polysaccharide present in nature. In plant stems, softwood hemicellulose is composed of glucomannans and hardwood is mostly composed of xylans [43]. Xylan is a heteropolysaccharide of β-1,4-linked D-xylopyranose motifs. Xylan also includes glucuronic acid and arabinose motifs along with xylopyranose. Based on plant species, composition of xylan differs in plants [136].

## 9.4.1.3  PECTIN

In plants, pectin (commonly termed as pectic polysaccharide) is the most vital component of primary cell wall and middle lamella. It is primarily composed of D-galacturonic acid [44]. Pectin is indispensable in deciphering plant morphology [137]. Three types of pectin namely, homogalacturonans, substituted galacturanan and rhamnogalacturonans, forms an important structural components of primary cell wall and middle lamella. Here, substituted galacturanan includes xylogalactouranan and arabinogalactouranans. Homogalacturonan (linear polymer of α-1,4 linked galacturonic acid residues) is the most important pectin found in cell wall and comprising approximately 60% of overall pectin. In plants, galactouronan helps in nutrients accumulation [45]. There are two types of rhamnogalacturonan namely rhamnogalacturonan-I and rhamnogalacturonan-II. Rhamnogalacturonan-I is composed of β-1,4-galactan, type-I arabinogalactan and α-1,5-L-arabinan which contribute pectin content of nearly 20–35%. Its percentage varies in cell wall of different plants as if dry weight of potato cell wall glycan contributes about 35% of polysaccharides [46]. Rhamnogalacturonan II comprises of four side chains having distinctive residues such as 3-deoxy-lyxo-2-heptulosaric acid, apiose, 3-deoxy-manno-2-octulosonic acid and aceric acid [47] which are linked to nine esterified molecules of homogalacturonan [48]. Rhamnogalacturonan-II can form a complex with boron via synthesizing borate-diol-ester which later on links to two molecules of homogalacturonan via cross-linking and thus provides increased cell wall stability [49]. If borate will fail to dimerise two homogalacturonan molecules, it leads to deleterious effect on reproductive and meristematic parts of plants. Xylogalactouronan, a substitutional galactouranan is composed of α-1,4-linked D-galacturonic acid polymers with substitution of β-D-xylose residue at the O-3 position. Various studies reported the presence of xylogalacturonan in storage tissues as well as in reproductive parts of plant. For example, an investigation conducted on duckweed plant reveal the presence of main macromolecules, galacturonan, xylogalacturonan, and rhamnogalacturonan constituting 20.3% pectin [50]. In *Arabidopsis thaliana*, it has been reported that pectin helps in deciphering the peptic structure [51, 52]. Another important substitutional galactouranan is arabinogalactan proteins which are crucial players of various cellular activities during development of plant including proliferation of cell, growth, and reproduction. Arabinogalactans binding reagents can effect pollen tube tips growth, cell expansion and somatic embryogenesis at developing stage [53, 54]. In plants, hydroxyproline-o glycosyltransferase enzyme plays vital character in the formation of arabinogalactan proteins [138]. Two types of arabinogalactan are present in plants like arabinogalactan I and arabinogalactan II. Arabinogalactan I is a heterogeneous polymer of arabinose and galactose residues where α-L-arabinofuranose residues are substituted at $3^{rd}$ oxygen of galactosyl residues [55]. These glycans are found in gums of plants and are also identified to link proteins synthesizing arabinogalactan protein. It has been also reported that β-galactose is substituted with arabinogalactan residues at position 6 of oxygen [56]. Arabinogalactan II is a polymer of β-D-galacturonic acid which is linked at C1 and C3 positions. It constitutes few linked chains of α-L-arabinofuranose [57, 139]. Arabinogalactan II binds to proteins synthesizing arabinogalactan protein. Enzymes like galactosyl transferases play an important role in the formation of arabinogalactan II [58]. In *Arabidopsis thaliana,* galactosyl transferases have been reported to play vital role in normal embryo development [59]. In cell culture method, pectin lyase helps in the degradation of pectin molecule which is utilized commercially for formation of protoplast and starch decomposition [60].

## 9.4.2   CELL MEMBRANE GLYCOME

The cell membrane is a semi-permeable membrane which encompasses the cell cytoplasm and acts as an obstacle between cell and its surrounding. It plays pivotal roles in the communication between external and internal of a cell. It aids in exchange of cellular metabolites and imparts constant environment for the reactions occurring within the cell.

Cell membrane is composed of about 20% carbohydrates with approximately 40% proteins and rest 50% lipids [140]. Carbohydrates in cell membrane differ than other macromolecules as they are very peculiar in their structure and are linked with each other in various forms. They are having α or β linkages with adjoining sugars at carbon 3rd or 4th position of hydroxyl groups. Carbohydrates are very diverse in contrast to other macromolecules and thus have unique feature of acting as a ligand which can be recognized by other molecules. In living organisms, carbohydrates are synthesized by gene products such as glycosyltransferases but not by genes directly. In tobacco plant, β-1-3 glucan (cell membrane glycan) can alter the cell structure which leads to defense response [61].

## 9.4.3   CELL SURFACE GLYCANS

Carbohydrates are the primary components of plant cell present on the exterior surface of cell. However, only unique carbohydrates are present on maturity. Glycosaminoglycans (GAGs) are the major compound polysaccharides occurring on the cell surface and extracellular matrix, which interact with different proteins. These compound polysaccharides are required in various important biological processes like growth and development of cell. These are unbranched repeating units of two disaccharide molecules viz. hexosamine and hexuronic acid which are coupled resulting in the formation of heterogeneous polysaccharides. The carbohydrate part of GAGs is manufactured in Golgi bodies bounded to protein by O-glycosylation synthesizing a complex molecule known as proteoglycan [62, 63].

## 9.4.4   MITOCHONDRIAL GLYCOME

Mitochondria, which is regarded as powerhouse of the cell is a double membrane-bound organelle found in all eukaryotic organisms. It is an important compartmentalized organelle which plays crucial roles in plants including transportation of vital components used in biological processes such as catabolism, cell signaling, defense mechanism, and apoptosis [64]. Mitochondrial membrane is mostly abundant in proteins and phospholipids but it also constitutes small amount of glycoconjugates in its structure. It has been also deduced that the process of glycosylation occurs at mitochondrial outer membrane where mannose and N-acetylglucosamine are the molecules which are glycosylated in acceptor proteins [65].

### 9.4.5   CHLOROPLAST GLYCOME

Chloroplast, a specialized organelle found in plant cell harnesses light energy for the synthesis of glucose and thus is a key site for conducting the process of photosynthesis. Like mitochondrion, chloroplast is also double membranous compartmentalized organelle where important chemical reactions occur leading ultimately to the synthesis of glucose. Envelope membranes of chloroplasts are further divided in two membranes-outer and inner. Its outer membrane is comprised of mainly free lipids and lipids conjugated with carbohydrates. However, inner membrane consists of galactolipids, phospholipids, and sulfolipids [66]. In chloroplast, concentration of galactolipid is more in outer than in inner membrane. Another important membrane of chloroplast is thylakoid membrane, where ATP is synthesized and is considered as light absorption site [66]. Thylakoid membrane is mainly composed of phospholipids, some sulfolipids, and galactolipids. Mono- and di-are the two types of galactolipids present in thylakoid membrane. In Mono-galactosyldiacylglycerol, residues of galactose are complexed with glycerol at carbon 3 position and on the contrary in digalactosyldiacylglycerol (DGD) glycerol is attached at terminal position [67]. About 15% of chloroplastic volume is comprised of starch granules in photosynthetic organisms including plants [68]. It is well known fact that chloroplast synthesizes sugar in the form of starch which is later on utilized by the plant during night.

### 9.5   TECHNIQUES USED FOR GLYCOME PROFILING

Cell which is regarded as the basic unit of life constitutes of various organelles like nucleus, GA, chloroplast, mitochondria, ER, etc. All these organelles are enclosed in a cell with a membrane called cell membrane. Many prokaryotic and eukaryotic cells are surrounded by an outer layer called cell wall. This cell wall protects the cell from outer environment by providing mechanical and chemical support. Cell wall is mainly composed of carbohydrates, proteins, and lipids. Carbohydrate (a major class of biological macromolecule) plays an important role in cell wall formation as they provide protection to the cell. Carbohydrates are also important in various cellular processes like adhesion, immune response, motility, and host-pathogen interactions [69–72]. Like genomics, proteomics, and metabolomics, glycomics is the new and emerging field which deals with the study of complex carbohydrates/glycan structures. Thus with the advent in science and knowledge of the genomics and proteomics, scientists have become more interested to unravel the functions of biological molecules. Glycomics find its role in almost all research fields including developmental biology, immunology, cancer, and signal transduction. Glycan structure is complex as it is made up of monosaccharide units which are differently linked to each other. The complex and diverse structures of limited monosaccharides units, isobaric, and isomeric structures, non-template-driven nature of these monosaccharides and various modifications like acetylation, phosphorylation, and sulfation may enhance the heterogeneity of glycans has made glycome profiling difficult [141, 142]. Various conventional and high throughput analytical techniques have been used for glycome profiling. In this section, we will discuss some of the conventional and recent techniques used in glycome profiling. The main focus will be on CE, liquid chromatography (LC), MS, and microarrays. The basic flow chart of the glycome profiling in an organism is detailed in Figure 9.1.

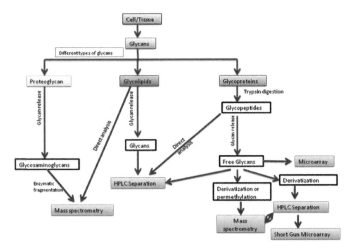

**FIGURE 9.1**   The basic flow chart of glycome profiling in an individual. Here, firstly the different types of glycans are isolated from the cell or tissue. Then the glycans are released as simpler form like glycoaminoglycans, glycans, and free glycans. The glycans are then analyzed by different methods like MS, HPLC, microarray.

## 9.5.1   CAPILLARY ELECTROPHORESIS (CE) BASED GLYCOME PROFILING

Capillary Electrophoresis (CE) is one of the most successful techniques used to separate biomolecules including DNA and small-molecule metabolites. Important parts of this technique are vials, electrodes, capillary, power supply unit, detector, data output and handling device. Three different types of vials are used in CE namely sample vial, source vial and destination vial. In this method an electrolyte (an aqueous buffer solution) is filled in capillary, source vial and destination vial. Firstly, the sample to be analyzed is injected in the capillary via capillary action which is followed by the return of capillary to the source. The electric field generated within source and destination vials initiates migration of the analytes. Here the electrodes are supplied with high-voltage power supply. In CE, analytes are separated by electro-osmotic flow in which positive or negative ions are dragged in same direction through the capillary. Analytes are migrated due to their electrophoretic mobility. These analytes are detected near the capillary outlet end. The results of the detector are sent as data output to the attached computer. Then obtained output is displayed as electropherogram on the computer screen. In an electropherogram, separated chemical compounds emerge as different retention times peaks. For separation of glycans via CE, pre fractionation of glycans is needed. For this, fractions of NLGs are generated by Peptide-N-Glycosidase F (PNGaseF) cleavage and then labeled by laser induced fluorescence probes. Glycan attached probes prevents sialyl and fucozyl units degradation by reductive amination. These probes also allow the detection of low concentration of glycans in neutral pH. After tagging, the ionized glycans are allowed to pass through a narrow channel with higher surface area to volume ratio and higher electrical resistance. In CE separation samples of even smaller volumes (1–50 nL) of structurally similar molecules are subjected to high electric fields (up to 750 V/cm or sometimes even higher) [73]. However, carbohydrate structure can be identified by comparing time between separation of sample from trace of CE and standard glycan. It has been found that among various CE methods, two analysis methods are commonly used for glycome analysis namely capillary zone electrophoresis (CZE) and micellar electro kinetic chromatography (MEKC) [143]. In CZE, Pre-column derivatization of samples with PMP (1-phenyl-3-methyl-5-pyrazolone) is used for eliminating purified polysaccharides (neutral and acidic sugars) simultaneously from Ephedra sinica [74]. However, CE with direct ultraviolet (UV) detection method is frequently used for studying neutral carbohydrates in alkali conditions and of cellulose fiber extracts [75]. Glycan analysis by CE is influenced by many parameters (e.g., nature of carbohydrate, capillary diameter, electrolyte pH and retention time) and it is worth to mention reverse electro osmotic flow using capillaries with polybrene-modification can perform the separation task much faster [76]. CE separation method can be used to identify N-linked carbohydrates on glycoproteins and various sugars like glucose, sucrose, lactose, and fructose can be determined in beverage, forensic, and pharmaceutical specimens [77–80]. There are certain limitations of CE such as requirement of new standard for glycan analysis every time, it requires pre fractionated and labeled glycans, in this each peak denotes single glycan and the difficulty of performing CE with MS [144].

## 9.5.2    CHROMATOGRAPHY GLYCOME PROFILING

### 9.5.2.1    HPLC

High-performance Liquid Chromatography (HPLC) is one of the most common analytical techniques utilized for separation, identification, and quantification of different compounds present in a solution. In this technique, pumps help in passing pressurized liquid solvent through column filled with adsorbent material along with sample mixture. Variation in the interaction between the specimen and adsorbent material creates differences in flow rates of different specimen components and finally, the specimen flows out of the column as separate components. In the late 1990s and early 21st century, HPLC methods for glycan profiling have been extensively developed [145–147]. For glycome profiling, HPLC is frequently used for examining charged and neutral carbohydrates. In HPLC separation, firstly the samples are anchored on 96-well plates, and after that glycans are enzymatically cleaved and fluorescently labeled at reducing ends. Finally, the quantitative analytical elements are generated. After the generation of analytical elements, these elements are correlated with the data present in database like: GlycoBase [81] or GlycoExtractor [82]. These databases have approximately 350 N-linked glycan standards which give structural information of the element and this data is confirmed via exoglycosidase digestions by using autoGU an automated program [81]. Within four to five days one can analyze 96 samples by HPLC technique [83]. For simple glycan separation normal-phase (NP) HPLC technique is used whereas for complex glycans reverse-phase (RP) analysis can be used only after derivatization of the glycans with a hydrophobic tag. In NP HPLC, for more sensitive fluorescence detection, amide, and amine based columns are used and the samples are labeled with 2-aminopyridine (AP) and 2-aminobenzamide (2-AB) tags [147, 84]. Whereas AP or 2-AB with derivatization or other hydrophobic labels can be used to retain glycans on RP-HPLC columns like octadecylsilane and porous graphitized carbon (PGC) [148]. Due to low hydrophobicity feature of 2-Aminopyridine it is considered as suitable tagging reagent for RP-HPLC [84]. By using HPLC separation method, higher recovery of the labeled oligosaccharides can be obtained as compared to conventional methods like size exclusion chromatography or solid-phase extraction. Exoglycosidase treatment induces mobility shift in monosaccharides and makes NP-HPLC analysis one of the important techniques for the separation of glycans. In contrast, RP-HPLC analysis is sometimes preferred over NP-HPLC for analyzing isomers as it can differentiate between bisected and non-bisected glycans [149]. In case of complex glycan mixture analysis two- (2D) or three-dimensional (3D) HPLC techniques can be used [150]. NP-HPLC separation technique is preferred due to the higher sensitivity, the diversity of available columns and the lower cost and complexity of the equipment. HPLC has greater resolving power and can be used for performing multiplexing with other analytic devices like chromatography or MS, both charged and neutral glycans can be analyzed at same time in NP-HPLC, labeling (stoichiometric and unselective) of oligosaccharides AP or 2-AB enables reliable quantification, generation of databases (Glyco-Base) with elution times of known structures and automatically assignment of structures to glycans via software auto GU [151]. One of the limitations of this technique is low throughput capability which is even lower than CE separation technique. This technique is limited to analysis of NLGs only.

### 9.5.2.2 HIGH-PERFORMANCE (HIGH-PH) ANION-EXCHANGE CHROMATOGRAPHY (HPAEC)

High-performance (high-pH) anion-exchange chromatography (HPAEC) is an advance chromatography technique which separates carbohydrates like alditols, monosaccharides, oligo saccharides, polysaccharides, and amino-sugars on the basis of its size, anomericity, linkage isomerism and composition. This technique is mainly used for analysis of carbohydrates falling in 12–14 pH range. At this pH range the hydroxyl group of carbohydrates is fully or partially converted into oxyanions and anion exchange columns (quaternary-ammonium-bonded pellicular) are utilized for separation of carbohydrates. Anion exchange chromatography enables selective carbohydrates elution at single run. Hydroxyl group number, positional isomerism, anomerism, and the polymerization degree are certain parameters which influence carbohydrates separation. HPAEC has illustrated that aldoses and ketoses has higher retention than their analogous alditols. In general, the retention period increases with increase in the numbers of carbon atoms [85]. HPAEC technique has great advantage over other techniques as it does not require fluorescent labeling for detection. The signal for analysis of glycan in HPAEC depends on the concentration of the glycan and standard redox potential of the oxidized group. However, for proper separation of glycans, carbohydrate standards should be properly calibrated and identify signals in HPAEC trace. The detector used in HPAEC method is insensitive to certain carbohydrates. The problem of detector insensitivity of HPAEC can be solved by HPAEC-PED (which means HPAEC coupled with pulsed amperometric detection). HPAEC-PED method is used for detection of aldehyde group of carbohydrates by electrochemical oxidation [152]. HPAEC-PED separation is greatly affected by oxyanions availability to anion exchanger column functional groups. However, this effect is only seen in disaccharides formed by two units of D-glucozyl residues with modified glycosidic bonds configuration. These disaccharides include maltose, isomaltose, trehalose, gentiobiose, glucobioses, and nigerose [153]. HPAEC technique can also be coupled with MS for separation, but the use of high salt concentrated flow buffer can affect the proper analysis of glycans. HPAEC coupled with MS can be used to study nucleotide sugars, monosaccharide composition, cellulose, and single-glycoprotein glycosylation [154].

### 9.5.3  MASS SPECTROMETRY (MS) GLYCOME PROFILING

Mass Spectrometry (MS) technique is used to analyze glycans elucidated from derivatized or ionized glycoconjugates like glycoproteins and glycolipids. These modified glycans generate MS spectra which are annotated later. Complex glycan structures can be identified by MS analysis due to its high selectivity and sensitivity. High-throughput approach can be used to study pre-fractionated glycan spectra generated from known biosynthetic pathways. Confirmation of these assignments is done through exo-glycosidase digestion. Preparation of sample in this approach is an important step as glycans pre-fractionation is needed before analysis due to complex nature of glycans. For minimum ion suppression in MS analysis, samples to be analyzed are separated into fractions of equal acidities [155–157]. Majority of NLGs are analyzed by MS technique. This method provides maximum information on glycan structure from small amount of samples in lesser time with low labor. MS is used for fast scanning and analysis of specimens in a high-throughput manner. In this method, normal or per-methylated glycans and glycopeptides are isolated and are enhanced by using lectin-based affinity chromatography. MS can be used to determine the size of the oligosaccharides and various components like hexoses, hexosamines, and sialic acids. MS method can be used to analyze stereo-chemical linkages of specific monosaccharides on the basis of known biochemical pathways [158, 159]. Two important MS approaches for analyzing glycans are Matrix-assisted laser desorption/ionization and electrospray ionization [160–163] as discussed in subsections.

### 9.5.3.1 MATRIX-ASSISTED LASER DESORPTION IONIZATION-MASS SPECTROMETRY (MALDI-MS)

MALDI-MS has been widely used in the glycan analysis. In this, firstly, the specimen to be analyzed is fused with a suitable matrix solution, and then the mixed solution is embedded on a target plate. After spotting, the specimen is allowed to dry which leads to the formation of matrix crystals incorporating the sample molecules. Then the matrix crystals are allowed to pass through a pulsed laser beam which causes desorption and ionization of the sample molecules which finally results in singly charged ions. Time-of-flight mass spectrometry (TOF-MS) has great importance as in this the velocity of different ions is based on mass and charge of ion. MALDI-MS can detect unmodified glycans. However, following parameters like restricted dynamic range, difference in ionization efficiencies, ion suppression, and heterogeneous distribution of specimen in matrix spot results in different spectra for the same spot that makes the accuracy of quantification of MALDI-TOF glycan profiling debatable [73]. To overcome these problems and to improve the ionization of carbohydrates, they are often permethylated [86, 164]. In MALDI technique, permethylation of glycan is done to increase detection stability and sensitivity for fragmentation of sulfated monosaccharides and sialosides [86, 87]. Glycans can be per-methylated in spin column by a high-throughput method (12–18 samples in less than 20 min) [165]. Neutral

N-glycans can be quantified accurately with MALDI-TOF MS by performing esterification step prior to Girard's reagent T derivatization which binds to the reducing end of the glycan, providing a permanent positive charge [166]. Sulfation modification is often lost during MALDI-MS. Sulfated glycans can be analyzed in negative ion mode as in positive ion mode its signal suppresses the neutral ones due to poor ionization efficiency. This creates a problem for analysis of neutral glycans in negative ion mode. The partial loss of desulfation is due to the permethylation according to the Ciucanu-Kerek protocol, using a highly alkaline solution which is done in samples prior to analysis. To overcome the partial loss, double-permethylation is done to analyze sulfated glycans in positive ion mode MALDI MS [167]. Here, the first permethylation stabilizes sialic acids and leaving the sulfate groups intact and second permethylation is subjected to acidic methanolysis, without disturbing the permethylated sialic acid residues. Then the sulfated hydroxyl groups are subsequently deuteromethylated. Thus the sulfate groups can be determined by the obtained mass shift. MALDITOF-MS is used to identify $N$-glycan complex including tri-antennary and tetra-antennary structures overexpressing in the serum specimen of hepatocellular carcinoma [168]. It can also identify O-linked glycans as shown in studies conducted in dendritic cells (DCs) [169]. MS-based glycomics does not provide comparative quantification results of glycans in comparable specimens. For comparative glycomic analysis in animal cell culture, a traditional proteomic based methodology SILAC (stable isotope labeling by amino acids in cell culture) is used. In this method, the conversion of monosaccharides during the oligosaccharides biosynthesis limits the use of metabolic tagging of isotope-labeled glycans for glycomic analysis [170]. Different carbohydrate samples can be compared directly in MS technique by using mass tags for glycans. In MALDI high salt concentrated samples can be analyzed without any problem as MALDI is less sensitive to salt contamination as compared to other ionization methods. Singly charged ions peak obtained in MALDI has reduced the complexity of the spectrum. Isomeric and isobaric structures cannot be analyzed with MS alone. For analyzing isomeric and isobaric glycans of differing spatial extension MS is coupled with ion mobility spectrometry [88, 126, 171]. Like other techniques MALDI-MS has also some limitations. Glycans with acid groups like multiple sialylation or heavily sulfated GAGs shows less sensitivity and stability when analyzed by MALDI-MS. Ionic liquid matrix which is composed of a conventional MALDI matrix (DHB and CHCA) with an organic base added (1-methylimidazole) can be used to overcome sensitivity and stability issue in acidic glycans [172]. Ionic liquid matrix can make homogeneous spots, which make the analysis of GAG-derived oligosaccharides containing up to 13 sulfate groups easy) [173]. The data interpretation in case of GAG-derived oligosaccharides is very difficult due to the multiple charged ions, differential cation complexation, and different arrangements of N- and O-sulfate groups. Thus to overcome this problem, a software tool was developed in 2008 [174]. Fourier transform ion cyclotron resonance (FTICR) technique can also be applied for glycomic analysis. In this, ions are trapped in a magnetic field and are subsequently excited to a larger cyclotron radius where mass to charge ratio of ion decides the frequency efficiency. FTICR technique is used to study serum N-glycome and mucin O-glycosylation [175, 176].

### 9.5.3.2 ELECTROSPRAY IONIZATION-MASS SPECTROMETRY (ESI-MS)

Electrospray ionization-mass spectrometry (ESI-MS) is an advance spectrometry technique used to analyze glycans with unstable modification like as sialylation and sulfation [89, 90]. In this technique, the liquid sample is introduced in a very narrow tip end capillary which is supplied with a high voltage which generates spray carrying charged droplets that later enters in mass spectrometer. The droplets are converted into gaseous charged molecules by using nitrogen gas which increases the temperature in the evaporation chamber. Different analyzes like quadrupoles (Q), TOF, and FTICR can be used in this technique to analyze the mass by charge ratio of ions [177]. This technique generates cool ions which nullify the complications with sialylated analytes. ESI generates multi charged proton adducts instead of singly charged ions as in MALDI-MS which leads in the deviation of spectra for these ion types in MS/MS analysis. ESI-MS results in ionization without fragmentation of glycans [87, 91, 92]. It is found that sialic acid is retained intact in ESI-MS. ESI-MS method uses volatile buffers as it is sensitive to salt concentration in samples. Glycoconjugates are not generally fragmented in ESI-MS as in this process low vibrational energy is given on analyte molecules. For better resolution of glycans, ESI-MS can be coupled either with CE or HPLC or PGC [77, 178, 179]. In ESI-MS coupled with CE, nonvolatile buffers create problem for analysis by MS and the flow rates of the two systems requires an interface to become compatible [77]. Thus to overcome this hitch, in recombinant antibiotics, APTS-labeled N-glycans are analyzed by using CE-ESI-MS with online LIF detection [180]. ESI-MS can be easily coupled with RP or NP-HPLC as in this the glycans are eluted in an aqueous solvent like acetonitrile which is suitable for electrospray ionization process. In HPLC-ESI-MS, ESI can be used directly by analyte solution infusion or by connecting to column of LC. To minimize the ion suppression effects it is necessary to quantify glycans by using chromatographic separation technique before performing MS analysis. In PGC-ESI-MS, acidic glycans can be detected by negative ion detection mode but positive ion mode is preferred to analyze glycans (acidic and neutral) as in negative ion phase acidic glycan suppresses the estimation of neutral glycan. PGC chromatography separates isomeric structures of O-linked glycans and NLGs. PGC-LC-ESI-MS is considered as most successful technique of LC combined with MS because of its high resolving power together with a detection capacity of fmol (femtomole). PGC-LS-ESI-MS is used to analyze reduced, labeled, and permethylated carbohydrates [181, 179]. Fourier transform ion cyclotron mass spectrometry (FT-ICR-MS), a highly sensitive approach, has been developed to define various glycans along with glycolipids [182–184]. For accurate glycan characterization, chip-based electrospray like nanoscale liquid delivery system along with tandem MS and FT-ICR-MS is used [185, 186].

### 9.5.4   LECTIN MICROARRAY GLYCOME PROFILING

Lectins are the class of proteins derived from diverse sources and binds to the carbohydrate residue of glycoproteins and glycolipids. Lectins are also regarded as a class of non-immune proteins having one non-catalytic domain which enables glycan recognition including glycans binding reversibly to specific free sugars or unaltered glycans [93, 187]. Plant seeds are considered as the great source of lectins and it can be obtained easily from plant seeds [94]. Lectins are of great importance as they help in proper physiology, development of plant and also help the plant in overcoming from stressed conditions. Lectins obtained from plants are nearly soluble and one can get purified lectins by immobilizing the lectins on carbohydrate matrix. Lectins can be used for glycoconjugates characterization. Lectins are classified based on small carbohydrate haptens that can identify galactose and N-Acetyl glucosamine binding lectins. This classification is further based on diverse carbohydrate-binding among plant lectins. To overcome the drawbacks of label-based approaches, various label-free methods like carbon nanowires and nanotubes, micro cantilevers and surface-plasmon resonance have been developed recently [188]. Previous research shows that glycan analysis uses lectin based techniques like enzyme linked lectin sorbent assay (ELLA), lectin affinity chromatography, lectin blots, lectin based flow cytometry and lectin histology [94]. Nature of interaction between glycan and lectin is very weak. Earlier, scanner present in microarray needed washing process. This washing step may result in reduced signal intensity as certain glycans linked to lectin on array may dissociate. With the advance in technology lectin microarray was developed based on the detection via evanescent field fluorescence, which has improved the array analysis methods [95]. In this method, lectins are placed individually on solid support as discrete spots using different chemical methods, optimum conditions like buffer, temperature, humidity, size, and morphology of spot. The proteins are covalently linked in random orientation to solid surface via various chemical methods like inserting carbine, forming biotin avidinbridge, attaching lysine protein backbone amine functional group to support via epoxyfunctionalized or N-hydroxysuccinimidyl derived esters, applying self-assembled thiols monolayers on three dimensional hydrogel or gold-coated surfaces [96–98]. All the above mentioned methods lack proper control in providing native multimeric quarternary structure, optimal orientation and multivalent carbohydrate recognition domains (CRD) cluster of lectins and required metal ions [189]. Thus to overrule the above problems, glycosylated antibody analysis can be performed by oxidizing glycan chains of lectin with periodate oxidation [98]. In this method the oxidized glycans chain of lectins are attached to solid support having hydrazine or hydroxylamine leaving lectins CRD unaltered. This method protects lectins by forming cross-reacting glycan portions and improves localization of lectins in native multimeric structures. Hydrazide compounds based approaches can be applied on glycoproteins and mammalian cells to obtain glycans [190]. In lectin microarray, lectins are linked to pin-print array or piezoelectric array. In human studies, this approach can be used for profiling biological and clinical samples glycosylation which results in the development of new biomarkers. It is used to study diversity in human immunodeficiency virus 1 (HIV-1), bacterial glycome and cancer glycome [99]. Certain plant lectins, specific antibody and recombinant lectins can be used for analyzing glycans (O- and N-linked) and glycolipids in lectin microarray. Integration of lectin microarray with real time PCR (qPCR) puts an end to the need of lengthy procedures used in glycan analyzes of glycoproteins and glycolipids from various sources and replaces it with a high-throughput approach based on lectin-sugar interactions [191]. Glycome profiling specificity in lectin microarray is based on carbohydrate-binding proteins array. This microarray technique has various advantages like throughput, speed, magnitude of glycolipids and linked glycans analysis. In this technique, glycan units that are not recognized by specific lectins cannot be analyzed [100]. However, the limitation of lectin microarray leads to the development of another microarray technique known as carbohydrate microarray. This technology has revolutionized the research in the field of carbohydrate protein interactions and helped in decoding carbohydrate ligands present in host-pathogen interactions and endogenous receptor systems [192–194]. Carbohydrate microarray approach is classified into two types: polysaccharide microarrays and oligosaccharide microarrays. In polysaccharide microarray method the samples derived from natural sources are randomly immobilized on solid support depending on hydrophobic physical absorption [195–197] or on the charge interaction [198] to generate polysaccharide microarrays whereas in case of oligosaccharide microarray technique, the samples are difficult to immobilize on the solid matrix due to their hydrophilic nature. Thus to overcome the immobilization problem of samples, samples are chemically conjugated with lipids to develop neoglycolipid (NGL) probes by using reductive amination process. Amphipathic properties of NGL probes are used in oligosaccharide microarray for analysis. Oligosaccharide microarray technique provides complete information on carbohydrate recognition events like structure-activity relationships of glycans [199, 200].

## 9.6  EXAMPLES OF GLYCOME PROFILING IN PLANTS

### 9.6.1  GLYCOME PROFILING IN CEREALS

Carbohydrates, one of the major components of food source and which can act as structural building units are abundant constituents of cereals. Earlier, scientists were giving more attention to study the biological roles of proteins than to the carbohydrates [101]. However, the unique complexity of carbohydrates is getting more attention now than nucleic acids and proteins. The quality of the rice grain can be determined by the amount of starch present in rice grain. Environmental factors like high temperature can modify the starch content in rice grains at the time of grain filling stage [102]. Amylose and amylopectin are two main components of starch which can disturb the texture as well as physiochemical properties of rice grains [103]. Amylose and amylopectin can be formed from both translucent and opaque parts of perfect and chalky grain. Chromatography technique like size-exclusion

HPLC can be used to determine the content of starch in rice grain. This technique revealed that there is similarity between the amylose content of translucent and opaque parts of rice grains [201]. However, a little difference between the long B chain of amylopectin and amylose was observed in rice grain which was confirmed by using fluorescence capillary electrophoresis (FCEP) technique. The difference in the chain-length of perfect grain and in opaque and translucent part of chalky grain was due to high temperature [202]. Further, high temperature can reduce the amylose contents and can decrease the weight ratio of short B chains to long B chains of amylopectin [104]. Addition of metal ions and gibberellic acid individually alters α- and β-amylase and dextrinase during sprouting in barley. However, combination of both metal ions and gibberellic acid increases the activity of enzymes during sprouting in barley [105]. Purity of β-glucan was studied in barley spike using β-glucan activity measurement method and it revealed that barley spike has 77.12% of β-glucan which mainly includes self-made and imported β-glucan [106]. In oats, gel permeation chromatography (GPC) technique was used to determine β-glucans [203]. Glycome profiling in wheat straw was done by using comprehensive microarray polymer profiling (CoMPP) which revealed that hydrothermal pretreatment can alter hemicellulose, xylan, arabinoxylans, xyloglucan, and mixed linkage glucan epitopes in wheat straw biomass [107]. Glycome profiling using CoMPP technique has been done in rice crop also. In this case, this analysis revealed that monoclonal antibodies against xylan can produce a weak signal on de-starched alcohol-insoluble residues (AIR) in rice plant. It is also reported that less xylan content in cell wall leads to the reduction of xylose content in rice culm cell walls and finally to recalcitrance reduction [108].

## 9.6.2   GLYCOME PROFILING IN HORTICULTURAL CROPS

Glycome profiling has been done in many horticultural crops. Few examples have been discussed here in this section. Corn stover, a graminaceous monocot which is commonly used as fodder, litter for livestock and as a fuel for bioenergy. Golden rod (*Solidago*), an herbaceous dicot used as food in America, has an anti-pyretic activity to hay fever caused by ragweed. Pretreatment of corn stover, golden rod with alkaline hydrogen peroxide (AHP) and then enzymatic hydrolysis can be used to identify various cell wall matrix features contributing to recalcitrance of cell wall matrix. This method can be used to determine the taxonomy and structural diversity based on mild alkaline oxidative pretreatment of diverse feedstock [204]. Glycome profiling can be used to determine alteration in non-cellulosic glucans extract of different crops followed up by pretreatment with monoclonal antibodies. Glycome analysis can further determine the variation of non-cellulosic glycan epitopes in crop extracts in terms of abundance and distribution, and it can also elucidate cell wall structural integrity by inducing pretreatment to the samples [204]. Glycome profiling can be used to study the difference in cell wall recalcitrance in different plants via AHP pretreatment mechanism in golden rod and corn stover [204]. Glycome profiling in tuber vegetable crop like potato was done to study the N-glycosylation of patatin proteins. Patatin, a type of glycoprotein present in potato has xylan and fructose oligosaccharides. Oligosaccharides are considered as plant allergens immunological cross-reaction source. For studying patatin in potato, PNAGaseA enzyme was used to digest oligosaccharides and then released oligosaccharides were analyzed by MALDI-MS [109]. Thus glycome analysis in potato helps in identifying new patatins and revealed that there is difference in glycan profile of different patatins. However, insights into the potato patatin glycosylation helps in selecting healthy potato with lesser amount of allergy causing specific glycan epitopes and for manufacturing of therapeutic proteins [109]. Recently, glycome profiling of corn stover with ammonia fiber expansion (AFEX) pretreatment was done to identify intact components of cell wall after enzymatic hydrolysis [117]. During enzymatic hydrolysis, glycome profiling is used to study carbohydrates cell wall recalcitrance. Glycan analysis of corn stover during enzyme hydrolysis showed that few glycans cannot be cleaved and enriched [117].

### 9.6.3   GLYCOME PROFILING IN TREES

Various techniques like glycome profiling, LC have been used for analysis of monosaccharides, determination of lignin present in cell wall and enzymatic hydrolysis in trees. Glycome profiling has been used to examine the xylem in cell walls of three *Eucalyptus* species (*E. globulus, E. grandis,* and *E. urophylla*). Based on the climate of tropic region, *Eucalyptus* species varies in their lignin content, cellulose productivity, and syringyl/guayacyl (S/G) ratio. The glycan diversity of 600 *Eucalyptus* species was determined by using diverse collection of glycan directed monoclonal antibodies [110]. With the advances in glycome profiling, monoclonal antibodies can be used to characterize and invigilate the structure of cell wall and its component extractability [111]. Glycome profiling of three species of *Eucalyptus* revealed that *E. urophylla* and *E. globulus* differ in their cell wall polymer linkages, with different phenotypes whereas *E. grandis* have characteristics of other two species. This research revealed that *E. europhylla* has strong connection between cell wall polymers and lignin which make it highly recalcitrance. Whereas cell wall polymers in *E. globulus* are loosely associated which leads to easily deconstruction of its cell wall. Cell wall component of *Eucalyptus* species varies both in terms of quality and quantity. These composition variations in species affect the lingo-cellulosic feedback process. Moreover, cell wall polymers association types and amount of polymer also affect the lingo-cellulosic feedback process. Glycome profiling has also been done in *Populus trichocarpa* to study the N-linked glycosylation mechanism. Recently, in *Populus trichocarpa* a unique workflow was developed to identify and quantify the glycans by using a novel approach: individuality normalization when labeling with isotopic glycan hydrazide tags (INLIGHT) [112]. INLIGHT helped the researchers to identify glycosylation sites for over 500 proteins including 12 enzymes and a peroxidase involved in lignin biosynthesis. Moreover, it also helped in identifying 27 glycans present in the *Populus trichocarpa* cell wall [112]. Xylan hydrolysis was studied in a cross between *Populus trichocarpa* × *Populus deltoids*. Glycome profiling of the *Populus* during hydrolysis revealed that there is polysaccharide fragmentation and removal of glycan in cell wall of *Populus*. Thus showing that hydrolysis of xylan influences the cell wall formation process [113].

### 9.6.4  GLYCOME PROFILING IN LEGUMES

Legumes are the major source of proteins, fats, carbohydrates along with vitamins, minerals, and certain polyphenolic compounds. Legumes are major food crop for developing and underdeveloped countries. Glycome profiling has also been carried out in legumes to study the cell wall composition and structure. Root hair is the extended part present only at the root tip of plants. The chemical characterization of cell wall of soybean root can be determined by its root hair. Glycome profiling in soybean elucidated structural difference in galactomannan, rhamnogalacturonan I and II, xyloglucan, 4-*O*-methyl glucuronoxylan and glucuronoxylan isolated from soybean roots and root hair cell walls [114]. Glycome profiling in soybean for analyzing the cell wall polysaccharides was done by using MALDI-TOF-MS [114]. Glycan analysis of soybean revealed that galacturonic acid-containing xyloglucan, mannose-rich polysaccharides was present only in soybean root hair cell walls. Glycome profiling was carried in alfalfa crop to analyze recalcitrance in cell wall. CAD1 (cinnamyl alcohol dehydrogenase 1) [205] and HCT (hydroxycinnamoyl CoA: shikimate hydroxycinnamoyl transferase) [206] mutants effect were studied in the recalcitrance of mutations in lignin biosynthesis in alfalfa. Recently SDS-PAGE (a proteomic approach) and LC-MS/MS (lectin affinity chromatography) approach was used to unravel the molecular mechanism of drought stress in leaves of common bean. Thirty-five glycoproteins involving cell wall processes, defense/stress-related proteins, and proteins related to proteolysis were found to be changed in abundance in leaves of common bean under drought [127].

## 9.7   CONCLUSION AND FUTURE PROSPECTS

Glycome profiling provides an insight to understand the structure and functional relationship of glycans, and it unfolds the mysteries of glycomics, which is of great benefit to the researchers as sugars play a vital role in processes like stress responses, signaling, and immunity. The finding from plant glycomics can be used in biopharming and biopharmaceutics. Thus there is a need to understand the nature and structure of glycans present in plants. These glycans can be analyzed by different techniques like electrophoresis, chromatography, spectrometry, and microarrays. All these techniques are high throughput, and there are certain limitations of these techniques. Thus, detailed knowledge about the method is a prerequisite for proper analysis of glycans. However, the integration of glycomics with other omic approaches like genomics, proteomics, epigenomics, metabolomics, and lipidomics can be used to study the mechanism of cellular pathways and their response in stress conditions.

## KEYWORDS

- **chromatography**
- **electrophoresis**
- **glycan**
- **glycome profiling**
- **mass spectrometry**
- **microarray**

# REFERENCES

1. Shahzad, M. A., et al., (2016). Plant glycomics. *Plant Omics: Trends and Applications* (pp. 445–476). Springer International Publishing.
2. Yadav, S., Yadav, D. K., Yadav, N., & Khurana, S. M. P., (2015). *Plant Omics: The Omics of Plant Science*, 299–329.
3. Faugeron, C., Lhernould, S., Lemoine, J., Costa, G., & Morvan, H., (1997a). Identification of unconjugated *N*-glycans in strawberry plants. *Plant Physiol. Biochem., 35*, 891–895.
4. Kimura, Y., Takagi, S., & Shiraishi, T., (1997). Occurrence of free *N*-glycans in pea (*Pisum sativum* L.) seedlings. *Biosci. Biotechnol. Biochem., 61*, 924–926. doi: 10.1271/bbb.61.924.
5. Kimura, Y., & Matsuo, S., (2000). Free *N*-glycans already occur at an early stage of seed development. *J. Biochem., 127*, 1013–1019. doi: 10.1093/oxfordjournals.jbchem.a022692.
6. Kimura, Y., & Kitahara, E., (2000). Structural analysis of free *N*-glycans occurring in soybean seedlings. *Biosci. Biotechnol. Biochem., 64*, 1847–1855. doi: 10.1271/bbb.64.1847.
7. Maeda, M., & Kimura, Y., (2014). Structural features of free N-glycans occurring in plants and functional features of de-N-glycosylation enzymes, ENGase, PNGase: The presence of unusual plant complex type N-glycans. *Front Plant Sci., 5*, 429. doi: 10.3389/fpls.2014.00429.
8. Wilson, I. B. H., Zeleny, R., Kolarich, D., Staudacher, E., Stroop, C. J. M., Kamerling, J. P., et al., (2001). Analysis of Asn linked glycans from vegetable foodstuffs: Widespread occurrence of Lewis a, core α 1,3-linked fucose and xylose substitutions. *Glycobiology, 11*, 261–274. doi: 10.1093/glycob/11.4.261.
9. Lis, H., & Sharon, N., (1993). Protein glycosylation: Structural and functional aspects. *Eur. J. Biochem., 218*(1), 1–27.
10. Varki, A., (1993). Biological roles of oligosaccharides: All of the theories are correct. *Glycobiology, 3*, 97–130.
11. Brockhausen, I., Schutzbach, J., & Kuhns, W., (1998). Glycoproteins and their relationship to human disease. *Acta Anat. (Basel), 161*, 36–78.
12. Dennis, J. W., Granovsky, M., & Warren, C. E., (1999). Protein glycosylation in development and disease *Bioessays, 21*, 412–421.
13. Haltiwanger, R. S., & Lowe, J. B., (2004). Role of glycosylation in development. *Annu. Rev. Biochem., 73*, 491–537.
14. Taniguchi, N., Miyoshi, E., Gu, J., Honke, K., & Matsumoto, A., (2006). Decoding sugar functions by identifying target glycoproteins. *Curr. Opin. Struct. Biol., 16*, 561–566.
15. Lerouge, P., Cabanes-Macheteau, M., Rayon, C., Fischette-Lainé, A. C., Gomord, V., & Faye, L., (1998). *N*-Glycoprotein biosynthesis in plants: Recent developments and future trends. *Plant Mol. Biol., 38*, 31–48. doi: 10.1023/A:1006012 005654.
16. Nguema-Ona, E., Vicré-Gibouin, M., Gotté, M., Plancot, B., Lerouge, P., Bardor, M., et al., (2014). Cell wall O-glycoproteins and N -glycoproteins: Aspects of biosynthesis and function. *Front. Plant Sci., 5*, 499. doi: 10.3389/fpls.2014.00499.
17. Kieliszewski, M. J., & Shpak, E., (2001). Synthetic genes for the elucidation of glycosylation codes for arabinogalactan-proteins and other hydroxyproline- rich glycoproteins. *Cell Mol. Life Sci., 58*, 1386–1398. doi: 10.1007/PL00000783.

18. Showalter, A. M., (2001). Arabinogalactan-proteins: Structure, expression and function. *Cell Mol. Life Sci., 58,* 1399–1417.

19. Schultz, C. J., Rumsewicz, M. P., Johnson, K. L., Jones, B. J., Gaspar, Y. M., & Bacic, A., (2002). Using genomic resources to guide research directions the arabinogalactan protein gene family as a test case. *Plant Physiol., 129,* 1448–1463. doi: 10.1104/pp.003459.

20. Showalter, A. M., Keppler, B., Lichtenberg, J., Gu, D., & Welch, L. R., (2010). A bioinformatics approach to the identification, classification, and analysis of hydroxyproline-rich glycoproteins. *Plant Physiol., 153,* 485–513. doi: 10.1104/pp.110.156554.

21. Kieliszewski, M. J., Lamport, D. T. A., & Cannon, M. C., (2011). Hydroxy proline rich glycoproteins: Form and function. *Annu. Plant Rev., 41,* 321–342.

22. Lamport, D. T. A., Kieliszewski, M. J., Chen, Y., & Cannon, M. C., (2011). Role of the extension superfamily in primary cell wall architecture. *Plant Physiol., 156,* 11–19. doi: 10.1104/pp.110.169011.

23. Nguema-Ona, E., Coimbra, S., Vicré-Gibouin, M., Mollet, J. C., & Driouich, A., (2012). Arabinogalactan proteins in root and pollen tube cells: Distribution and functional properties. *Ann. Bot., 110,* 383–404. doi: 10.1093/aob/mcs143.

24. Nguema-Ona, E., Vicré-Gibouin, M., Cannesan, M. A., & Driouich, A., (2013). Arabinoga-lactan-proteins in root microbe interactions. *Trends Plant Sci., 18,* 440–449. doi: 10.1016/j.t plants. 2013.03.006.

25. Zachara, N. E., & Hart, G. W., (2006). Cell signaling, the essential role of O-GlcNAc. *Biochim. Biophys. Acta, 1761,* 599–617.

26. Mitsui, T., Kimura, S., & Igaue, L., (1990). Isolation and characterization of Golgi membranes from suspension cultured cells of rice. *Plant Cell Physiol., 31,* 15–25.

27. Kishimoto, T., Watanabe, M., Mitsui, T., & Mori, H., (1999). Glutelin basic subunits have a mammalian mucin type O-linked disaccharide side chain. *Arch Biochem. Biophys., 370,* 271–277.

28. Hall, Q., & Cannon, M. C., (2002). The cell wall hydroxyl proline-rich glycoprotein RSH is essential for normal embryo development in *Arabidopsis. Plant Cell, 14,* 1161–1172. doi: 10.1105/tpc.010477.

29. Motose, H., Sugiyama, M., & Fukuda, H., (2004). A proteoglycan mediates inductive interaction during plant vascular development. *Nature, 429,* 873–878.

30. Lee, K. J. D., Sakata, Y., Mau, S. L., Pettolino, F., Bacic, A., Quatrano, R. S., et al., (2005). Arabinogalactan proteins are required for apical cell extension in the moss *Physcomitrella patens. Plant Cell, 17,* 3051–3065. doi: 10.1105/tpc.105.034413.

31. Nguema-Ona, E., Bannigan, A., Chevalier, L., Baskin, T. I., & Driouich, A., (2007). Disruption of arabinogalactan proteins disorganizes cortical microtubules in the root of *Arabidopsis thaliana. Plant J., 52,* 240–251. doi: 10.1111/j.1365- 313X.2007.03224.

32. Seifert, G. J., & Roberts, K., (2007). The biology of arabinogalactan proteins. *Annu. Rev. Plant Biol., 58,* 137–161.

33. Cannon, M. C., Terneus, K., Hall, Q., Tan, L., Wang, Y., Wegenhart, B. L., et al., (2008). Self-assembly of the plant cell wall requires an extension scaffold. *Proc. Natl. Acad. Sci. U.S.A, 105,* 2226–2231. doi: 10.1073/pnas.0711980105.

34. Ellis, M., Egelund, J., Schultz, C. J., & Bacic, A., (2010). Arabinogalactan protein: Key regulator s at the cell surface. *Plant Physiol., 153,* 403–419. doi: 10.1104/pp.110.156000.

35. Velasquez, S. M., Ricardi, M. M., Dorosz, J. G., Fernandez, P. V., Nadra, A. D., Pol-Fachin, L., Egelund, J., et al., (2011). O-Glycosylated cell wall proteins are essential in root hair growth. *Science, 332*, 1401–1403.

36. Cannesan, M. A., Durand, C., Burel, C., Gangneux, C., Lerouge, P., Ishii, T., et al., (2012). Effect of arabinogalactan proteins from the root caps of pea and *Brassica napus* on *Aphanomyces euteiches* zoospore chemotaxis and germination. *Plant Physiol., 159*, 1658–1670. doi: 10.1104/pp.112.198507.

37. Moore, J. P., Fangel, J. U., Willats, W. G. T., & Vivier, M. A., (2014a). Pectic-$\beta$(1,4)-galactan, extension and arabinogalactan-protein epitopes differentiate ripening stages in wine and table grape cell walls. *Ann Bot.* doi: 10.1093/aob/mcu053.

38. Dormann, P., & Benning, C., (2002). Galactolipids rule in seed plants. *Trends Plant Sci., 7*, 112–118.

39. Grunwald, C., (1978). Steryl glycoside biosynthesis. *Lipids, 13*, 697–703.

40. Terao, A., Hyodo, H., Satoh, S., & Iwai, H., (2013). Changes in the distribution of cell wall polysaccharides in early fruit pericarp and ovule, from fruit set to early fruit development, in tomato (*Solanum lycopersicum*). *J. Plant Res., 126*, 719–728.

41. Gough, C., & Cullimore, J., (2011). Lipo-chitooligosaccharide signaling in endosymbiotic plant-microbe interactions. *Mol. Plant Microbe. Interact., 24*(8), 867–878.

42. Crawford, R. L., (1981). *Lignin Biodegradation and Transformation*. John Wiley and Sons, New York, NY. ISBN: 0-471-05743-6.

43. McMillan, J. D., (1993). Pretreatment of lignocellulosic biomass. In: Himmel, M. E., Baker, J. O., & Overend, R. P., (eds.), *Enzymatic Conversion of Biomass for Fuel Production* (pp. 292–323). American Chemical Society, Washington DC.

44. Pornsak, S., (2003). Chemistry of pectin and its pharmaceutical uses: A review. *Silpakorn Univ. Int. J., 3*(1/2), 206.

45. Camejo, D., Martí, M. C., Jiménez, A., Cabrera, J. C., Olmos, E., & Sevilla, F., (2011). Effect of oligogalacturonides on root length, extracellular alkalinization and $O_2$-accumulation in alfalfa. *J. Plant Physiol., 168*, 566–575.

46. Obro, J., Harholt, J., Scheller, H. V., & Orfila, C., (2004). *Phytochemistry, 65*, 1429–1438.

47. Pabst, M., Fischl, R. M., Brecker, L., Morelle, W., Fauland, A., Köfeler, H., Altmann, F., & Léonard, R., (2013). Rhamnogalacturonan II structure shows variation in the side chains monosaccharide composition and methylation status within and across different plant species. *Plant J., 76*, 61–72.

48. O'Neill, M. A., Eberhard, S., Albersheim, P., & Darvill, A. G., (2001). *Science, 294*, 846.

49. Ishii, T., Matsunaga, T., Pellerin, P., O'Neill, M. A., Darvill, A., & Albersheim, P., (1999). The plant cell wall polysaccharide rhamnogalacturonan II self-assembles into a covalently cross-linked dimer. *J. Biol. Chem., 274*(19), 13098. doi: 10.1074/jbc.274.19.13098.

50. Zhao, X., Moates, G. K., Wellner, N., Collins, S. R. A., Coleman, M. J., & Waldron, K. W., (2014). Chemical characterization and analysis of the cell wall polysaccharides of duckweed (*Lemna minor*). *Carbohydr. Polym., 111*, 410–418.

51. Gardner, S., Burrell, M. M., & Fry, S. C., (2002). Screening of Arabidopsis thaliana stems for variation in cell wall polysaccharides. *Phytochemistry, 60*, 241–254.

52. Dilokpimol, A., & Geshi, N., (2014). Arabinogalactan proteins: Focus on carbohydrate active enzymes. *Front Plant Sci., 5*, 198.

53. Majewska-Sawka, A., & Nothnagel, E. A., (2000). The multiple roles of arabinogalactan proteins in plant development. *Plant Physiol., 122*, 3–10.

54. Knoch, E., Dilokpimol, A., & Geshi, N., (2014). Arabinogalactan proteins: Focus on carbohydrate active enzymes. *Front Plant Sci., 5*, 198.

55. Mohnen, D., (1999). In: Barton, D., Nakanishi, K., & Meth-Cohn, O., (eds.), *Comprehensive Natural Products Chemistry* (pp. 497–527). Elsevier, Dordrecht.

56. Van, D. V. J. W., (1994). *Characterization and Mode of Action of Enzymes Degrading Galactan Structures of Arabinogalactans.* Doctoral thesis, Wageningen University, Wageningen.

57. Ridley, B. L., O'Neill, M. A., & Mohnen, D., (2001a). Pectins: Structure, biosynthesis, and oligogalacturonide related signaling. *Phytochemistry, 57*, 929. doi: 10.1016/S0031-9422 (01) 00113-3.

58. Dilokpimol, A., Poulsen, C. P., Vereb, G., Kaneko, S., Schulz, A., & Geshi, N., (2014). Galactosyl transferases from *Arabidopsis thaliana* in the biosynthesis of type II arabinogalactan: Molecular interaction enhances enzyme activity. *BMC Plant Biol., 14*, 90.

59. Geshi, N., Johansen, J. N., Dilokpimol, A., Rolland, A., Belcram, K., Verger, S., et al., (2013). A galactosyltransferase acting on arabinogalactan protein glycans is essential for embryo development in *Arabidopsis. Plant J., 76*, 128–137.

60. Cao, J., (2012). The pectin lyases in Arabidopsis thaliana: Evolution, selection and expression profiles. *PLoS One, 7*, e46944.

61. Fu, Y., Yin, H., Wanga, W., Wang, M., Zhang, H., Zhao, X., et al., (2011). β-1, 3-glucan with different degree of polymerization induced different defense responses in tobacco. *Carbohydr. Polym., 86*, 774–782.

62. Silbert, J. E., & Sugumaran, G., (2002). Biosynthesis of chondroitin/dermatan sulfate. *IUBMB Life, 54*, 177–186.

63. Sugahara, K., & Kitagawa, H., (2002). Heparin and heparan sulfate biosynthesis. *IUBMB Life, 54*, 163–175.

64. McBride, H. M., Neuspiel, M., & Wasiak, S., (2006). Mitochondria: More than just a powerhouse. *Curr. Biol., 16*(14), 551–560.

65. Hubbard, C. S., & Ivatt, R. J., (1981). *Annu. Rev. Biochem., 50*, 555–583.

66. Block, M. A., Dorne, A. J., Joyard, J., & Douce, R., (1983). Preparation and Characterization of Membrane Fractions Enriched in Outer and Inner Envelope Membranes from Spinach Chloroplasts, *J. Biol. Chem. 258*, 13273–13280.

67. Kelly, A. A., & Dormann, P., (2004). Green light for galactolipid trafficking. *Curr. Opin. Plant Biol., 7*, 262–269.

68. Austin, J. R., Frost, E., Vidi, P. A., Kessler, F., & Staehelin, L. A., (2006). Plastoglobules are lipoprotein sub-compartments of the chloroplast that are permanently coupled to thylakoid membranes and contain biosynthetic enzymes. *Plant Cell Online, 18*(7), 1693–1703.

69. Dube, D. H., & Bertozzi, C. R., (2005). Glycans in cancer and inflammation potential for therapeutics and diagnostics. *Nat. Rev. Drug Discov., 4*, 477–488.

70. Ohtsubo, K., & Marth, J. D., (2006). Glycosylation in cellular mechanisms of health and disease. *Cell, 126*, 855–867.

71. Sharon, N., (2006). Carbohydrates as future anti-adhesion drugs for infectious diseases. *Biochim. Biophys. Acta, 1760*, 527–537.

72. Marth, J. D., & Grewal, P. K., (2008). Mammalian glycosylation in immunity. *Nat. Rev. Immunol., 8*, 874–887.

73. Vanderschaeghe, D., Festjens, N., Delanghe, J., & Callewaert, N., (2010). Glycome profiling using modern glycomics technology: Technical aspects and applications. *Biol. Chem., 391*, 149–161.

74. Xia, Y. G., Wang, Q. H., Liang, J., Yang, B. Y., Li, G. Y., & Kuang, H. X., (2011). Development and application of a rapid and efficient CZE method coupled with correction factors for determination of monosaccharide composition of acidic hetero-polysaccharides from *Ephedra sinica*. *Phytochem. Anal., 22*, 103–111.

75. Rovio, S., Simolin, H., Koljonen, K., & Siren, H., (2008). Determination of monosaccharide composition in plant fiber materials by capillary zone electrophoresis. *J. Chromatogr., 1185*, 139–144.

76. Sarazin, C., Delaunay, N., Costanza, C., Eudes, V., Mallet, J. M., & Gareil, P., (2011). New avenue for mid-UV-range detection of underivatized carbohydrates and amino acids in capillary electrophoresis. *Anal. Chem., 83*, 7381–7387.

77. Mechref, Y., & Novotny, M., (2009). Glycomic analysis by capillary electrophoresis-mass spectrometry. *Mass Spectrom. Rev., 28*, 207–222.

78. Tseng, H. M., Gattollin, S., Pritchard, J., Newbury, H. J., & Barrett, D. A., (2009). Analysis of mono, di- and oligosaccharides by CE using a two-stage derivatization method and LIF detection. *Electrophoresis, 30*, 1399–1405.

79. Gotti, R., (2011). Capillary electrophoresis of phytochemical substances in herbal drugs and medicinal plants. *J. Pharm. Biomed. Anal., 55*, 775–801.

80. Sarazin, C., Delaunay, N., Costanza, C., Eudes, V., & Gareil, P., (2012). Application of a new capillary electrophoretic method for the determination of carbohydrates in forensic, pharmaceutical, and beverage samples. *Talanta, 99*, 202–206.

81. Campbell, M. P., Royle, L., & Rudd, P. M., (2015). Glyco base and auto GU: Resources for interpreting HPLC-glycan data. In: Lütteke, T., & Frank, M., (eds.), *Glycoinformatics: Methods in Molecular Biology* (Vol. 1273). Humana Press, New York, NY.

82. Artemenko, N. V., Campbell, M. P., & Rudd, P. M., (2010). Glyco extractor: A web-based interface for high throughput processing of HPLC-glycan data. *Journal of Proteome Research, 9*(4), 2037–2041. doi: 10.1021/pr901213u.

83. Royle, L., Campbell, M. P., Radcliffe, C. M., White, D. M., Harvey, D. J., et al., (2008). HPLC-based analysis of serum N-glycans on a 96-well plate platform with dedicated database software. *Anal. Biochem., 376*, 1–12.

84. Pabst, M., Kolarich, D., Poltl, G., Dalik, T., et al., (2009). Comparison of fluorescent labels for oligosaccharides and introduction of a new post labeling purification method. *Anal. Biochem., 384*, 263–273.

85. Paskach, T. J., Lieker, H. P., Reilly, P. J., & Thielecke, K., (1991). High-performance anion-exchange chromatography of sugars and sugar alcohols on quaternary ammonium resins under alkaline conditions. *Carbohydrate Research, 215*(1), 1–14.

86. Ciucanu, I., & Kerek, F., (1984). A simple rapid method for the permethylation of carbohydrates. *Carbohydr. Res., 131*, 209–217.

87. Zaia, J., (2010). Mass spectrometry and glycomics. *OMICS, 14*, 401–418.

88. Vakhrushev, S. Y., Langridge, J., Campuzano, I., Hughes, C., & Peter-Katalinic, J., (2008). Ion mobility mass spectrometry analysis of human glycourinome. *Anal. Chem., 80*, 2506–2513.

89. Itoh, S., Kawasaki, N., Hashii, N., Harazono, A., et al., (2006). N-linked oligosaccharide analysis of rat brain Thy-1 by liquid chromatography with graphitized carbon column/ion trap-Fourier transform ion cyclotron resonance mass spectrometry in positive and negative ion modes. *J. Chromatogr. A, 1103*, 296–306.

90. Pabst, M., & Altmann, F., (2008). Influence of electro sorption, solvent, temperature, and ion polarity on the performance of LC-ESI-MS using graphitic carbon for acidic oligosaccharides. *Anal. Chem., 80,* 7534–7542.

91. Harvey, D. J., Mattu, T. S., Wormald, M. R., Royle, L., et al., (2002). Internal residue loss: Rearrangements occurring during the fragmentation of carbohydrates derivatized at the reducing terminus. *Anal. Chem., 74*, 734–740.

92. Deguchi, K., Ito, H., Takegawa, Y., Shinji, N., et al., (2006). Complementary structural information of positive- and negative-ion MSn spectra of glycopeptides with neutral and sialylated N-glycans. *Rapid Commun. Mass Spectrom., 20*, 741–746.

93. Peumans, W. J., & Vandamme, E. J. M., (1995). Lectins as plant defense proteins. *Plant Physiol., 109*, 347–352.

94. Rudiger, H., & Gabius, H. J., (2001). Plant lectins: Occurrence, biochemistry, functions and applications. *Glycoconj. J., 18*, 589–613.

95. Narimatsu, H., Sawaki, H., Kuno, A., Kaji, H., Ito, H., & Ikehara, Y., (2010). A strategy for discovery of cancer glyco-biomarkers in serum using newly developed technologies for glycoproteomics. *FEBS J., 277*, 95–105.

96. Pilobello, K. T., Krishnamoorthy, L., Slawek, D., & Mahal, L. K., (2005). Development of a lectin microarray for the rapid analysis of protein glycopatterns. *Chem. Bio. Chem., 6*, 985–989.

97. Hsu, K. L., & Mahal, L. K., (2006). A lectin microarray approach for the rapid analysis of bacterial glycans. *Nat. Protoc., 1*, 543–549.

98. Chen, P., Liu, Y., Kang, X., Sun, L., Yang, P., et al., (2008). Identification of N-glycan of α-fetoprotein by lectin affinity microarray. *J. Cancer Res. Clin. Oncol., 134*, 851–860.

99. Krishnamoorthy, L., Bess, J., Preston, A. B., Nagashima, K., & Mahal, L. K., (2009). HIV-1 and microvesicles from T cells share a common glycome, arguing for a common origin. *Nat. Chem. Biol., 5*, 244–250.

100. Gupta, G., Surolia, A., & Sampathkumar, S. G., (2010). Lectin microarrays for glycomic analysis. *OMICS, 14*, 419–436.

101. Ernst, B., & Magnani, J. L., (2009). From carbohydrate leads to glycomimetic drugs. *Nat. Rev. Drug Disco., 8*, 661–677.

102. Mitsui, T., Yamakawa, H., & Kobata, T., (2016). Molecular physiological aspects of chalking mechanism in rice grains under high-temperature stress *Plant Prod. Sci., 19*, 22–29.

103. Muench, D. G., Wu, Y., Zhang, Y., Li, X., Boston, R. S., & Okita, T. W., (1997). Molecular cloning, expression and subcellular localization of a BiP homolog from rice endosperm tissue. *Plant Cell Physiol., 38*, 404–412.

104. Inouchi, N., Ando, H., Asaoka, M., Okuno, K., & Fuwa, H., (2000). The effect of environmental temperature on distribution of unit chains of rice amylopectin. *Starch/Starke, 52*, 8–12.

105. Shan, L., Bin, G., Juan, X., Qing, K., Junhong, Y., Jianjun, D., Lianju, S., Shuli, H., & Liu, J., (2009). Study on the effection of ions and GA3 on the carbohydrases during barley malt production. *Food and Fermentation Industries* TS262.5.

106. Liu, X., Shen, S., Song, J., Hu, B., Zhang, J., & Cai, J., (2007). Preliminary study on the extracting of β-glucan from barley spike's waste. *Barley and Cereal Sciences*. S512.3.

107. Alonso-Simón, A., Kristensen, J. B., Obro, J., Felby, C., Willats, W. G. T., & Jorgensen, H., (2010). High-throughput microarray profiling of cell wall polymers during hydrothermal pretreatment of wheat straw. *Biotechnol. Bioeng.*, *105*, 509–514. doi: 10.1002/bit.22546.

108. Chen, X., Vega-Sánchez, M. E., Verhertbruggen, Y., Chiniquy, D., Canlas, P. E., Fagerström, A., et al., (2013). Inactivation of OsIRX10 leads to decreased xylan content in rice culm cell walls and improved biomass saccharification. *Mol. Plant, 6*, 570–573. doi: 10.1093/mp/sss135.

109. Lattova, E., Brabcová, A., Bartová, V., Potesil, D., Bárta, J., & Zdráhal, Z., (2015). N-glycome profiling of patatins from different potato species of solanum genus. *J. Agric. Food Chem.* doi: 10.1021/acs.jafc.5b00426.

110. Salazar, M. M., Pattathil, S., Camargo, E. L. O., Grandis, A., Gonalves, D. C., Nascimento, L. C., Marques, W. L., et al., (2012). Transcriptome analysis and glycome profiling of three eucalyptus species. *Abstract in Plant and Animal Genome XX*.

111. Salazar, M. M., Grandis, A., Pattathil, S., Neto, J. L., Camargo, E. L. O., Alves, A., Rodrigues, J. C., et al., (2016). Eucalyptus cell wall architecture: Clues for lignocellulosic biomass deconstruction. *Bioenerg. Res.* doi: 10.1007/s12155-016-9770-y.

112. Loziuk, P. L., Hecht, E. S., & Muddiman, D. C., (2016). N-linked glycosite profiling and use of skyline as a platform for characterization and relative quantification of glycans in differentiating xylem of *Populus trichocarpa*. *Anal Bioanal. Chem.* doi: 10.1007/s00216-016-9776-9775.

113. Trajano, H. L., Pattathil, S., Tomkins, B. A., Tschaplinski, T. J., Hahn, M. G., Van, B. G. J., & Wyman, C. E., (2015). Xylan hydrolysis in *Populus trichocarpa × P. deltoides* and model substrates during hydrothermal pretreatment. *Bioresource Technology, 179*, 202–210.

114. Muszynski, A., Neill, M. A. O., Ramasamy, E., Pattathil, S., Avci, U., Pena, M. J., Libault, M., et al., (2015). Xyloglucan, galactomannan, glucuronoxylan, and rhamnogalacturonan I do not have identical structures in soyabean root and root hair cell walls. *Planta*. doi: 10.1007/s00425-015-2344-y.

115. Faugeron, C., Lhernould, S., Maes, E., Lerouge, P., Strecker, G., & Morvan, H., (1997). Tomato plant leaves also contain unconjugated *N*-glycans. *Plant Physiol. Biochem., 35*, 73–79.

116. Guillard, M., Gloerich, J., Wessels, H. J., Morava, E., Wevers, R. A., et al., (2009). Automated measurement of permethylated serumN-glycans by MALDI-linear ion trap mass spectrometry. *Carbohydr. Res., 344*, 1550–1557.

117. Gunawan, C., Xue, S., Pattathi, S., Da Costa, S. L., Dale, B. E., & Balan, V., (2017). Comprehensive characterization of non-cellulosic recalcitrant cell wall carbohydrates in unhydrolyzed solids from AFEX-pretreated corn stover. *Biotechnol., Biofuels, 10*, 82. doi: 10.1186/s13068-017-0757-5.

118. Majewska-sawka, A., & Nothnagel, E. A., (2000). The multiple roles of arabinogalactan proteins in plant development. *Plant Physiol., 122*, 3–9. doi: 10.1104/pp.122.1.3.

119. Moore, J. P., Fangel, J. U., Willats, W. G., & Vivier, M. A., (2014). Pectic-b (1,4)-galactan, extensin and arabinogalactan-protein epitopes differentiate ripening stages in wine and table grape cell walls. *Ann. Bot., 114*(6), 1279–1294. doi: 10.1093/aob/mcu053.

120. Muyang, L., Pattathil, S., Hahn, M. S., & Hodge, D. B., (2014). Identification of features associated with plant cell wall recalcitrance to pretreatment by alkaline hydrogen peroxide in diverse bioenergy feedstocks using glycome profiling. *Royal Society of Chemistry, 4*, 17282–17292.

121. Pan, Y., & Wu, H., (2009). Optimization of oat β-glucan extraction technology and determination of molecular characteristic by using GPC. *Journal of Beijing Technology and Business University* (Natural Science Edition).

122. Priem, B., Solokwan, J., Wieruszeski, J. M., Strecker, G., Nazih, H., & Morvan, H., (1990b). Isolation and characterization of free glycans of the oligomannoside type from the extra cellular medium of a plant cell suspension. *Glycoconjugate J., 7*, 121–132. doi: 10.1007/BF01050375.

123. Showalter, A. M., (2001). Arabinogalactan proteins: Structure, expression and function. *Cell Mol. Life Sci., 58*, 1399–1417. doi: 10.1007/PL00000784.

124. Snovida, S. I., & Perreault, H., (2007). A 2, 5-dihydroxybenzoic acid/N,N-dimethylaniline matrix for the analysis of oligosaccharides by matrix-assisted laser desorption/ionization mass spectrometry. *Rapid Commun. Mass Spectrom., 21*, 3711–3715.

125. Tadege, M., Wen, J., Tu, H., Kwak, Y., Eschstruth, A., Cayrel, A., Endre, G., et al., (2008). Large scale insertional mutagenesis using the Tnt1 retrotransposon in the model legume *Medicago truncatula. Plant J., 54*, 335–347.

126. Yamagaki, T., & Sato, A., (2009). Isomeric oligosaccharides analyses using negative-ion electrospray ionization ion mobility spectrometry combined with collision-induced dissociation MS/MS. *Anal. Sci., 25*, 985–988.

127. Zadraznik, T., Moen, A., Egge-Jacobsen, W., Meglic, V., & Sustar-Vozlic, J., (2017). Towards a better understanding of protein changes in common bean under drought: A case study of N-glycoproteins. *Plant Physiology and Biochemistry, 118*, 400–412.

128. Lannoo, N., Van, D. E. J. M., Albenne, C., & Jamet, E., (2014). Plant glycobiology—a diverse world of lectins, glycoproteins, glycolipids and glycans. *Front. Plant Sci., 5*, 604. doi: 10.3389/fpls.2014.00604.

129. Priem, B., Morvan, H., Hafez, A. M. A., & Morvan, C., (1990a). Influenced 'un xylmannoside d' origin evegetalesur l'elongationdel' hypocotylede Lin. *C. R. Acad. Sci. Paris., 311*, 411–416.

130. Priem, B., Gitti, R., Bush, C. A., & Gross, K. C., (1993). Structure of ten free N-glycans in ripening tomato fruit. Arabinose is a constituent of a plant N-glycan. *Plant Physiol., 102*, 445–458. doi: 10.1104/pp.102.2.445.

131. Faugeron, C., Lhernould, S., Maes, E., Lerouge, P., Strecker, G., & Morvan, H., (1997b). Tomato plant leaves also contain unconjugated *N*-glycans. *Plant Physiol. Biochem., 35*, 73–79.

132. Kornfeld, R., & Kornfeld, S., (1985). Assembly of asparagine linked oligosaccharides. *Annu. Rev. Biochem., 54*, 631–664.

133. Tan, L., Showalter, A. M., Egelund, J., Hernandez-Sanchez, A., Doblin, M. S., & Bacic, A., (2012). Arabino galactan-proteins and the research challenges for these enigmatic plant cell surface proteo glycans. *Front. Plant Sci., 3*, 140. doi: 10.3389/fpls.2012.00140.

134. Velasquez, S. M., SalgadoSalter, J., Petersen, B. L., & Estevez, J. M., (2012). Recent advances on the post-translational modifications of EXTs and their roles in plant cell walls. *Front Plant Sci., 3*, 93. doi: 10.3389/fpls.2012.00093.

135. Moore, J. P., Fangel, J. U., Willats, W. G., & Vivier, M. A., (2014b). Pectic-b(1,4)-galactan, extensin and arabinogalactan - protein epitopes differentiate ripening stages

in wine and table grape cell walls. *Ann. Bot., 114*(6), 1279–1294. doi: 10.1093/aob/mcu053.

136. Shibuya, N., & Iwasaki, T., (1985). Structural features of rice bran hemicellulose. *Phytochemistry, 24*, 285–289.

137. Palin, R., & Geitmann, A., (2012). The role of pectin in plant morphogenesis. *Biosystem, 109*, 397–402.

138. Basu, D., Liang, Y., Liu, X., Himmeldirk, K., Faik, A., Kieliszewski, M., et al., (2013). Functional identification of a hydroxyproline-o-galactosyltransferase specific for arabinogalactan protein biosynthesis in Arabidopsis. *J. Biol. Chem., 288*, 10132–10143.

139. Ridley, B. L., O'Neill, M. A., & Mohnen, D., (2001b). Complex carbohydrate research center and department of biochemistry and molecular biology. *Phytochemistry, 57*, 929. doi: 10.1016/ S0031-9422(01)00113-3.

140. Kjellbom, P., & Larsson, C., (1984). Preparation and polypeptide composition of chlorophyll-free plasma membranes from leaves of light-grown spinach and barley. *Physiol. Plant, 62*, 501–509.

141. Gabius, H. J., Siebert, H. C., Andre, S., Jimenez-Barbero, J., & Rudiger, H., (2004). Chemical biology of the sugar code. *Chembiochem., 5*, 740–764.

142. Cummings, R. D., (2009). The repertoire of glycan determinants in the human glycome. *Mol. Biosyst., 5*, 1087–1104.

143. Li, S. P., Wu, D. T., Lv, G. P., & Zhao, J., (2013). Carbohydrates analysis in herbal glycomics. *Trends in Analytical Chemistry, 52*, 155–169. http://dx.doi.org/10.1016/j.trac.2013.05.020.

144. Rakus, J. F., & Mahal, L. K., (2011). New technologies for glycomic analysis: Toward a systematic understanding of the glycome. *Annu. Rev. Anal. Chem., 4*, 367–392.

145. Lo-Guidice, J. M., & Lhermitte, M., (1996). HPLC of oligosaccharides in glycobiology. *Biomed. Chromatogr., 10*, 290–296.

146. Rudd, P. M., Guile, G. R., Kuster, B., Harvey, D. J., Opdenakker, G., et al., (1997). Oligosaccharide sequencing technology. *Nature, 388*, 205–207.

147. Anumula, K. R., (2006). Advances in fluorescence derivatization methods for high-performance liquid chromatographic analysis of glycoprotein carbohydrates. *Anal. Biochem., 350*, 1–23.

148. Koizumi, K., (1996). High-performance liquid chromatographic separation of carbohydrates on graphitized carbon columns. *J. Chromatogr. A, 720*, 119–126.

149. Kuroda, Y., Shikata, K., Takeuchi, F., Akazawa, T., Kojima, N., Nakata, M., Mizuochi, T., & Goto, M., (2007). Structural alterations in outer arms of IgG oligosaccharides in patients with Werner syndrome. *Exp. Gerontol., 42*, 545–553.

150. Takahashi, N., (1996). Three-dimensional mapping of N-linked oligosaccharides using anion-exchange, hydrophobic and hydrophilic interaction modes of high-performance liquid chromatography. *J. Chromatogr. A, 720*, 217–225.

151. Campbell, M. P., Royle, L., Radcliffe, C. M., Dwek, R. A., & Rudd, P. M., (2008). GlycoBase and autoGU: Tools for HPLC-based glycan analysis. *Bioinformatics, 24*, 1214–1216.

152. Cataldi, T. R., Campa, C., & De Benedetto, G. E., (2000). Carbohydrate analysis by high-performance anion-exchange chromatography with pulsed amperometric detection: The potential is still growing. *Fresenius J. Anal. Chem., 368*, 739–758.

153. Corradini, C., Corradini, D., Huber, C. G., & Bonn, G. K., (1994). Synthesis of a polymeric-based stationary phase for carbohydrate separation by high-pH

anion-exchange chromatography with pulsed amperometric detection. *Journal of Chromatography A, 685*(2), 213–220.

154. Behan, J. L., & Smith, K. D., (2011). The analysis of glycosylation: A continued need for high pH anion exchange chromatography. *Biomed. Chromatogr., 25*, 39–46.

155. Cooper, C. A., Gasteiger, E., & Packer, N. H., (2001). GlycoMod — a software tool for determining glycosylation compositions from mass spectrometric data. *Proteomics, 1*, 340–349.

156. Goldberg, D., Sutton-Smith, M., Paulson, J., & Dell, A., (2005). Automatic annotation of matrix-assisted laser desorption/ionization N-glycan spectra. *Proteomics, 5*, 865–875.

157. McDonald, C. A., Yang, J. Y., Marathe, V., Yen, T. Y., & Macher, B. A., (2009). Combining results from lectin affinity chromatography and glycocapture approaches substantially improves the coverage of the glycoproteome. *Mol. Cell Proteomics, 8*, 287–301.

158. Von, D. L. C. W., Lutteke, T., & Frank, M., (2006). The role of informatics in glycobiology research with special emphasis on automatic interpretation of MS spectra. *Biochim Biophys Acta, 1760*, 568–577.

159. North, S. J., Hitchen, P. G., Haslam, S. M., & Dell, A., (2009). Mass spectrometry in the analysis of N-linked and O-linked glycans. *Curr. Opin. Struct. Biol., 19*, 498–506.

160. Whitehouse, C. M., Dreyer, R. N., Yamashita, M., & Fenn, J. B., (1985). Electrospray interface for liquid chromatographs and mass spectrometers. *Anal. Chem., 57*, 675–679.

161. Karas, M., Bachmann, D., Bahr, U., & Hillenkamp, F., (1987). Matrix-assisted ultraviolet-laser desorption of nonvolatile compounds. *Int. J. Mass Spectrom. Ion Process, 78*, 53–68.

162. Karas, M., & Hillenkamp, F., (1988). Laser desorption ionization of proteins with molecular masses exceeding 10,000 Daltons. *Anal Chem., 60*, 2299–2301.

163. Meng, C. K., Mann, M., & Fenn, J. B., (1988). *Of Protons or Proteins, Z Phys* D – Atoms, Molecules and Clusters *10*, 361–368.

164. Dell, A., Khoo, K. H., Panico, M., McDowell, R. A., Etienne, A. T., Reason, A. J., & Morris, H. R., (1993). FAB-MS and ES-MS of glycoproteins. In: Fukuda, M., & Kobata, A., (eds.), *Glycobiology: A Practical Approach* (pp. 187–222). Oxford, UK: IRL Press.

165. Kang, P., Mechref, Y., & Novotny, M. V., (2008). High-throughput solid-phase permethylation of glycans prior to mass spectrometry. *Rapid Commun. Mass Spectrom., 22*, 721–734.

166. Jang, K. S., Kim, Y. G., Gil, G. C., Park, S. H., & Kim, B. G., (2009). Mass spectrometric quantification of neutral and sialylated nglycans from a recombinant therapeutic glycoprotein produced in the two Chinese hamster ovary cell lines. *Anal. Biochem., 386*, 228–236.

167. Lei, M., Mechref, Y., & Novotny, M. V., (2009). Structural analysis of sulfated glycans by sequential double-permethylation using methyl iodide and deuteromethyl iodide. *J. Am. Soc. Mass Spectrom., 20*, 1660–1671.

168. Goldman, R., Ressom, H. W., Varghese, R. S., Goldman, L., et al., (2009). Detection of hepatocellular carcinoma using glycomic analysis. *Clin. Cancer Res., 15*, 1808–1813.

169. Bax, M., Garcia-Vallejo, J. J., Jang-Lee, J., North, S. J., Gilmartin, T. J., Hernandez, G., Crocker, P. R., Leffler, H., Head, S. R., Haslam, S. M., et al., (2007). Dendritic cell maturation results in pronounced changes in glycan expression affecting recognition by siglecs and galectins. *J. Immunol., 179*, 8216–8224.

170. Ong, S. E., Kratchmarova, I., & Mann, M., (2003). Properties of 13C-substituted arginine in stable isotope labeling by amino acids in cell culture (SILAC). *J. Proteome Res., 2,* 173–181.

171. Zhu, M., Bendiak, B., Clowers, B., & Hill, H. H. Jr., (2009). Ion mobility-mass spectrometry analysis of isomeric carbohydrate precursor ions. *Anal. Bioanal. Chem., 394,* 1853–1867.

172. Fukuyama, Y., Nakaya, S., Yamazaki, Y., & Tanaka, K., (2008). Ionic liquid matrixes optimized for MALDI-MS of sulfated/sialylated/ neutral oligosaccharides and glycopeptides. *Anal. Chem., 80,* 2171–2179.

173. Tissot, B., Gasiunas, N., Powell, A. K., Ahmed, Y., Zhi, Z. L., Haslam, S. M., Morris, H. R., et al., (2007). Towards GAG glycomics: Analysis of highly sulfated heparins by MALDI-TOF mass spectrometry. *Glycobiology, 17,* 972–982.

174. Tissot, B., Ceroni, A., Powell, A. K., Morris, H. R., Yates, E. A., Turnbull, J. E., Gallagher, J. T., et al., (2008). Software tool for the structural determination of glycosaminoglycans by mass spectrometry. *Anal. Chem., 80,* 9204–9212.

175. Williams, T. I., Saggese, D. A., Toups, K. L., Frahm, J. L., An, H. J., Li, B., Lebrilla, C. B., & Muddiman, D. C., (2008). Investigations with O-linked protein glycosylations by matrix-assisted laser desorption/ionization Fourier transform ion cyclotron resonance mass spectrometry. *J. Mass Spectrom., 43,* 1215–1223.

176. Barkauskas, D. A., An, H. J., Kronewitter, S. R., De Leoz, M. L., Chew, H. K., De Vere, W. R. W., Leiserowitz, G. S., et al., (2009). Detecting glycan cancer biomarkers in serum samples using MALDI FT-ICR mass spectrometry data. *Bioinformatics, 25,* 251–257.

177. Reinhold, V. N., Reinhold, B. B., & Chan, S., (1996). Carbohydrate sequence analysis by electrospray ionization-mass spectrometry. *Methods Enzymol., 271,* 377–402.

178. Zamfir, A., & Peter-Katalinic, J., (2004). Capillary electrophoresis mass spectrometry for glycoscreening in biomedical research. *Electrophoresis, 25,* 1949–1963.

179. Ruhaak, L. R., Deelder, A. M., & Wuhrer, M., (2009). Oligosaccharide analysis by graphitized carbon liquid chromatography-mass spectrometry. *Anal. Bioanal. Chem., 394,* 163–174.

180. Gennaro, L. A., & Salas-Solano, O., (2008). On-line CE-LIF-MS technology for the direct characterization of N-linked glycans from therapeutic antibodies. *Anal. Chem., 80,* 3838–3845.

181. Costello, C. E., Contado-Miller, J. M., & Cipollo, J. F., (2007). A glycomics platform for the analysis of permethylated oligosaccharide alditols. *J. Am. Soc. Mass Spectrom., 18,* 1799–1812.

182. Hakanssan, K., Cooper, H. J., Emmett, M. R., Costello, C. E., Marshall, A. G., & Nilsson, C. L., (2001). Electron capture dissociation and infrared multiphoton dissociation MS/ MS of an N-glycosylated tryptic peptic to yield complementary sequence information. *Anal. Chem., 73,* 4530–4536.

183. McFarland, M. A., Marshall, A. G., Hendrickson, C. L., Nilsson, C. L., Fredman, P., & Mansson, J. E., (2005). Structural characterization of GM1 ganglioside by infrared multiphoton dissociation: Electron capture dissociation and electron detachment dissociation electrospray ionization FT-ICR-MS/MS. *J. Am. Soc. Mass Spectrom., 16,* 752–762.

184. Park, Y., & Labriella, C. B., (2005). Applications of Fourier transform ion cyclotron mass spectrometry to oligosaccharides. *Mass Spectrom. Rev., 24,* 232–264.

185. Froesch, M., Bindila, L., Baykut, G., Allen, M., Peter-Katalinić, J., & Zamfi, A. D., (2004). Coupling of fully automated chip electrospray to Fourier transform ion cyclotron mass spectrometry for high-performance glycoscreening and sequencing. *Rapid Commun. Mass Spectrom., 18*, 3084–3092.

186. Zamfir, A., Vakhrushev, S., Sterling, A., Niebel, H. J., Allen, M., & Peter-Katalinic, J., (2004). Fully automated chip-based mass spectrometry for complex carbohydrate system analysis. *Anal. Chem., 76*, 2046–2054.

187. Goldstein, I. J., Hughes, R. C., Monsigny, M., Osawa, T., & Sharon, N., (1980). What should be called a lectin? *Nature, 285*, 66.

188. Chandra, H., Reddy, P. J., & Srivastava, S., (2011). Protein microarrays and novel detection platforms. *Expert Rev. Proteomics, 8*(1), 61–79. doi: 10.1586/epr.10.99. PMID: 21329428.

189. Pilobello, K. T., Agarwal, P., Rouse, R., & Mahal, L. K., (2013). Advances in lectin microarray technology: Optimized protocols for piezoelectric print conditions. *Curr. Protoc. Chem. Biol., 5*(1), 1–23.

190. Wollscheid, B., Bausch-Fluck, D., Henderson, C., O'Brien, R., Bibel, M., Schiess, R., et al., (2009). Mass-spectrometric identification and relative quantification of N-linked cell surface glycoproteins. *Nat. Biotechnol., 27*, 378–386.

191. Narimatsu, H., Sawaki, H., Kuno, A., Kaji, H., Ito, H., & Ikehara, Y., (2009). A strategy for discovery of cancer glyco-biomarkers in serum using newly developed technologies for glycoproteomics. *FEBS J., 277*, 95–105.

192. Paulson, J. C., Blixt, O., & Collins, B. E., (2006). Sweet spots in functional glycomics. *Nat. Chem. Biol., 2*, 238–248.

193. Liu, Y., & Feizi, T., (2008). Microarrays - key technologies and tools for glycobiology. In: Fraser-Reid, B. O., Tatsuta, K., & Thiem, J., (eds.), *Glycoscience - Chemistry and Chemical Biology* (2nd edn., pp. 2121–2132) Springer, Berlin/Heidelberg/New York.

194. Horlacher, T., & Seeberger, P. H., (2008). Carbohydrate arrays as tools for research and diagnostics. *Chem. Soc. Rev., 37*, 1414–1422.

195. Willats, W., Rasmussen, S., Kristensen, T., Mikkelsen, J., & Knox, J., (2002). Sugar-coated microarrays: A novel slide surface for the high-throughput analysis of glycans. *Proteomics, 2*, 1666–1671. [PubMed: 12469336].

196. Wang, D., Liu, S., Trummer, B., Deng, C., & Wang, A., (2002). Carbohydrate microarrays for the recognition of cross-reactive molecular markers of microbes and host cells. *Nat. Biotechnol., 20*, 275–281. [PubMed: 11875429].

197. Moller, I., Sørensen, I., Bernal, A. J., Blaukopf, C., Lee, K., Øbro, J., Pettolino, F., et al., (2007). High- throughput mapping of cell-wall polymer within and between plants using novel microarrays. *Plant J., 50*, 1118–1128.

198. Shipp, E., & Hsieh-Wilson, L., (2007). Profiling the sulfation specificities of glycosaminoglycan interactions with growth factors and chemotactic proteins using microarrays. *Chem. Biol., 14*, 195–208. [PubMed: 17317573].

199. Fukui, S., Feizi, T., Galustian, C., Lawson, A., & Chai, W., (2002). Oligosaccharide microarrays for high throughput detection and specificity assignments of carbohydrate-protein interactions. *Nat. Biotechnol., 20*, 1011–1017. [PubMed: 12219077].

200. Feizi, T., & Chai, W., (2004). Oligosaccharide microarrays to decipher the glyco code. *Nat. Rev. Mol. Cell Biol., 5*, 582–588.

201. Kaneko, et al., (2016). Proteomic and glycomic characterization of rice chalky grains produced under moderate and high-temperature *Conditions in Field System Rice, 9*, 26. doi: 10.1186/s12284-016-0100-y.

202. Tsutsui, K., Kaneko, K., Hanashiro, I., Nishinari, K., & Mitsui, T., (2013). Characteristics of opaque and translucent parts of high temperature stressed grains of rice. *J. Appl. Glycosci.* doi: 10.5458/jag.jag.JAG-2012014.

203. Yan, P., Hao, W., Jing-jie, L., Cong-fen, H. E., Yin-mao, D., & Chang-Tao, W., (2009). GPC determination of molecular weight and extraction of beta-glucan from oat bran [J]. *Food Science, 30*(20), 49–52.

204. Li, M., Pattathil, S. K., Hahn, M. G., & Hodge, D. B., (2014). Identification of features associated with plant cell wall recalcitrance to pretreatment by alkaline hydrogen peroxide in diverse bioenergy feedstocks using glycome profiling. *RSC Adv., 4*, 17282.

205. Zhao, Q., Tobimatsu, Y., Zhou, R., Pattathil, S., Gallego-Giraldo, L., et al., (2013). Loss of function of cinnamyl alcohol dehydrogenase 1 leads to unconventional lignin and a temperature-sensitive growth defect in *Medicago truncatula. PNAS, 110*(33) 13660–13665. https://doi.org/10.1073/pnas.1312234110.

206. Pattathil, S., Avci, U., Miller, J. S., & Hahn, M. G., (2012). Immunological approaches to plant cell wall and biomass characterization: Glycome profiling. *Methods Mol. Biol., 908*, 61–72.

# CHAPTER 10

# Sugars: Coping the Stress in Plants

RITIKA RAJPOOT

*Department of Crop Physiology, University of Agricultural Sciences, Bangalore – 560056, Karnataka, India, E-mail: ritikaa15@gmail.com*

## ABSTRACT

Plants being autotrophic in nature, produce sugars by the process of photosynthesis and use it in various metabolic processes. Soluble sugars like sucrose, glucose, and fructose not only act as metabolic precursors and cell structure constituents but do also act as signal molecules for regulating various processes involved in plant development and growth. Apart from these primary roles, these soluble sugars also play an important role in counteracting various types of stresses including biotic as well as abiotic. These sugars actually act as signaling molecules that initiate various stress response pathways by modifying the expression of various stress-specific genes and proteins involved in stress pathways. Actually, under stress conditions, plants observe an increased production of reactive oxygen species (ROS), which in turn leads to stress. Soluble sugars do have a close association with various metabolic enzymes, which are strongly interrelated to stress induced ROS accumulation in plants. The main goal of this chapter is to provide updated knowledge about the role of various sugars in helping the plants to cope with stress. Future insights about the other possible sugars to be explored for such activities have also been documented.

## 10.1 INTRODUCTION

Any condition which alters the normal growth and development of plants is stress. Being a sessile organism, plant has to face various types of adverse conditions causing stress either biotic (e.g., temperature, pathogen, etc.), or abiotic (e.g., metal toxicity). Electron transport reactions associated with respiration and photosynthesis often lead to leakage of electrons onto molecular oxygen ($O_2$). Consequent reductions of $O_2$ by the leaked electrons lead to production of various reactive oxygen species (ROS) in the cellular compartments like mitochondria, chloroplasts, and peroxisomes (Figure 10.1) [1]. In normal pathway of oxygen reduction, leads to production of different types of ROS include the superoxide radical ($O_2^-$), hydrogen peroxide ($H_2O_2$), and the hydroxyl radical ($\cdot OH$) respectively and among these $H_2O_2$ is most common. In plant cells the production and scavenging of ROS is tightly regulated and monitored. However, certain conditions, such as pathogen defense and exposure to environmental stresses often lead to overproduction of ROS [2–4]. This uncontrolled and sudden change in ROS concentration is known as "oxidative burst" [5]. Increase in the level of ROS in cells may cause a significant damage to biomolecules. These oxidized biomolecules are known to act as signaling molecules for oxidative stress in plants [6–8]. To cope up with oxidative damage, plants have their own defense system known as antioxidant (enzymatic and non-enzymatic) system [9].

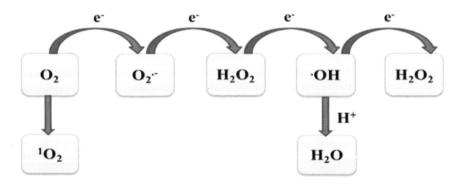

**FIGURE 10.1**   Production of the reactive oxygen species.

In plants, sugars play an integral role in nutrient, structure, and as signaling molecules. Soluble sugars play a central role in balancing ROS through its important relationship with photosynthesis, mitochondrial respiration and fatty acid b-oxidation [10]. Further, it is anticipated that sugars and phenols interact with each other to form a complex redox system and help the plant against stress by quenching the excessively produced ROS. Hexokinase in mitochondria regulates ROS level by stimulating antioxidative defense system in plants [11]. Soluble sugar is also known to play a dual role by involving in ROS producing pathways and also ROS scavenging pathways. They are involved in many metabolic processes linked with the formation of ROS (photosynthesis and mitochondrial respiration) equally. It plays a role in carotenoid and oxidative pentose phosphate (OPP) pathway acting as an antioxidative molecule [10]. Various sugars like Disaccharides, RFOs, fructans, and trehalose function as antioxidants [12, 13, 121]. Moreover, they scavenge OH radical more efficiently than $O_2^-$ [122]. Stress is known to alter sugar levels in plants [123]. Some of the widely studied sugars like glucose and sucrose control the expression of various genes of sugar metabolism by acting as primary messengers. During the stress in plants, these sugars interact with various stress regulatory pathways and modulate the various mechanisms involved in defense [14]. Studies have also shown that stress stimuli and sugars also trigger different signaling pathways independently (Table 10.1) [15, 16].

**TABLE 10.1**   Types of Sugars Involved in Different Stresses

| SL. No. | Stress | Accumulated Sugar | References |
|---------|--------|-------------------|------------|
| 1. | Drought | Raffinose, polyols, proline, and glycine betaine, glucose, fructose, sucrose, sinapoyl malate, trehalose | [121, 124, 125, 126] |
| 2. | Heat | Trehalose | [12] |
| 3. | Cold/chilling | Raffinose, galactose, fructans | [17, 127] |
| 4. | Salinity | trehalose, glucose, and mannitol, sorbitol | [18, 127] |

Carbon, hydrogen, and oxygen form the typical constituent of carbohydrates having the empirical formula Cm ($H_2O$) n (m and n≥3, m≥n). They are further characterized into monosaccharides, disaccharides, oligosaccharides, and polysaccharides on the basis of different degree of polymerization. Carbohydrates are also characterized as structural and nonstructural. Structural carbohydrates including hemicellulose and cellulose, mainly play a role in the structure of plants; however, nonstructural carbohydrates, which are further characterized as reducing and non-reducing sugars, serve as the major carbohydrate reserves [19]. Different carbohydrates are stored in different tissues in different plants. Maltose is reported to be stored in cytosol in Arabidopsis [12], whereas glucose, fructose, sucrose, and fructans are known to be accumulated in stem internodes [128], etc. Plants are unable to utilize all the carbon skeletons formed from the process of photosynthesis. Hence, they store carbon skeletons as short or longer-term reserve carbohydrates. One of the many mechanisms used by plants to minimize the deleterious effects caused by abiotic stresses is the synthesis of compatible solutes. These include amino acids, quaternary compounds, amines, and several sugars. However, the concept of 'sugar as antioxidant' is now widely accepted and studied.

## 10.2   THE PROTECTIVE ROLE OF SUGARS UNDER STRESS

The primary role of carbohydrates is to provide energy to the growing plant. During the process of photosynthesis, sugars are transported from the leaves (source) to different plant parts (sink) like roots, tubers, etc. This source-sink regulation in plants plays a vital role in plant development and defense against stress. Drought has been known to modulate source-sink balance in plants mainly due to decrease in the performance of photosynthetic organs [20]. Decrease in photosynthesis leads to decrease export to the sink, ultimately compromising the growth of plants. Drought stress was found to inhibit starch synthesis by increasing discs in tuber of potato [21]. Studies show that salinity also influences the source-sink relations. Salt stress is known to alter the type of sugars synthesized in the source [22]. However, there is increased evidence suggesting that different types of stresses modulates the source-sink concentration of sugars but still the mechanism is not fully known.

Apart from that, it is also known to act as signaling molecules, controlling transcriptional, post-transcriptional, and post-translational processes [23, 24]. Soluble sugars also protect the plants against stress by maintaining the osmotic condition [25]. RFOs and fructans are the common carbohydrates that accumulate in mild stress in plants [25]. It was also reported that increase in proline, anthocyanin, and other soluble sugars plays an important role in maintaining the antioxidant defense against the drought stress [26]. Besides this, sugars play a major role in maintaining the integrity of cellular membrane which is an important factor during stress [27]. Pathogen also modulates the carbohydrate metabolism in plants suggesting the role of sugars in biotic stress (Figure 10.2) [129]. Furthermore, extracellular invertase was found to be increased during salinity suggesting the overlapping role of sugar in biotic and abiotic stress [28].

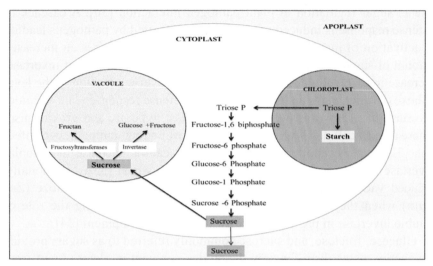

**FIGURE 10.2** Carbohydrate metabolism in plants.
*Source:* Redrawn from Irma Vijn, and Sjef [29].

## 10.3 TYPES OF CARBOHYDRATES INVOLVED IN STRESS

Sucrose, trehalose, raffinose family oligosaccharides (RFOs), and fructans (collectively referred to as 'sugars' from here on) are the major types of water-soluble carbohydrates that are essentially involved in plant stress responses.

### 10.3.1 SUCROSE

The carbohydrate which is mainly transported in higher plants is sucrose. In the process of photosynthesis sucrose is formed in the leaves and transported to various plant organs like root, stem, vegetative, and reproductive organs via phloem. It further breaks down into glucose and fructose by an important enzyme called invertase. Different types of invertases are present in plants based on their localization, apoplast, the cytoplasm, and the vacuole, invertases. These invertases play critical roles in regulating the level of sucrose and ratio of sucrose and hexose. Invertases have vital role in defense and resistance in plants. During biotic stress, different pathogens uses plant sugars for their own need by modulating the plants sugar content and activating its defense system. Invertase plays a complex role in sugar regulation in plant pathogen interaction [30]. A cascade of defense response is induced when plants are affected by pathogens leading to activation of many stress responsive genes. This requires an increased amount of sugar which can be provided by increased activity of invertase. Increase in extracellular invertase leads to the over-expression in the level of hexose which plays an important role in defense response [31]. Vacoular invertase has shown a role in drought stress, hypoxia, and gravitropism whereas extracellular invertase has role in wounding and pathogen infection [32]. It was reported that water deficit causes decrease in vacuolar invertase leading induction of male sterility in wheat [33]. Additionally, reduced vacuolar invertase was found in ovaries of young maize (*Zea mays*) when they were imposed by drought stress showing the role of soluble invertase in providing hexose sugar for development [34].

Glucose, fructose, and sucrose commonly referred to as sugars provide the energy metabolism for growth and development in plants. Apart from this glucose and sucrose are also known as important signaling molecules controlling the expression of genes related to their metabolism, stress, and survival [11]. Glucose was known to regulate wide range of genes associated with stress showing their importance in environmental responses [14]. It is a precursor of various important compounds involved in antioxidative defense like ascorbate [130] and carotenoids [131] and various amino acids like Cys, Glu, and Gly which form glutathione [132]. Ascorbate and glutathione is known to be involved in ascorbate-glutathione cycles and detoxification of peroxides protecting the plants against oxidative stress [133].

Plants utilize hexose sugar after their phosphorylation by hexokinases (HXKs) and fructokinases (FRKs). Apart from this, HXKs work as a glucose sensor, which has separate catalytic and signaling activities [35]. Glucose plays an important role in hexokinase pathway [36] while fructose is associated with abscissic acid (ABA) and ethylene pathway [37].

Sucrose balances its concentration by promoting carbohydrate metabolism and down regulating photosynthesis when present in excess and promoting photosynthesis when its concentration is low in plants [38]. High level of sugar increases the expression of various genes which plays an important role in sink function, such as growth, storage of proteins and biosynthesis of starch and fructans [39]. Plants have various proteins which sense the concentration of sugars, and the interaction of these proteins initiates the signaling transaction cascade [40]. Sucrose plays a vital role in plant defense mechanism. It was found that the expression of PRms (pathogenesis-related genes) in transgenic rice accumulates higher level of sucrose indicating its importance in biotic stress [41]. Sucrose is known to increase the formation of isoflavonoids by stimulating the phenylpropanoid metabolism. Isoflavonoid is a significant element in the defense system of legumes [42]. Sucrose stimulates the production of both fructans and anthocyanins under stress in plants [43]. It was found that decline in sucrose concentration in *Brassica oleracea L. var. italica* florets results into considerable damage to hydrogen peroxide-scavenging system [134]. HXKs are known to involved plant resistance against biotic [44] and abiotic stress. Upregulation of hexokinase gene NbHxk1 was noted by pathogen infection in *Nicotiana benthamiana* [44].

Starch is stored in the chloroplast during daytime by an ADP-glucose pyrophosphorylase (AGPase) and by sugar-induced initiation of gene expression [45]. Starch is also degraded by an invertase. Invertases and other enzymes which have role in sucrose splitting, for example, sucrose synthase have great effect on different sugar signaling pathways [43]. Proline has an important effect as an osmoprotectant in plants under stress, e.g., drought. Increase in the level of proline under stress is also associated with increase in sucrose level indicating it depends on Suc-specific signaling events [46].

## 10.3.2   FRUCTANS

Fructans are the polymers of fructose and synthesized by fructosyltransferase (FT) enzymes using sucrose. Fructans are the major storage carbohydrates in temperate grasses and cereals, while starch is often only present at low levels [47]. They are widely distributed in stems, leaves, inflorescences, and seeds of plants. The vacuole is the main site for the synthesis; however, studies have also shown the presence of fructans in the apoplast, phloem, and xylem tissues [48, 49]. Fructans are further characterized on the basis of linkage-type between the fructosyl residues and the position of the glucose residue [50]. Fructans having a terminal glucose residue include the β(2,1) type fructans (inulin), and the linear β(2,6) (levan) or branched type fructans (graminan) with both β(2,6) and β(2,1) linkages (Table 10.2).

**TABLE 10.2**   Types of Fructans

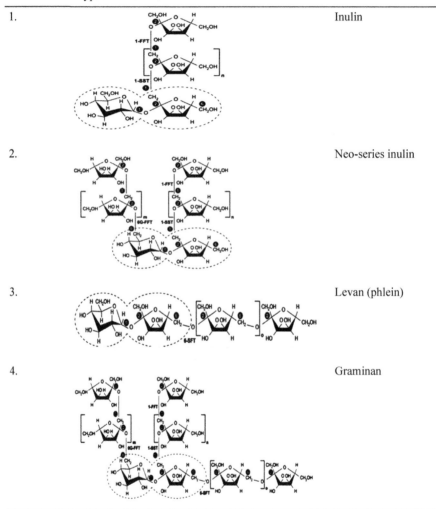

| 1. | Inulin |
| 2. | Neo-series inulin |
| 3. | Levan (phlein) |
| 4. | Graminan |

*Source:* Reprinted with permission from Ritsema and Smeekens Ref. [51]. © 2003 Elsevier.

It has a high solubility in water, but it does not crystallize under extremely cold conditions, and its biosynthetic machinery functions under sub-zero temperatures. This property of fructans helps the plant to sustain under cold stress. Under cold conditions, fructans accumulate in plants, and due to this reason, it has been correlated with an increase in freezing tolerance [52]. Moreover, the existence of fructans in the apoplast significantly indicates its role in abiotic stress tolerance by directly interacting with membrane lipids

[53]. Since the major cause of abiotic stress is biomembranes distortion and rupture resulting into cell death, fructan is known to stabilize the membranes through inserting part of the polysaccharide into the lipid headgroup region thereby preventing the leakage during freezing or drought condition [54]. A significant increase in fructans and other soluble sugars was observed in *Aloe barbadensis Miller* when subjected to water deficit. Further, it was found that plant synthesizes more branched fructans, the neofructans under stressful conditions contrary to inulin, neo-inulin, and neo-levan fructans which were major in control plants [55]. Increase in degree of polymerization under drought and chilling structure has also been reported in wheat [56]. Under drought conditions, plants protect the cellular structure by increasing fructans branching leading to increased interaction with hydrophilic phospholipid heads and amino acids by hydrogen bonding. Plants tend to accumulate fructans, i.e., fructose-based oligo-and polysaccharides in their vacuoles against the biotic or abiotic stress. It was found that when spring wheat (*Triticum aestivum L. cv. Marvdasht*) were grown under prolong dehydration, the plant showed increase in fructans along with sucrose, proline, and certain soluble proteins indicating that during dehydration plants tends to survive by decreasing expression of membrane damaging enzymes and increasing the production of osmoprotectants [57]. Moreover, He et al. showed that transgenic tobacco (*Nicotiana tabacum* L.) containing 6-SFT gene from *Psathyrostachys huashanica* showed increased accumulation of fructans with higher tolerance against drought, cold, and salinity stress [58].

Apart from this, fructans are also known to maintain ROS homeostasis by playing an important role in vacuole antioxidant mechanisms [13, 59]. During stress the concentration of ROS mainly $H_2O_2$ increases in cell. $H_2O_2$ move to the vacuole through diffusion or aquaporins. Fructans which are present between the head group of vacuolar membrane stabilize excessively produced $H_2O_2$. Peroxidases present in vacuolar membranes catalyzes the excessively produced $H_2O_2$ into •OH and •OOH with the help of various electron donor molecules like phenolic compounds •OH radicals produced from the peroxidases are also scavenged by fructans producing less harmful fructan radical and water. Therefore, it can be concluded that sugars present in the vicinity of vacuoles plays an important role in scavenging various ROS and protecting the plants against stress. Interestingly, in immature wheat kernels, it was found that different concentrations of fructans showed significant correlation with ASC and GSH concentrations showing its strong correlation with antioxidative defense in cytoplasm. Sequence alignments and enzymatic properties indicate that FTs are evolved from vacuolar invertases [60]. Fundamentally, the FT enzyme uses sucrose or fructans as an alternative of water as an acceptor substrate. This could be possibly due to lower water content and higher sucrose content in cells by desiccation or low temperatures, which might have been the driving force during this evolutionary process [27]. Therefore, it is strongly suggested that in fructans accumulating plants, fructans play an integral part of a ROS scavenging antio-xidative defense system.

### 10.3.3 TREHALOSE

Trehalose being a non-reducing disaccharide is highly soluble but non-reactive in nature. This property of trehalose makes it possible to accumulate it at higher concentration in plants and works as an osmoprotectant. Trehalose plays an important role as it initiates defense responses by the plants under stress [61]. Trehalose activated plant defense response, i.e., increased papilla deposition, phenylalanine ammonia-lyase (PAL) and peroxidase activities in wheat against *B. graminis* infection [62]. Foliar application of trehalose protects the rice plants against salt stress [63]. Transgenic rice having overexpressed *E. coli* trehalose biosynthetic genes (otsA and otsB) showed increased tolerance against abiotic stress. Moreover, increase in accumulation of trehalose also caused increased photosynthesis suggesting its role in carbohydrate metabolism [64]. Increased accumulation of trehalose in chickpea was related with better growth of the plant when subjected to chilling stress by providing the protection against oxidative damage and maintaining the carbon assimilation [65].

It is also known to reduce accumulation of protein during stressful conditions [66]. Trehalose is synthesized by dephosphorylating T6P (trehalose-6-phosphate phosphate) by an enzyme T6P phosphatase. T6P is formed from glucose-6-phosphate and UDP-glucose by trehalose-6-phosphate synthase (TPS) [67] (Figure 10.3).

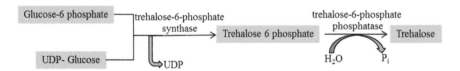

**FIGURE 10.3**    Schematic diagram showing synthesis of trehalose in cell cytoplasm.

Further studies showed trehalose especially T6P is found to be a powerful signaling metabolite which controls carbohydrate metabolism and plays an important role in plant growth and reproduction [68, 69]. T6P is known to play an important role in utilization of sucrose in plants. The increased level of T6P in the immature embryo has been studied to prevent stress-induced kernel abortion [70].

Apart from this, trehalose has special property of stabilizing membranes and molecules by two mechanisms, i.e., water replacement and the glass formation mechanisms. During stressful conditions (desiccation or chilling) trehalose forms hydrogen bonds with membranes and macromolecules by replacing water molecules hence protecting the plant by water replacement mechanism [71]. Trehalose under dehydration conditions, changes into solid glassy state. This property helps trehalose to prevent denaturing of biomolecules and retaining their functional property during drought or chilling stress [72]. The importance of trehalose was first recognized in desiccation tolerant plants where these plants can tolerate extreme dehydration and become viable after rehydration. It has been reported that drought tolerant plants tend to accumulate higher amount of trehalose and sucrose in them to sustain against drought stress [73]. Further, it was found that trehalose was also associated with other stresses, for example, the level of trehalose was found to be doubled when imposed with heat stress (40°C) and was elevated to eight times under cold exposure (4°C) [74]. It is reported that *Arabidopsis thaliana* mutant, sweetie, was characterized to accumulate higher levels of trehalose. This mutant shows higher constitutive expressions of genes involved in various abiotic stresses [17].

Expression of trehalose gene involved in its biosynthesis was reported in different organs in different plants during stress. Expression of trehalose gene was found in shoots and roots when imposed with chilling stress in rice (Os*TPP1*) and wheat [75, 76], whereas in cotton the (*TPS*) expression was observed in leaves and roots under osmotic stress [77] and in maize the expression of TPS1 gene was found to be in ears [78]. There are few reports also suggesting that the exogenous application of trehalose also protects the plants against stress [119]. It is postulated that trehalose prevents senescence and programmed cell death by inducing autophagy during desiccation in *T. loliiformis* [79].

## 10.3.4   *RAFFINOSE FAMILY OLIGOSACCHARIDES (RFOS)*

The RFOs (raffinose, stachyose, and verbascose) are soluble, non-reducing $\alpha(1,6)$ galactosyl (Gal) extensions of sucrose. They are ubiquitous in the plant kingdom and known to be found in seeds of a large variety of plant species and have tendency to get stored in other tissues during stressful conditions [80]. Galactinol and myo-inositol are the main metabolites of the RFOs where UDP-galactose and myo-inositol forms galactinol via the important enzyme known as galactinol synthase (GolS). GolS is a key enzyme which is known to regulate the accumulation of galactinol and RFOs [81]. RFOs are reported to play a vital role in drought conditions where they act as an osmoprotectant and maintain the turgor pressure inside the cell. It also protects the metabolically important proteins [82]. RFOs also protect the plants by acting as an antioxidant and scavenging the excessively produced ROS during stressful conditions. It has been reported that RFOs decreases the inactivation of electron transport (2,6-dichlorophenolindophenol reduction) and cyclic photophosphorylation in photosynthesis in thylakoid membrane chloroplast of *Spinacia oleracea* [83].

Though RFOs are known to play an important role in stress tolerance [84], they are also known to participate in various others cellular functions, including carbon storage and transport [85], RNA export from the nucleus to the cytoplasm [86], signaling [87] membrane stability during drought stress [88], unilamellar liposomes protection from dehydration by interacting with sugar-membrane [89]. Several studies showed the increased expression and accumulation of GolS in plants when subjected to various environmental stresses. The activity of GolS was found to be increased when seeds of kidney beans (*Phaseolus vulgaris*) and vegetative tissues of *Arabidopsis thaliana* were subjected to cold stress indicating its role in cold stress [90]. When Arabidopsis seeds were subjected to drought/salt/heat stress, GolS1 and GolS2 mRNAs were found to be expressed [81]. Kidney bean seeds also showed increased activity of GolS when exposed to cold stress [90].

## 10.4   CONCLUSION

In conclusion, sugars or sugar-like compounds present in the plants play an integral role by acting as signaling molecules and ROS scavenging. Sugars work as an antioxidative defense system by regulating a complex network of modulating the osmoregulation in the plants or by scavenging the excessively produced ROS or by participating in stress-specific hormone signaling.

## KEYWORDS

- **abscisic acid**
- **fructokinases**
- **oxidative pentose phosphate**
- **raffinose family oligosaccharides**
- **reactive oxygen species**
- **trehalose-6-phosphate synthase**

## REFERENCES

1. Apel, K., & Hirt, H., (2004). Reactive oxygen species: Metabolism, oxidative stress, and signal transduction. *Annu. Rev. Plant Biol., 55*, 373–399.
2. Dat, J., Vandenabeele, S., Vranová, E., Van, M. M., Inzé, D., & Van, B. F., (2000). Dual action of the active oxygen species during plant stress responses. *Cellular and Molecular Life Sciences CMLS, 57*(5), 779–795.
3. Hammond-Kosack, K. E., & Jones, J. D., (1996). Resistance gene-dependent plant defense responses. *The Plant Cell, 8*(10), 1773.
4. Grant, J. J., & Loake, G. J., (2000). Role of reactive oxygen intermediates and cognate redox signaling in disease resistance. *Plant Physiology, 124*(1), 21–30.
5. Apostol, I., Heinstein, P. F., & Low, P. S., (1989). Rapid stimulation of an oxidative burst during elicitation of cultured plant cells role in defense and signal transduction. *Plant Physiology, 90*(1), 109–116.
6. Mittler, R., (2002). Oxidative stress, antioxidants, and stress tolerance. *Trends in Plant Science, 7*(9), 405–410.
7. Pasqualini, S., Piccioni, C., Reale, L., Ederli, L., Della, T. G., & Ferranti, F., (2003). Ozone-induced cell death in tobacco cultivar bel W3 plants. The role of programmed cell death in lesion formation. *Plant Physiology, 133*(3), 1122–1134.
8. Verma, S., & Dubey, R. S., (2003). Lead toxicity induces lipid peroxidation and alters the activities of antioxidant enzymes in growing rice plants. *Plant Science, 164*(4), 645–655.
9. Gill, S. S., & Tuteja, N., (2010). Reactive oxygen species and antioxidant machinery in abiotic stress tolerance in crop plants. *Plant Physiology and Biochemistry, 48*(12), 909–930.
10. Couée, I., Sulmon, C., Gouesbet, G., & El Amrani, A., (2006). Involvement of soluble sugars in reactive oxygen species balance and responses to oxidative stress in plants. *Journal of Experimental Botany, 57*(3), 449–459.
11. Bolouri-Moghaddam, M. R., Le Roy, K., Xiang, L., Rolland, F., & Van, D. E. W., (2010). Sugar signaling and antioxidant network connections in plant cells. *The FEBS Journal, 277*(9), 2022–2037.
12. Kaplan, F., & Guy, C. L., (2004). β-amylase induction and the protective role of maltose during temperature shock. *Plant Physiology, 135*(3), 1674–1684.

13. Peshev, D., Vergauwen, R., Moglia, A., Hideg, É., & Van, D. E. W., (2013). Towards understanding vacuolar antioxidant mechanisms: A role for fructans. *Journal of Experimental Botany, 64*(4), 1025–1038.

14. Price, J., Laxmi, A., Martin, S. K. S., & Jang, J. C., (2004). Global transcription profiling reveals multiple sugar signal transduction mechanisms in *Arabidopsis. The Plant Cell, 16*(8), 2128–2150.

15. Ehness, R., Ecker, M., Godt, D. E., & Roitsch, T., (1997). Glucose and stress independently regulate source and sink metabolism and defense mechanisms via signal transduction pathways involving protein phosphorylation. *The Plant Cell, 9*(10), 1825–1841.

16. Roitsch, T., (1999). Source-sink regulation by sugar and stress. *Current Opinion in Plant Biology, 2*(3), 198–206.

17. Veyres, N., Danon, A., Aono, M., Galliot, S., Karibasappa, Y. B., Diet, A., & Boitel-Conti, M., (2008). The arabidopsis sweetie mutant is affected in carbohydrate metabolism and defective in the control of growth, development and senescence. *The Plant Journal, 55*(4), 665–686.

18. Zhifang, G., & Loescher, W. H., (2003). Expression of a celery mannose 6-phosphate reductase in Arabidopsis thaliana enhances salt tolerance and induces biosynthesis of both mannitol and a glucosyl-mannitol dimer. *Plant, Cell and Environment 26*, 275–283.

19. McCarty, E. C., (1938). *The Relation of Growth to the Varying Carbohydrate Content in Mountain Brome (No. 165913).* United States Department of Agriculture, Economic Research Service.

20. Berman, M. E., & DeJong, T. M., (1996). Water stress and crop load effects on fruit fresh and dry weights in peach (*Prunus persica*). *Tree Physiology, 16*(10), 859–864.

21. Geigenberger, P., Reimholz, R., Geiger, M., Merlo, L., Canale, V., & Stitt, M., (1997). Regulation of sucrose and starch metabolism in potato tubers in response to short-term water deficit. *Planta, 201*(4), 502–518.

22. Gucci, R., Moing, A., Gravano, E., & Gaudillère, J. P., (1998). Partitioning of photosynthetic carbohydrates in leaves of salt-stressed olive plants. *Functional Plant Biology, 25*(5), 571–579.

23. Koch, K. E., Wu, Y., & Xu, J., (1996). Sugar and metabolic regulation of genes for sucrose metabolism: Potential influence of maize sucrose synthase and soluble invertase responses on carbon partitioning and sugar sensing. *Journal of Experimental Botany,* 1179–1185.

24. Rolland, F., Baena-Gonzalez, E., & Sheen, J., (2006). Sugar sensing and signaling in plants: Conserved and novel mechanisms. *Annu. Rev. Plant Biol., 57*, 675–709.

25. Peshev, D., & Van, D. E. W., (2013). Sugars as antioxidants in plants. In: Tuteja, N., & Gill, S. S., (eds.), *Crop Improvement UNDER Adverse Conditions* (pp. 285–308). Springer-Verlag, Berlin, Heidelberg, Germany.

26. Sperdouli, I., & Moustakas, M., (2012). Interaction of proline, sugars, and anthocyanins during photosynthetic acclimation of Arabidopsis thaliana to drought stress. *Journal of Plant Physiology, 169*, 577–585.

27. Valluru, R., & Van, D. E. W., (2008). Plant fructans in stress environments: Emerging concepts and future prospects. *Journal of Experimental Botany, 59*(11), 2905–2916.

28. Roitsch, T., Balibrea, M. E., Hofmann, M., Proels, R., & Sinha, A. K., (2003). Extracellular invertase: Key metabolic enzyme and PR protein. *Journal of Experimental Botany, 54*(382), 513–524.

29. Vijn, I., & Smeekens, S., (1999). Fructan: More than a reserve carbohydrate. *Plant Physiology, 120*(2), 351–360.
30. Tauzin, A. S., & Giardina, T., (2014). Sucrose and invertases, a part of the plant defense response to the biotic stresses. *Frontiers in Plant Science, 5.*
31. Xiang, L., Le, R. K., Bolouri-Moghaddam, M. R., Vanhaecke, M., Lammens, W., Rolland, F., & Van, D. E. W., (2011). Exploring the neutral invertase–oxidative stress defense connection in Arabidopsis thaliana. *Journal of Experimental Botany, 62*(11), 3849–3862.
32. Roitsch, T., & González, M. C., (2004). Function and regulation of plant invertases: Sweet sensations. *Trends in Plant Science, 9*(12), 606–613.
33. Dorion, S., Lalonde, S., & Saini, H. S., (1996). Induction of male sterility in wheat by meiotic-stage water deficit is preceded by a decline in invertase activity and changes in carbohydrate metabolism in anthers. *Plant Physiology, 111*(1), 137–145.
34. Andersen, M. N., Asch, F., Wu, Y., Jensen, C. R., Næsted, H., Mogensen, V. O., & Koch, K. E., (2002). Soluble invertase expression is an early target of drought stress during the critical, abortion-sensitive phase of young ovary development in maize. *Plant Physiology, 130*(2), 591–604.
35. Moore, B., Zhou, L., Rolland, F., Hall, Q., Cheng, W. H., Liu, Y. X., & Sheen, J., (2003). Role of the Arabidopsis glucose sensor HXK1 in nutrient, light, and hormonal signaling. *Science, 300*(5617), 332–336.
36. Cho, J. I., Ryoo, N., Eom, J. S., Lee, D. W., Kim, H. B., Jeong, S. W., & Hahn, T. R., (2009). Role of the rice hexokinases OsHXK5 and OsHXK6 as glucose sensors. *Plant Physiology, 149*(2), 745–759.
37. Cho, Y. H., & Yoo, S. D., (2011). Signaling role of fructose mediated by FINS1/FBP in *Arabidopsis thaliana. PLoS Genetics, 7*(1), e1001263.
38. O'Hara, L. E., Paul, M. J., & Wingler, A., (2013). How do sugars regulate plant growth and development? New insight into the role of trehalose-6-phosphate. *Molecular Plant, 6*(2), 261–274.
39. Gupta, A. K., & Kaur, N., (2005). Sugar signaling and gene expression in relation to carbohydrate metabolism under abiotic stresses in plants. *Journal of Biosciences, 30*(5), 761–776.
40. Horacio, P., & Martinez-Noel, G., (2013). Sucrose signaling in plants: A world yet to be explored. *Plant Signaling and Behavior, 8*(3), e23316.
41. Gómez-Ariza, J., Campo, S., Rufat, M., Estopà, M., Messeguer, J., Segundo, B. S., & Coca, M., (2007). Sucrose-mediated priming of plant defense responses and broad-spectrum disease resistance by overexpression of the maize pathogenesis-related PRms protein in rice plants. *Molecular Plant-Microbe Interactions, 20*(7), 832–842.
42. Morkunas, I., Marczak, Ł., Stachowiak, J., & Stobiecki, M., (2005). Sucrose-induced lupine defense against *Fusarium oxysporum*: Sucrose-stimulated accumulation of isoflavonoids as a defense response of lupine to *Fusarium oxysporum. Plant Physiology and Biochemistry, 43*(4), 363–373.
43. Van, D. E. W., & El-Esawe, S. K., (2014). Sucrose signaling pathways leading to fructan and anthocyanin accumulation: A dual function in abiotic and biotic stress responses? *Environmental and Experimental Botany, 108*, 4–13.
44. Sarowar, S., Lee, J. Y., Ahn, E. R., & Pai, H. S., (2008). A role of hexokinases in plant resistance to oxidative stress and pathogen infection. *Journal of Plant Biology, 51*(5), 341–346.

45. Hendriks, J. H., Kolbe, A., Gibon, Y., Stitt, M., & Geigenberger, P., (2003). ADP-glucose pyrophosphorylase is activated by posttranslational redox-modification in response to light and to sugars in leaves of Arabidopsis and other plant species. *Plant Physiology, 133*(2), 838–849.

46. Hanson, J., Hanssen, M., Wiese, A., Hendriks, M. M., & Smeekens, S., (2008). The sucrose regulated transcription factor bZIP11 affects amino acid metabolism by regulating the expression of asparagine synthetase1 and proline dehydrogenase 2. *The Plant Journal, 53*(6), 935–949.

47. Chalmers, J., Lidgett, A., Cummings, N., Cao, Y., Forster, J., & Spangenberg, G., (2005). Molecular genetics of fructan metabolism in perennial ryegrass. *Plant Biotechnology Journal, 3*(5), 459–474.

48. Ernst, M., & Pfenning, J., (2000). Fructan in stem exudates of *Helianthus tuberosus* L. In: *Proceedings of the Eighth Seminar on Inulin* (pp. 56–58). Stuttgart, Germany: EFA.

49. Van, D. E. W., Michiels, A., Van, W. D., Vergauwen, R., & Van, L. A., (2000). Cloning, developmental, and tissue-specific expression of sucrose: Sucrose 1-fructosyl transferase from *Taraxacum officinale*. Fructan localization in roots. *Plant Physiology, 123*(1), 71–80.

50. Lewis, D. H., (1993). Nomenclature and diagrammatic representation of oligomeric fructans: A paper for discussion. *New Phytologist, 124*(4), 583–594.

51. Ritsema, T., & Smeekens, S., (2003). Fructans: Beneficial for plants and humans. *Current Opinion in Plant Biology, 6*(3), 223–230.

52. Pontis, H. G., (1989). Fructans and cold stress. *Journal of Plant Physiology, 134*(2), 148–150.

53. Van, D. E. W., Yoshida, M., Clerens, S., Vergauwen, R., & Kawakami, A., (2005). Cloning, characterization, and functional analysis of novel 6-kestose exohydrolases (6-KEHs) from wheat (*Triticum aestivum* L.). *New Phytologist, 166*, 917–932.

54. Livingston, D. P., Hincha, D. K., & Heyer, A. G., (2009). Fructan and its relationship to abiotic stress tolerance in plants. *Cellular and Molecular Life Sciences, 66*(13), 2007–2023.

55. Salinas, C., Handford, M., Pauly, M., Dupree, P., & Cardemil, L., (2016). Structural Modifications of fructans in aloe barbadense miller (Aloe Vera) grown under water stress. *PloS One, 11*(7), e0159819.

56. Santoiani, C. S., Tognetti, J. A., Pontis, H. G., & Salerno, G. L., (1993). Sucrose and fructan metabolism in wheat roots at chilling temperatures. *Physiologia. Plantarum, 87*(1), 84–88.

57. Koobaz, P., Ghanati, F., Hosseini, S. G., Moradi, F., & Hadavand, H., (2017). Drought tolerance in four-day-old seedlings of a drought-sensitive cultivar of wheat. *Journal of Plant Nutrition, 40*(4), 574–583.

58. He, X., Chen, Z., Wang, J., Li, W., Zhao, J., Wu, J., & Chen, X., (2015). Sucrose: Fructan-6-fructosyltransferase (6-SFT) gene from *Psathyrostachys* huashanica confers abiotic stress tolerance in tobacco. *Gene., 570*(2), 239–247.

59. Keunen, E. L. S., Peshev, D., Vangronsveld, J., Van, D. E. W. I. M., & Cuypers, A. N. N., (2013). Plant sugars are crucial players in the oxidative challenge during abiotic stress: Extending the traditional concept. *Plant, Cell and Environment, 36*(7), 1242–1255.

60. Ritsema, T., Hernandez, L., Verhaar, A., Altenbach, D., Boller, T., Wiemken, A., & Smeekens, S., (2006). Developing fructan-synthesizing capability in a plant invertase via mutations in the sucrose-binding box. *The Plant Journal, 48*, 228–237.

61. Piazza, A., Zimaro, T., Garavaglia, B. S., Ficarra, F. A., Thomas, L., Marondedze, C., Feil, R., et al., (2015). The dual nature of trehalose in citrus canker disease: A virulence factor for *Xanthomonas* citri Subsp. citri and a trigger for plant defense responses. *J. Exp. Bot., 66*, 2795–2811.

62. Reignault, P. H., Cogan, A., Muchembled, J., Lounes-Hadj, S. A., Durand, R., & Sancholle, M., (2001). Trehalose induces resistance to powdery mildew in wheat. *New Phytologist, 149*(3), 519–529.

63. Shahbaz, M., Abid, A., Masood, A., & Waraich, E. A., (2017). Foliar-applied trehalose modulates growth, mineral nutrition, photosynthetic ability, and oxidative defense system of rice (*Oryza sativa* L.) under saline stress. *Journal of Plant Nutrition, 40*(4), 584–599.

64. Garg, A. K., Kim, J. K., Owens, T. G., Ranwala, A. P., Do Choi, Y., Kochian, L. V., & Wu, R. J., (2002). Trehalose accumulation in rice plants confers high tolerance levels to different abiotic stresses. *Proceedings of the National Academy of Sciences, 99*(25), 15898–15903.

65. Farooq, M., Hussain, M., Nawaz, A., Lee, D. J., Alghamdi, S. S., & Siddique, K. H., (2017). Seed priming improves chilling tolerance in chickpea by modulating germination metabolism, trehalose accumulation, and carbon assimilation. *Plant Physiology and Biochemistry, 111*, 274–283.

66. Jain, N. K., & Roy, I., (2010). Trehalose and protein stability. *Current Protocols in Protein Science*, 4–9.

67. Cabib, E., & Leloir, L. F., (1958). The biosynthesis of trehalose phosphate. *Journal of Biological Chemistry, 231*(1), 259–275.

68. Van, D. A. J., Schluepmann, H., & Smeekens, S. C., (2004). Arabidopsis trehalose-6-phosphate synthase 1 is essential for normal vegetative growth and transition to flowering. *Plant Physiology, 135*(2), 969–977.

69. Wahl, V., Ponnu, J., Schlereth, A., Arrivault, S., Langenecker, T., Franke, A., & Schmid, M., (2013). Regulation of flowering by trehalose-6-phosphate signaling in *Arabidopsis thaliana. Science, 339*(6120), 704–707.

70. Bledsoe, S. W., Henry, C., Griffiths, C. A., Paul, M. J., Feil, R., Lunn, J. E., & Lagrimini, L. M., (2017). The role of Tre6P and SnRK1 in maize early kernel development and events leading to stress-induced kernel abortion. *BMC Plant Biology, 17*(1), 74.

71. Crowe, J. H., (2007). Trehalose as a chemical chaperone. In: *Molecular Aspects of the Stress Response: Chaperones, Membranes, and Networks* (pp. 143–158). Springer New York.

72. Richards, A. B., Krakowka, S., Dexter, L. B., Schmid, H., Wolterbeek, A. P. M., Waalkens-Berendsen, D. H., & Kurimoto, M., (2002). Trehalose: A review of properties, history of use and human tolerance, and results of multiple safety studies. *Food and Chemical Toxicology, 40*(7), 871–898.

73. Iturriaga, G., Gaff, D. F., & Zentella, R., (2000). New desiccation-tolerant plants, including a grass, in the central highlands of Mexico, accumulate trehalose. *Australian Journal of Botany, 48*(2), 153–158.

74. Kaplan, F., Kopka, J., Haskell, D. W., Zhao, W., Schiller, K. C., Gatzke, N., & Guy, C. L., (2004). Exploring the temperature-stress metabolome of *Arabidopsis. Plant Physiology, 136*(4), 4159–4168.

75. Pramanik, M. H. R., & Imai, R., (2005). Functional identification of a trehalose 6-phosphate phosphatase gene that is involved in transient induction of trehalose biosynthesis during chilling stress in rice. *Plant Molecular Biology, 58*(6), 751–762.

76. El-Bashiti, T., Hamamcı, H., Öktem, H. A., & Yücel, M., (2005). Biochemical analysis of trehalose and its metabolizing enzymes in wheat under abiotic stress conditions. *Plant Science, 169*(1), 47–54.

77. Kosmas, S. A., Argyrokastritis, A., Loukas, M. G., Eliopoulos, E., Tsakas, S., & Kaltsikes, P. J., (2006). Isolation and characterization of drought-related trehalose 6-phosphate-synthase gene from cultivated cotton (*Gossypium hirsutum* L.). *Planta, 223*(2), 329–339.

78. Zhuang, Y., Ren, G., Yue, G., Li, Z., Qu, X., Hou, G., & Zhang, J., (2007). Effects of water-deficit stress on the transcriptomes of developing immature ear and tassel in maize. *Plant Cell Reports, 26*(12), 2137–2147.

79. Williams, B., Njaci, I., Moghaddam, L., Long, H., Dickman, M. B., Zhang, X., & Mundree, S., (2015). Trehalose accumulation triggers autophagy during plant desiccation. *PLoS Genet., 11*(12), e1005705.

80. Keller, F., & Pharr, D. M., (1996). Metabolism of carbohydrates in sinks and sources: Galactosyl-sucrose oligosaccharides. *Photoassimilate Distribution in Plants and Crops: Source-Sink Relationships*, 157–183.

81. Panikulangara, T. J., Eggers-Schumacher, G., Wunderlich, M., Stransky, H., & Schöffl, F., (2004). Galactinol synthase 1: A novel heat shock factor target gene responsible for heat-induced synthesis of raffinose family oligosaccharides in *Arabidopsis*. *Plant Physiology, 136*, 3148–3158.

82. Bartels, D., & Sunkar, R., (2005). Drought and salt tolerance in plants. *Critical Reviews in Plant Sciences, 24*(1), 23–58.

83. Santarius, K. A., (1973). The protective effect of sugars on chloroplast membranes during temperature and water stress and its relationship to frost, desiccation, and heat resistance. *Planta, 113*, 105–114.

84. Sengupta, S., Mukherjee, S., Basak, P., & Majumder, A. L., (2015). Significance of galactinol and raffinose family oligosaccharide synthesis in plants. *Frontiers in Plant Science, 6*.

85. Perera, I. Y., Hung, C. Y., Moore, C. D., Stevenson-Paulik, J., & Boss, W. F., (2008). Transgenic Arabidopsis plants expressing the type 1 inositol 5-phosphatase exhibit increased drought tolerance and altered abscisic acid signaling. *The Plant Cell, 20*(10), 2876–2893.

86. Okada, M., & Ye, K., (2009). Nuclear phosphoinositide signaling regulates messenger RNA export. *RNA Biology, 6*(1), 12–16.

87. Xue, H., Chen, X., & Li, G., (2007). Involvement of phospholipid signaling in plant growth and hormone effects. *Current Opinion in Plant Biology, 10*, 483–489.

88. Hoekstra, F. A., Wolkers, W. F., Buitink, J., Golovina, E. A., Crowe, J. H., & Crowe, L. M., (1997). Membrane stabilization in the dry state. *Comparative Biochemistry and Physiology, 117A*, 335–341.

89. Hincha, D. K., Zuther, E., & Heyer, A. G., (2003). The preservation of liposomes by raffinose family oligosaccharides during drying is mediated by effects on fusion and lipid phase transitions. *Biochim. Biophys. Acta, 1612*, 172–177.

90. Liu, J. J., Krenz, D. C., Galvez, A. F., & De Lumen, B. O., (1998). Galactinol synthase (GS): Increased enzyme activity and levels of mRNA due to cold and desiccation. *Plant Sci., 134*, 11–20.

91. Bledsoe, S. W., Henry, C., Griffiths, C. A., Paul, M. J., Feil, R., Lunn, J. E., & Lagrimini, L. M., (2017). The role of Tre6P and SnRK1 in maize early kernel development and events leading to stress-induced kernel abortion. *BMC Plant Biology, 17*(1), 74.

92. Bonfig, K. B., Schreiber, U., Gabler, A., Roitsch, T., & Berger, S., (2006). Infection with virulent and a virulent *P. syringae* strains differentially affects photosynthesis and sink metabolism in *Arabidopsis* leaves. *Planta, 225*(1), 1–12.

93. Eleutherio, E. C., Araujo, P. S., & Panek, A. D., (1993). Protective role of trehalose during heat stress in *Saccharomyces cerevisiae*. *Cryobiology, 30*(6), 591–596.

94. Kaur, S., Gupta, A. K., & Kaur, N., (2002). Effect of osmo-and hydropriming of chickpea seeds on seedling growth and carbohydrate metabolism under water deficit stress. *Plant Growth Regulation, 37*(1), 17–22.

95. Moore, B., Zhou, L., Rolland, F., Hall, Q., Cheng, W. H., Liu, Y. X., & Sheen, J., (2003). Role of the Arabidopsis glucose sensor HXK1 in nutrient, light, and hormonal signaling. *Science, 300*(5617), 332–336.

96. Ritsema, T., Hernández, L., Verhaar, A., Altenbach, D., Boller, T., Wiemken, A., & Smeekens, S., (2006). Developing fructan-synthesizing capability in a plant invertase via mutations in the sucrose-binding box. *The Plant Journal, 48*(2), 228–237.

97. Sengupta, S., Patra, B., Ray, S., & Majumder, A. L., (2008). Inositol methyl transferase from a halophytic wild rice, *Porteresia coarctata* roxb. (Tateoka): Regulation of pinitol synthesis under abiotic stress. *Plant, Cell and Environment, 31*, 1442–1459.

98. Shafiq, S., Akram, N. A., & Ashraf, M., (2015). Does exogenously-applied trehalose alter oxidative defense system in the edible part of radish (*Raphanus sativus* L.) under water-deficit conditions? *Scientia Horticulturae, 185*, 68–75.

99. Sheveleva, E., Chmara, W., Bohnert, H. J., & Jensen, R. G., (1997). Increased salt and drought tolerance by d-ononitol production in transgenic *Nicotiana tabacum* L. *Plant Physiology, 115*, 1211–1219.

100. Stevenson, J. M., Perera, I. Y., Heilmann, I., Persson, S., & Boss, W. F., (2000). Inositol signaling and plant growth. *Trends in Plant Science, 5*, 252–258.

101. Van, D. E. W., Michiels, A., De Roover, J., Verhaert, P., & Van, L. A., (2000). Cloning and functional analysis of chicory root fructan 1-exohydrolase I (1-FEH): A vacuolar enzyme derived from a cell wall invertase ancestor? Mass fingerprint of the 1-FEH I enzyme. *The Plant Journal, 24*, 447–456.

102. Zhuang, Y., Guijie, R., Guidong, Y., Zhaoxia, L., Xun, Q., Guihua, H., Yun, Z., & Juren, Z., (2007). Effects of water-deficit stress on the transcriptomes of developing immature ear and tassel in maize. *Plant Cell Reports, 26*(12), 2137–2147.

103. Santarius, K. A., (1973). The protective effect of sugars on chloroplast membranes during temperature and water stress and its relationship to frost, desiccation, and heat resistance. *Planta, 113*(2), 105–114.

104. Hincha, D. K., Zuther, E., & Heyer, A. G., (2003). The preservation of liposomes by raffinose family oligosaccharides during drying is mediated by effects on fusion and lipid phase transitions. *Biochimica et Biophysica Acta (BBA)-Biomembranes, 1612*(2), 172–177.

105. Nemeskéri, E., Sárdi, É., Remenyik, J., Kőszegi, B., & Nagy, P., (2010). Study of the defensive mechanism against drought in French bean (*Phaseolus vulgaris* L.) varieties. *Acta Physiologiae Plantarum, 32*(6), 1125–1134.

106. Stoyanova, S., Geuns, J., Hideg, E., & Van, D. E. W., (2011). The food additives inulin and stevioside counteract oxidative stress. *International Journal of Food Sciences and Nutrition, 62,* 207e214.

107. Kaur, S., Gupta, A. K., & Kaur, N., (2002). Effect of osmo-and hydropriming of chickpea seeds on seedling growth and carbohydrate metabolism under water deficit stress. *Plant Growth Regulation, 37*(1), 17–22.

108. Legay, N., Piton, G., Arnoldi, C., Bernard, L., Binet, M. N., Mouhamadou, B., & Clément, J. C., (2018). Soil legacy effects of climatic stress, management and plant functional composition on microbial communities influence the response of *Lolium perenne* to a new drought event. *Plant and Soil, 424*(1, 2), 233–254. https://doi.org/10.1007/s11104-017-3403-x.

109. Mattana, M., Biazzi, E., Consonni, R., Locatelli, F., Vannini, C., Provera, S., & Coraggio, I., (2005). Overexpression of Osmyb4 enhances compatible solute accumulation and increases stress tolerance of Arabidopsis thaliana. *Physiologia Plantarum, 125*(2), 212–223.

110. Wang, L. F., (2014). Physiological and molecular responses to drought stress in rubber tree (Hevea brasiliensis Muell. Arg.). *Plant Physiology and Biochemistry, 83,* 243–249.

111. Morsy, M. R., Jouve, L., Hausman, J. F., Hoffmann, L., & Stewart, J. M., (2007). Alteration of oxidative and carbohydrate metabolism under abiotic stress in two rice (*Oryza sativa* L.) genotypes contrasting in chilling tolerance. *Journal of Plant Physiology, 164*(2), 157–167.

112. Shiomi, N., Benkeblia, N., Onodera, S., Yoshihira, T., Kosaka, S., & Osaki, M., (2006). Fructan accumulation in wheat stems during kernel filling under varying nitrogen fertilization. *Canadian Journal of Plant Science, 86*(4), 1027–1035.

113. Bonfig, K. B., Gabler, A., Simon, U. K., Luschin-Ebengreuth, N., Hatz, M., Berger, S., & Roitsch, T., (2010). Post-translational derepression of invertase activity in source leaves via down-regulation of invertase inhibitor expression is part of the plant defense response. *Molecular Plant, 3*(6), 1037–1048.

114. Smirnoff, N., (2001). L-ascorbic acid biosynthesis. *Vitamins & Hormones, 61,* 241–266.

115. Pallett, K. E., & Young, A. Y. A., (1993). Carotenoids. *Antioxidants in Higher Plants* (pp. 91–110). Boca Raton.

116. Noctor, G., & Foyer, C. H., (1998). Ascorbate and glutathione: Keeping active oxygen under control. *Annual Review of Plant Biology, 49*(1), 249–279.

117. Smirnoff, N., & Pallanca, J. E., (1996). *Ascorbate Metabolism in Relation to Oxidative Stress. 24*(2), 472–478.

118. Nishikawa, F., Kato, M., Hyodo, H., Ikoma, Y., Sugiura, M., & Yano, M., (2005). Effect of sucrose on ascorbate level and expression of genes involved in the ascorbate biosynthesis and recycling pathway in harvested broccoli florets. *Journal of Experimental Botany, 56*(409), 65–72.

119. Shafiq, S., Akram, N. A., & Ashraf, M., (2015). Does exogenously-applied trehalose alter oxidative defense system in the edible part of radish (*Raphanus sativus* L.) under water-deficit conditions? *Scientia Horticulturae, 185,* 68–75.

120. Moore, B., Zhou, L., Rolland, F., Hall, Q., Cheng, W. H., Liu, Y. X., & Sheen, J., (2003). Role of the Arabidopsis glucose sensor HXK1 in nutrient, light, and hormonal signaling. *Science, 300*(5617), 332–336.

121. Nemeskéri, E., Sárdi, É., Remenyik, J., Kőszegi, B., & Nagy, P., (2010). Study of the defensive mechanism against drought in French bean (*Phaseolus vulgaris* L.) varieties. *Acta Physiologiae Plantarum, 32*(6), 1125–1134.

122. Stoyanova, S., Geuns, J., Hideg, E., & Van, D. E. W., (2011). The food additives inulin and stevioside counteract oxidative stress. *International Journal of Food Sciences and Nutrition, 62*, 207e214.

123. Kaur, S., Gupta, A. K., & Kaur, N., (2002). Effect of osmo-and hydropriming of chickpea seeds on seedling growth and carbohydrate metabolism under water deficit stress. *Plant Growth Regulation, 37*(1), 17–22.

124. Legay, N., Piton, G., Arnoldi, C., Bernard, L., Binet, M. N., Mouhamadou, B., & Clément, J. C., (2018). Soil legacy effects of climatic stress, management and plant functional composition on microbial communities influence the response of *Lolium perenne* to a new drought event. *Plant and Soil, 424*(1, 2), 233–254. https://doi.org/10.1007/s11104-017-3403-x.

125. Mattana, M., Biazzi, E., Consonni, R., Locatelli, F., Vannini, C., Provera, S., & Coraggio, I., (2005). Overexpression of Osmyb4 enhances compatible solute accumulation and increases stress tolerance of Arabidopsis thaliana. *Physiologia Plantarum, 125*(2), 212–223.

126. Wang, L. F., (2014). Physiological and molecular responses to drought stress in rubber tree (Hevea brasiliensis Muell. Arg.). *Plant Physiology and Biochemistry, 83*, 243–249.

127. Morsy, M. R., Jouve, L., Hausman, J. F., Hoffmann, L., & Stewart, J. M., (2007). Alteration of oxidative and carbohydrate metabolism under abiotic stress in two rice (*Oryza sativa* L.) genotypes contrasting in chilling tolerance. *Journal of Plant Physiology, 164*(2), 157–167.

128. Shiomi, N., Benkeblia, N., Onodera, S., Yoshihira, T., Kosaka, S., & Osaki, M., (2006). Fructan accumulation in wheat stems during kernel filling under varying nitrogen fertilization. *Canadian Journal of Plant Science, 86*(4), 1027–1035.

129. Bonfig, K. B., Gabler, A., Simon, U. K., Luschin-Ebengreuth, N., Hatz, M., Berger, S., & Roitsch, T., (2010). Post-translational derepression of invertase activity in source leaves via down-regulation of invertase inhibitor expression is part of the plant defense response. *Molecular Plant, 3*(6), 1037–1048.

130. Smirnoff, N., (2001). L-ascorbic acid biosynthesis. *Vitamins & Hormones, 61*, 241–266.

131. Pallett, K. E., & Young, A. Y. A., (1993). Carotenoids. *Antioxidants in Higher Plants* (pp. 91–110). Boca Raton.

132. Noctor, G., & Foyer, C. H., (1998). Ascorbate and glutathione: Keeping active oxygen under control. *Annual Review of Plant Biology, 49*(1), 249–279.

133. Smirnoff, N., & Pallanca, J. E., (1996). *Ascorbate Metabolism in Relation to Oxidative Stress, 24*(2).

134. Nishikawa, F., Kato, M., Hyodo, H., Ikoma, Y., Sugiura, M., & Yano, M., (2005). Effect of sucrose on ascorbate level and expression of genes involved in the ascorbate biosynthesis and recycling pathway in harvested broccoli florets. *Journal of Experimental Botany, 56*(409), 65–72.

# Index

Printed and bound by CPI Group (UK) Ltd, Croydon, CR0 4YY

23/10/2024

01777702-0012